W9-ANR-260

Second Edition

WILDLIFE
ISSUES
in a
CHANGING
WORLD

Second Edition

WILDLIFE ISSUES
in a
CHANGING WORLD

Michael P. Moulton
James Sanderson

Lewis Publishers

Boca Raton   London   New York   Washington, D.C.

Library of Congress Cataloging-in-Publication Data

Catalog information may be obtained from the Library of Congress

This book contains information obtained from authentic and highly regarded sources. Reprinted material is quoted with permission, and sources are indicated. A wide variety of references are listed. Reasonable efforts have been made to publish reliable data and information, but the author and the publisher cannot assume responsibility for the validity of all materials or for the consequences of their use.

Neither this book nor any part may be reproduced or transmitted in any form or by any means, electronic or mechanical, including photocopying, microfilming, and recording, or by any information storage and retrieval system, without prior permission in writing from the publisher.

The consent of CRC Press LLC does not extend to copying for general distribution, for promotion, for creating new works, or for resale. Specific permission must be obtained in writing from CRC Press LLC for such copying.

Direct all inquiries to CRC Press LLC, 2000 Corporate Blvd., N.W., Boca Raton, Florida 33431.

Trademark Notice: Product or corporate names may be trademarks or registered trademarks, and are used only for identification and explanation, without intent to infringe.

© 1999 by CRC Press LLC
Lewis Publishers is an imprint of CRC Press LLC

No claim to original U.S. Government works
International Standard Book Number 0-56670-351-4
Printed in the United States of America 1 2 3 4 5 6 7 8 9 0
Printed on acid-free paper

CONTENTS

PREFACE

n the spring of 1979, Dr. Larry D. Harris at the University of Florida created a class called Man and Wildlife Resources. In the fall of 1987, he changed the name to Wildlife Issues in a Changing World. Dr. Harris, a lifelong conservationist and co-founder of the Florida Greenways program, was frustrated because he felt he was "preaching to the choir" in his graduate-level classes. He needed a forum where he could reach many people at once regarding conservation of our natural resources. His solution was to create an undergraduate class called Wildlife Issues and open it up to anyone interested in wildlife and conservation. The class was moved from a small classroom to an auditorium.

In the spring semester of 1998, class enrollment reached 1000 students, and more than 3000 students are expected to take the class for all of 1998. Two classes per semester are planned. Are students today more interested in wildlife issues? We suggest that most people, young and old, are interested in issues concerning wildlife.

When we began investigating general issues, we quickly discovered that some wildlife issue was covered in nearly every edition of every major newspaper. The problem was to make sense of all the separate and apparently isolated issues. Early in 1996, record numbers of dead or dying manatees (an endangered species) were turning up in southwest Florida. In New Mexico, the winter was unusually dry and bears were wandering into Albuquerque neighborhoods in search of food. In Chicago, a trial involving smuggled parrots was wrapping up. In Miami, Fly River turtles from Papua New Guinea were turning up in the pet trade. U.S. Fish and Wildlife Service officials arrested a footwear manufacturer for using endangered species skins from South America to make boots. In the North Atlantic, harp seal hunts resumed, and one of the greatest offshore fisheries was so badly overharvested that it was closed. Walrus had been hunted to extinction there 250 years ago.

Several remaining monarch butterfly populations died as the result of a freak snowfall in the mountains outside Mexico City. In China, tiger penises were sold as aphrodisiacs. In Kenya, only ten dugongs remain. While the U.S. Fish and Wildlife Service tries to protect 11 endangered piping plovers by poisoning some of the more than 5000 gulls that feast on the plovers' eggs and young, the Humane Society of the United States tries to protect the gulls. In the Democratic Republic of the Congo and Rwanda, gorilla habitat was under immense pressure from refugees. In Cameroon, forestry practices produced an unexpected benefit for bush meat hunters. Estimates are that three gorillas and three chimpanzees are consumed for food each day. Wildlife issues, some humorous and some sad, were with us wherever we went.

When early humans depended on animals for food, there were rarely any wildlife issues. The only issues probably concerned ownership of the kill. Today, wildlife is threatened for many reasons. Often wildlife populations cannot regenerate, and the survival of many species depends on active protection.

Issues arise when at least two people disagree. A wildlife issue arises when two or more people disagree over some aspect involving wildlife. Here, we urge that resolution of the issue must occur proactively. That is, resolving an issue does not include a default option. Issues must be decided upon and a course of action followed, and changed if necessary. In no instance is doing nothing an option.

We urge readers to take a historical perspective and use the scientific method to resolve wildlife issues. That is, we discourage participants from basing their decisions on "belief," however common such belief is. For instance, one often hears statements such as "there are more manatees in Florida today than when Columbus arrived." A counterargument is that there are more warm-water refugia provided by nuclear power plants than when the Spanish arrived, and so there are more manatees now than ever. The scientific method requires testable hypotheses to decide issues. Neither of these previous statements is testable today.

With all the national laws and international agreements to protect wildlife, it's a wonder there are any wildlife issues at all. It seems like the number of wildlife laws has grown in proportion to the number of wildlife issues. Protection laws and agreements generally are a response to wildlife issues and do not foresee them. For instance, the African Elephant Protection Act passed by Congress and signed by the president of the United States became law well after Africa had lost most of its elephants. Other laws appear to serve local research projects. The Endangered Species Act has "teeth" because it provides for stiff penalties, but the act can be circumvented by filing for a "take" permit, which is often granted. Another law protects wild horses and burros and demands that their habitat be managed in a natural way. However, because horses and burros are introduced species, they are generally harmful to the very habitat that must also be protected. Humans can enact laws but they cannot change the laws of nature, however stiff the penalties are made.

No number of laws will protect wildlife species, just as no number of parks and reserves will save all species. Worldwide habitat destruction is continuing, and the developed world continues to consume vast quantities of global resources. Introduced species are homogenizing the world, creating widespread problems. Can we address these issues and resolve them? In Chapter 20, we present several novel solutions that have been developed and applied with success.

During a recent trip to Papua New Guinea, a group of foreign tourists urged their native guide to break off discussions with the logging companies. "Preserve your forests and wildlife," they told him. "Do not sacrifice them for money." He responded by saying his people wanted access to modern healthcare, schools for their children, and motorboats and fuel. "You would put us in zoos and stare at us," he said. Then he asked a critical question that silenced the tourists: "Will you pay me for my trees and wildlife to save them?" He knew they would not.

People all over the world want nothing more than a better life for themselves and their families. Somehow we must redefine our quest for a better life. A better life, a high-quality life, means living with nature, not apart from nature. As the Papua New Guinea guide said, "You can try to explain conservation to my people, but they will not understand. They were born conservationists and their lives depend on conserving their resources. They do not understand how one could act otherwise." Certainly, issues depend on one's perspective.

Although most people will never see mountain gorillas in the wild, we are certain that most people favor protecting them through preserves and laws. By understanding some of the worldwide wildlife issues we discuss, we hope that readers will gain a better understanding of issues in their own backyard as well as others'. As educators, we believe that one way to resolve wildlife issues is through understanding and not emotion. Quality of life is synonymous with living as a part of nature and accepting wildlife for what it is. It is a depressing thought to imagine living in a world without free-roaming gorillas, elephants, or manatees. Therefore, we are prepared to deal with and resolve difficult wildlife issues and make hard choices. With greater appreciation for wildlife issues, our readers will gain the confidence to deal with them.

The growing use of the World Wide Web has enabled users to explore globally wildlife issues most never knew existed. We have included exercises designed to expand our readers' horizon so as to broaden their perspective and encourage their own thinking. Since we could not cover all wildlife issues in this text, other issues are explored in the exercises. Certainly, there is as yet no substitute for a good library, and we frequently suggest that readers seek one out for further exploration of issues we discuss. The act of discovery yields reward in and of itself.

In this edition, we have tried to organize wildlife issues around a common theme. Jared Diamond's Evil Quartet is the organizing principle. Easily remembered by even casual readers, the members of the quartet are useful in categorizing wildlife issues so that causes of the issues can be addressed.

Many students suggested that they wished there was as much wildlife around today as there was at the turn of the century. This led us to expand this edition to include a history of wildlife in the United States. We present an in-depth discussion of wildlife issues in eight states. The profound consequences of the loss of much wildlife in the late 19th and early 20th centuries and passage of the Pittman–Robertson Act, one of the most important wildlife laws ever enacted, are discussed in detail.

ACKNOWLEDGMENTS

We wish to thank Dr. Larry D. Harris for creating the Wildlife Issues class and for enthusiastically encouraging this book. We wish to thank the following people for sending information on the various state wildlife programs: Mark McCollough and Karen Morris, Maine Department of Inland Fisheries and Wildlife; James Sciascia and Susan Predl, The New Jersey Division of Fish, Game and Wildlife; Calvin DuBrock, Pennsylvania Game Commission, Bureau of Wildlife Management; Karen Bates and John Carlson, Jr., Department of Fish and Game, California; James A. Bailey and Darrel L. Weybright, New Mexico Department of Game and Fish; and Dr. Steven J. Bissell, Colorado Division of Wildlife. We also wish to thank Nicole Sanderson for additional information on wildlife programs in New Jersey.

Mike Moulton wishes to thank Becky, Rachel, and Linda for support and encouragement and access to the computer. He also thanks Deborah M. Epperson, Drs. Jon D. Johnson, John M. Davis, Kathryn E. Sieving, and Mel Sunquist for stimulating conversation about various wildlife issues.

Jim Sanderson wishes to thank his spouse, Joan Morrison, for support and encouragement. He also thanks Manuel Molles, University of New Mexico, and his son Anders for additional research material. Dr. Estaban Sarmiento and Dr. Dieter Steklis added much to the discussion of gorillas. Dr. Katie Sieving also provoked several interesting wildlife issue discussions. He thanks Karl Amman for the information on the gorilla bush meat trade.

We also reserve special thanks for our friends at Lewis Publishers. Dennis Buda provided us the opportunity to organize our thoughts and present them to others. Our editor, Ms. Sandy Pearlman, deserves thanks for all the hard work she did—and made us do—to make our efforts a reality. That picture of the medicinal shop in China is perfect, Sandy.

ABOUT THE AUTHORS

Michael P. Moulton is an Associate Professor of Wildlife Ecology and Conservation at the University of Florida. He spent his under graduate years at the University of Colorado and later received his Ph.D. from the University of Tennessee. His principal area of research deals with introduced species of birds.

Jim Sanderson is currently working for Los Alamos National Laboratory in New Mexico. He is also Director of Conservation for Mountain View Farms Breeding and Conservation Centre, Vancouver, Canada. He received his Ph.D. from the University of New Mexico in 1976. An avid traveler, Dr. Sanderson has collected and synthesized wildlife issues from around the world. His interests include quantitative ecology, community ecology, and landscape ecology.

INTRODUCTION

With increasing frequency, we hear and read about mysterious decreases in or disappearances of living organisms. Often these organisms are reptiles, fish, birds, or mammals—what people commonly call wildlife. This book addresses current wildlife issues. Exactly what does the term *wildlife* mean and what do we mean when we suggest there is an *issue* at stake? For the purposes of this book, we will accept the definition of "issue" as given in *Webster's New Collegiate Dictionary*: an issue is "a matter in dispute between two or more parties, a point of debate or controversy." Certainly "two or more parties" refers to two or more humans. Thus, without two or more humans involved, there can be no issue.

What does the term wildlife mean? Basically, all living wild animals are wildlife. We will spend much time discussing this catchall term. For instance, are elephants in a highly managed African park really wildlife? Are zoo animals wildlife? Are insects wildlife? How about bacteria? Humans create parks for endangered species. Do we create parks for endangered or rare malaria-causing tsetse flies? The term wildlife is often synonymous with benign wildlife. That is, animals are wildlife as long as they pose no direct threat to humans. When something poses a direct threat to humans, it becomes something other than wildlife, such as a pest or disease-causing agent that must be searched out and destroyed. We generally do not deliberately seek out and destroy a wildlife species. Thus, wildlife is a relative term with respect to human experience.

WILDLIFE ISSUES

The phrase *wildlife issue* is by definition a wildlife controversy that pits at least two and usually more human participants against one another over a problem relating to wildlife. Nearly everyone in North America is familiar with timber harvesting in the Pacific Northwest and the need to preserve northern spotted owl (*Strix occidentalis*) habitat. This is clearly a wildlife issue involving the continued harvesting of timber in forests that are essential habitat for the northern spotted owl. Some humans derive material gain from harvesting timber. Others want to have spotted owls around for their viewing pleasure or perhaps want to look at forested hillsides and resent the harvesting of timber on public lands.

Other wildlife issues are equally controversial. The manatee is a large endangered aquatic mammal that lives in Florida's streams, rivers, and lakes. Many boaters treat manatees as if they were nothing more than "speed bumps" and fail to exercise appropriate caution to avoid collisions with them. Many manatees show linear scars on their backs from boat propellers. Are manatees worth saving or are they doomed because they are an inconvenience to boaters? As with any issue, there are strong proponents on both sides. Each side frequently offers arguments to try to sway those who are neutral or undecided. For instance, a boating industry spokesperson suggested that increased manatee deaths were the result of there being more manatees than the environment can support. Statements such as "there are more manatees now than when Columbus was here" are offered as evidence.

Many wildlife issues span a broader context, where wildlife is a subset of a broader issue. For example, it has been suggested that the human-altered hydrological cycle in Everglades National Park is responsible for declining wading bird populations. Because much water is diverted to nearby human population centers such as Miami, the issue of water in the Everglades becomes a wildlife issue insofar as it impacts wildlife populations. Certainly, there is an issue here. Human control of water resources directly impacts the Everglades landscape and hence its living organisms, including wildlife.

In many areas of the United States, raccoon populations are increasing. An injured raccoon found by a sympathetic human is often taken to a wildlife rehabilitation center where it is treated and released. Yet raccoon populations are increasing in part because of a lack of predators to keep populations in check. Raccoon themselves are predators of birds' eggs and chicks, and in some areas raccoons now consume a large percentage of gopher tortoise eggs. Often, the raccoon does not wait for the tortoise to lay the eggs but rather scoops them out of the living tortoise, killing it. Alligators lose eggs to raccoons. In towns and rural areas everywhere, raccoons raid garbage cans, and their populations are increasing. When the population density becomes too high, rabies outbreaks are common and spread rapidly throughout the regional population. Humans are susceptible to rabies, and thus an issue is raised. Should humans indeed act as predators of raccoons, keeping their populations in check?

Should humans shoot raccoons on sight? Should wildlife rehabilitation centers treat raccoons or euthanize them? Indeed, are raccoons really wildlife or a human-adapted nuisance that is no longer wild?

Some wildlife issues seem bizarre or almost unbelievable. For example, on February 17, 1994, the *Gainesville Sun* reported, "Toke a toad, go to jail." A California couple was arrested on charges of possessing the illegal chemical bufotenine, obtained from a toad (*Bufo alvarius*). The toad secretes a venom from a gland on its back. When smoked, users say it produces a high that eclipses the psychedelic properties of LSD. Laws against the use of bufotenine date back to the 1960s.

As reported in the *Gainesville Sun* on July 12, 1994, "A chemical used for greener lawns is killing fish, birds, otters" and is contaminating drinking water. According to the article, a certain chemical had caused ten documented massive fish kills in Florida. The chemical (Nemacur) is used to control microscopic worms that are especially dangerous to golf course turf. In May 1994, after the chemical had been applied on 100 acres, thousands of fish turned up dead in a nearby canal after a heavy rain. The U.S. Golf Association studied the chemical, and the association's research director issued a statement to the public that there was no threat to groundwater. The study did not look at runoff however, which the state of Florida says caused ten other incidents of fish, bird, and otter kills. The company that makes the lawn chemical wants to cooperate with Florida, but golf courses are major customers for the chemical. Surely there is a testable hypothesis somewhere here. What is it?

"Locust threaten continent's food" was yet another wildlife issue. Swarms of three-inch-long desert locusts capable of devouring 100 tons of vegetation per day were eating their way across Africa. The locust infestation posed a serious threat to the food supply. Some 250 locust species cause formidable problems for agricultural crops. Is it the locust or the intensity of agriculture that is the problem here?

How about this attention-grabber from the January 13, 1994 *St. Petersburg Times*: "Caterpillar kills 5 in Brazil." A venomous, hairy species of caterpillar whose sting causes burns and internal bleeding killed five people in Brazil. The latest victim was 57 years old and the cause of death was officially "death by caterpillar sting." The fire caterpillar (*Lonomia obliqua*) can grow to three inches, is green, and is covered with a dense mat of fine hairs. The hairs act like hypodermic needles which, when contacted, inject a venom that interferes with coagulation of the blood. Rapid deforestation and the disappearance of predators appear to have led to an increased presence of the stinging caterpillars near towns and cities in Brazil's southernmost state, Rio Grande do Sul.

What is the basis for deciding whether an issue exists? Issues are human constructs and would not be issues if humans were not involved. That is, the basis of all issues lies in the different perceptions or beliefs among humans. Two humans with identical beliefs would never have any issues to decide. They would simply agree on a certain course of action and proceed. This is equivalent to suggesting that a person has never had a disagreement or difference with anyone else—and we all know how probable that is!

Consider the following example. Nearshore commercial fishing in Florida using gill nets was a profitable enterprise. Indeed, many people who had enjoyed recreational fishing blamed the great success of nearshore commercial fishing for depleting fish stocks. In 1994, Florida voters passed an amendment to the state constitution banning commercial netting of fish with gill nets within three miles on the Atlantic coast and within nine miles on the Gulf coast. On July 1, 1995, the amendment became law. How did this issue arise?

Recreational fishing in Florida is one of the state's top tourist attractions. Evidently fishermen (and fisherwomen) had perceived a decline in the number and size of fish caught for several years. Numerous causes, such as pollution, commercial gill netting, development of beachfront hotels, and global warming, were discussed in the press and on local radio shows. Notice that recreational fishing was not considered to be a cause for concern. This is likely because the recreational fishing industry was behind the campaign in the first place. After months of debate, the presumed cause was identified—commercial gill netting. Once a cause was found, it was targeted for solution. The recreational fishing industry had to convince voters statewide to pass a law prohibiting gill netting.

A grass roots campaign was organized statewide, and recreational and weekend fishermen were armed with petitions. Over 400,000 signatures were collected—enough to place a proposed amendment on the statewide ballot during the general election. The amendment passed, and Governor Lawton Chiles threatened to call out the National Guard to enforce the net ban. The grass roots campaign succeeded. People from all professions who enjoyed fishing took the day off—to go fishing!

Of course, there were two sides to this issue. The opposition consisted of commercial fishing interests that had difficulty getting organized. Their arguments progressed in stages. Initially, they denied there were fewer fish. How could commercial fishing alone deplete huge fish stocks? Surely, this was impossible. This was the denial phase. Such statements as "there are no data to support the conclusion that stocks are depleted," "natural fluctuations in fish populations occur all the time," and even "we don't believe you" appeared in the media. When these arguments failed to reverse the general opinion, the campaign moved to phase two—acceptance of "facts" without supporting "data." The commercial interests accepted that fish stocks were reduced but continued to deny responsibility.

When the grass roots petition drive spread throughout Florida towns and cities, the third and final phase—total confusion—set in and the opposition faltered. A statement typical of this phase was, "Now what will we do if we are unable to make a living fishing?" Note that this statement is no longer a wildlife issue. The wildlife issue would be decided in the voting booth. Lawmakers at the state level could now endlessly debate the commercial fishing business. Commercial fishing companies failed to grasp the widespread interest in sport fishing. Indeed, had they decided to help increase fish stocks through alternative stocking programs or to simply move farther offshore, they might have saved the day. Today, gill netting is illegal within the offshore distances set by law. Curiously, these limits are beyond where recreationists with boats most often fish. The number and size of the fish have not yet increased.

THE PROBLEM WITH "BELIEF"

Notice the role of "beliefs" on both sides of the Florida fishing issue. One side "believed" fishing stocks were being depleted by commercial fishing interests. The opposing side "believed" it was not responsible and that some other factor was causing the decline. Note also that commercial interests did not blame recreational fishing businesses. In the absence of good, solid scientific data, these "beliefs" became the foundation for an open public debate based not on fact but on faith (Lakatos 1978).

At one time, people believed the earth was flat. Ptolemy, the greatest scientist of his time, believed the earth was the center of the universe and the sun and all heavenly bodies revolved around the earth. This belief persisted for 1500 years, until Copernicus whispered to his friends that the sun was the body around which all planets, including earth, revolved and that he had accumulated scientific data to support this conclusion. Copernicus died before seeing his disciples held up to public ridicule by the church. Belief is a powerful force.

Belief is not limited to medieval times. Early in the 20th century, passenger pigeons (*Ectopistes migratorius*) formed flocks that darkened the sky and blocked out the sun for hours at a time. Who would have believed the few bird-watchers who foresaw the demise and eventual complete extinction of the ubiquitous pigeon? Philosopher Imre Lakatos (1978) reminds us that "no degree of commitment to beliefs makes them knowledge." What then is the difference between accepted knowledge and belief? Even Copernicus's view of the solar system was changed by Kepler and Kepler's by Newton. Were these views beliefs or scientific facts? How does a series of facts lead to a theory that permits predictions which can be tested?

One goal of this book is to help you sort out the issues that will be independent of people's beliefs. A trend or pattern may exist even if no one believes it and may not exist even if everyone believes it. The solution to this conundrum is to always use the scientific method to decide an issue.

RESOLVING ISSUES

There is often more than one way to resolve an issue. We might, for instance, want to decide an issue on the basis of economics. Cattle grazing on public lands is a wildlife issue because cattle frequently destroy habitat that could otherwise be used by deer and elk. The Bureau of Land Management is responsible for collecting grazing fees paid by ranchers to use public lands. In 1997, this fee was approximately $1.35 a head per month. That is, a rancher with grazing rights pays $1.35 per animal unit per month for the right to graze these cattle. Suppose the Wilderness Society or the Nature Conservancy would be willing to pay $2.00 per head per month to the U.S. government to keep a like number of cattle off public lands. Economically, it would be a good business decision by the government to accept the higher offer.

Sadly, some issues resolve themselves before humans can decide to act responsibly. The passenger pigeon became extinct before laws could be enacted to protect it. The Carolina parakeet left us similarly, its joyous melody denied to us before our forebears could introduce a law and bring it to a vote. The scientific method is supposed to prevent such atrocities as extinction from happening without the majority of the voting public deciding that extinction is an acceptable alternative instead of deciding by default.

THE SCIENTIFIC METHOD

The scientific method works in a series of steps. First, a pattern or trend is observed. This could correspond to our belief that a trend exists, for instance. An explanation is offered as to why the trend or pattern exists. We are led to the creation of a scientific hypothesis. Predictions are made based upon the hypothesis. Finally, tests of the predictions are conducted using controlled laboratory experiments or statistical tests. If a prediction fails, we can return to the creation of a new hypothesis from which to make predictions. If a prediction is true, then we draw the indicated conclusion.

Unfortunately, when dealing with emotional issues, hypotheses are not clearly stated or are untestable. For instance, someone might suggest that there are more manatees now than when Columbus arrived in the New World. This statement is untestable, and the scientific method cannot be used to prove or disprove it. The most important characteristic of hypotheses is that they must be mutually exclusive alternatives. In other words, two opposing hypotheses must span the space of all possibilities and lead to opposite conclusions. One hypothesis is referred to as the null hypothesis, or no difference hypothesis, and is labeled H_0. The opposing hypothesis, labeled H_1, is called the alternative hypothesis and states that there is a difference. Some examples will help clarify these hypotheses:

H_0 There is no difference in frog abundance between areas where frog harvesting is allowed versus protected areas.

H_1 There is a difference in frog abundance between the two areas.

Another example is:

H_0 Agricultural practices in the Everglades Agricultural District have no effect on concentrations of phosphorus in the water.

H_1 Agricultural practices in the Everglades Agricultural District alter the phosphorus concentration in the water leaving the district.

Notice that in both cases, one of the two hypotheses must be correct. This is not the case for hypotheses that are not mutually exclusive, such as:

H_0 The boss's new car is a Ford.

H_1 The boss's new car is a Chevy.

TESTING HYPOTHESES WITH EXPERIMENTS

Experiments are powerful ways to test hypotheses. In an experiment, one: (1) applies some treatment and (2) compares the effects of the treatment on some variable that could be influenced by the treatment to the effects on the variable when the treatment is not applied. A controlled experiment requires a *control* and a series of *experimental treatments* or manipulations. In ecology, control plots are simply plots of ground, for instance, that do not receive the treatment. The aim is, of course, to see what happens if the treatments were not applied. We also need to specify some variable of interest that the experimental treatments might influence. There is no point in measuring variables that are (1) not likely to be affected by the experimental treatment or (2) not measurable.

For example, if you were interested in comparing the effects of a new fuel on the performance of your car, there is no sense in checking your tire pressure before the new fuel is used and then after the new fuel has been tried. A better measure of performance would be mileage or acceleration.

Now recall our first pair of hypotheses above which dealt with frogs. A prediction of our hypothesis is that frog hunters reduce populations of frogs in areas where frogs are hunted. We could survey areas where frogs are hunted to see if their numbers are reduced each evening, but this alone would really show nothing because other factors might be involved. For instance, wading birds eat frogs, as do fish, so the frog count might go down anyway. Similarly, the frog population might go up because frogs invaded the area when the bigger hunted frogs were killed out.

For these reasons, we need to conduct a controlled experiment. We could establish ten replicates of two experimental plots, say 30 m^2 on a side (called a quadrat), and survey the quadrats for frogs. In ten of the quadrats, there will be no frog hunting. In the other ten quadrats, hunting of frogs will be allowed. The quadrats will be chosen randomly along the shoreline of a lake where frogs are hunted.

On successive mornings after an evening of hunting, we will catch and release as many frogs as we can from the 20 quadrats, making certain not to release frogs into any of the other quadrats. Ideally, the survey would be done simultaneously in all the quadrats and for the same length of time in each quadrat. We will count every frog caught in each quadrat. Note that it would be unlikely that we could catch every frog in every quadrat, nor is it essential to do so.

After ten successive evenings, we could compare the number of frogs in each quadrat. However, what is important is not the number of frogs in each quadrat but the change in the number of frogs in each quadrat. For instance, let's say that in one of the unhunted quadrats, the number of frogs went from 20 down to 10 over the course of ten days. In one of the hunted quadrats, the number of frogs might have increased from 15 to 30. However, it is important to compare all the quadrats simultaneously. We could sum up all the frogs caught in the unhunted quadrats each day and compare the change to that found in the hunted quadrats. The changes can be compared statistically.

If there was a decrease in the number of frogs over the course of our study in both sets of quadrats, and the difference in the two decreases was large, then we could conclude that hunting negatively impacted the frog population. If there was a slight increase in the unhunted quadrats and a slight decrease in the hunted quadrats, however, the difference might not be significant, and we would be led to conclude that hunting did not impact the frog population. The important parameter is the difference in the changes in population in each set of quadrats and not the absolute number of frogs.

Now consider the second set of hypotheses above:

H_0 Agricultural practices in the Everglades Agricultural District have no effect on concentrations of phosphorus in the water.

H_1 Agricultural practices in the Everglades Agricultural District alter the phosphorus concentration in the water leaving the district.

Here we have a problem setting up a controlled experiment because we cannot specify any kind of control. That is, we have no way of comparing test plots to plots where nothing was done. Certainly, we could measure the phosphorus concentration in the water exiting the Everglades Agricultural District, but the result would not tell us anything about whether the level of phosphorus is higher or lower because of agricultural practices. Furthermore, we would be unlikely to pinpoint the source of the phosphorus because of all the other inputs to the water supply. Unfortunately, in most wildlife issues, we face a similar situation: we cannot set up controlled experiments. The scale is simply too grand. What we end up doing is collecting a great deal of circumstantial evidence. Remember, however, that even though the evidence is circumstantial, it is still evidence!

The value of the scientific method is that often new tests are generated that lead to the prediction of novel facts, and one well-formed sequence of experiments can lead to volumes of new information. Of course, there are always errors associated with experiments and hypothesis testing. These errors are referred to as *Type I* or *Type II* errors. To understand these errors, we need to understand how hypotheses are tested. The probability used to reject the null hypothesis is called the *significance level* and is denoted by α. The significance level is chosen before any experiments are done. Typically, $\alpha = 5\%$ is used in ecological experiments. The value of the test statistic is called the *critical value* and is found in a table of whatever statistical procedure we are using.

A *Type I* error occurs when the null hypothesis is rejected when it is in fact true. That is, suppose the null hypothesis H_0 is true and we reject it based on the value of α. We have made a Type I error. Now suppose H_0 is false. If we accept H_0, then a *Type II* error has occurred. The only way to minimize both types of errors is to have many replications of the experiment. Table 1.1 summarizes the two types of errors.

In a recent article in *Science*, P.K. Dayton (1998) argued that in resource management issues, a Type II error is far more serious than a Type I error. Suppose we conduct a study to see if there will be a negative effect from

TABLE 1.1 The Two Kinds of Errors in Statistical Hypothesis Testing

	H_0 is true	H_0 is false
H_0 is rejected	Type I	No error
H_0 is accepted	No error	Type II

harvesting on a population of fish. There either is an effect or there is not. If we conclude that there is an effect when there really is not, we commit a Type I error. However, if we conclude that there is no effect when there really is one, we commit a Type II error.

Throughout this book, we will use the scientific method to evaluate wildlife issues. We will approach these issues as scientists, as biologists and ecologists, and not as environmentalists or activists. We will take positions based on science and not belief, no matter how widespread such beliefs are. Our focus will be current threats to wildlife around the world. We will use the World Wide Web to explore wildlife issues globally, from the clubbing of harp seals in Canada to the introduction of exotic birds in Hawaii, the creation of parks in Africa, and the consumption of tiger bones in Asia. To understand current wildlife issues, we must understand and appreciate the origin and distribution of global biodiversity and threats to this diversity, wildlife ecology, and wildlife conservation in a world dominated by humans. You will learn that nothing less than the survival of our own species is at stake.

THE IMPORTANCE OF HISTORY

Wildlife issues do not exist apart from our everyday life experiences. Indeed, a wildlife issue cannot be fully understood and appreciated unless we understand how the issue arose in the first place and why it came to be an issue. In other words, we must always consider how the issue came into being. Fair questions are: How did this become an issue? What is the history of this issue? Why is the history important? Perhaps no other single factor is more important than history in helping us to reach a fair and equitable solution to a wildlife issue. Both the short- and long-term history of events are important elements in our scientific quest to understand wildlife issues. Consider the following example.

One of the American bison's strongholds is Yellowstone National Park in the northwest corner of Wyoming. Within the park, all wildlife, including bison, are protected. The Yellowstone fires of 1986 showed that park managers are managing Yellowstone as naturally as possible, letting fires started by nature burn nearly uncontrolled. Wildlife is also managed as naturally as possible. In principle, such a policy is easy to execute—or is it?

One winter, a bison attempted to cross Yellowstone Lake, which was, as it turned out, not thickly covered with ice. The bison broke through the ice and was unable to secure its footing to lift itself from an icy death. Three snowmobilers witnessed the incident and reported it. Park rangers had to decide what to do. Without human intervention, the bison would be unable to free itself, but human intervention in a naturally occurring phenomenon was strictly prohibited. Park rangers decided to do nothing, and during the frigid night the bison succumbed. The issue was whether to attempt a rescue of the bison or to do nothing. If you were a park ranger, what would you have done?

First, you might have determined the sequence of events. Unknown to the park rangers, the bison was walking down the snowmobile trail that ran along the lake. Yellowstone was under deep snow, and bison often use snowmobile paths in winter because the snow is compacted and hence less deep. As the bison slowly walked down the path, the snowmobilers approached quickly from around the bend. When the snowmobilers blocked the bison's path, the bison abandoned its course and headed down an embankment that was the bank of Yellowstone Lake. Bison rarely walk across a frozen lake because their hooves are not well adapted to crossing ice. However, this particular bison was left with little choice. Within five minutes, the bison had crashed through the ice as the snowmobilers watched. Given these new facts, would your course of action change?

The immediate wildlife issue was whether to rescue the bison or to let nature take its most certain course. However, no wildlife issue is ever immediate, and few rarely surprise us from behind. Most often, there is a sequence of events that precipitates what we refer to as a wildlife issue (i.e., an issue that arises because two or more people disagree on a course of action involving wildlife). No decision can be made until the sequence of events is understood as best as possible under the time constraints involved. In our example of the bison, no questions were asked of the snowmobilers. The recreationists reported the incident immediately and were dismissed. Policy was followed and nature did the rest. There was no "discovery" process, no history, no sequence of events. There was simply a bison struggling to free itself, its front legs slipping, pitifully equipped to secure footing on wet ice.

The above events took place in a relatively short period of time. Other events take longer to materialize. For instance, there are two species of African rhinoceros, and both are endangered because of poaching by humans. Rhinos have few natural predators except for lions and hyenas. Ngorongoro Crater in northern Tanzania, East Africa, is a wonderful park to enjoy viewing African white rhino. There are about 20 rhino living in the crater, and most are seen each day by the many tourists driving throughout the park. At night, park rangers protect road access to the crater, so the rhino are relatively safe, at least from humans.

Rhinos are notoriously nearsighted, and a rhino calf rarely wanders more than a short distance from its mother's side. If it does, it is likely to become lost in spite of the fact that it may be only 5 m away. One afternoon in the crater, as the sun set and the tourist trucks drove the steep walls to exit, a rhino calf became "separated" from its mother. Hyenas are opportunistic predators and always seem to be in the right place at the right time. Not surprisingly, one approached the confused calf. A few meters away, the rhino mother grunted and snorted and jerked her head up and down. Leaning backward, with its hindquarters low to the ground, the hyena approached the calf. Slowly the hyena crept forward, and then suddenly, quickly, its jaws secured the rhino calf by the nose. The jaws of the hyena are among the most powerful of any predator, able to crush bone for the marrow contained within it. The calf resisted and broke free but did not flee. These events were witnessed by a professional wildlife photographer who had spent a lifetime in Africa. Had you been a park ranger on the scene, what would you have done? Would a factor in your decision be that white rhinos are endangered because humans have poached them to the brink of extinction? Or would you follow the stated wildlife policy, which is to let the animals work out their own struggles? The photographer filmed the gruesome sequence of events.

Unrelenting human poaching has led to the mass extirpation of the white rhino across Africa. Unless humans intervene on behalf of the rhino, extinction will be the result. The history of both species of African rhino is well documented. Intervention by humans on behalf of the species must take place inside as well as outside of parks and preserves if the rhino is to survive. Very few African parks are even remotely "wild" today, and most are extensively managed. Deciding on a form of "natural" management policy is certainly one of the most complex wildlife issues of our day. Each conservation biologist must decide when to intervene and when to be an observer—well before the situation arises.

In both examples, the importance of history and understanding the sequence of events should have carried some weight in helping us reach a decision. That is, the importance of history cannot be ignored, and an understanding of the historical events that preceded the wildlife issue frequently leads to a more comprehensive understanding and deeper appreciation of the issue at hand. Indeed, history, even ancient history, is of profound importance in resolving many wildlife issues. Without a historical perspective, many issues are impossible to resolve. Thus, a historical understanding of life on earth is essential, and understanding how we got to where we are today is crucial.

THE EVIL QUARTET: A FRAMEWORK FOR WILDLIFE ISSUES

In 1984, Jared M. Diamond published a chapter in a book on extinctions. The title of his chapter was "'Normal' Extinctions of Isolated Populations (Diamond 1984). Diamond reviewed the recent history of extinctions and produced a synthesis of their causes. He labeled the four principal causes of recent extinctions the "Evil Quartet." In 1996, Stuart Pimm (1996) analyzed Hawaiian extinctions and concluded that once one of the quartet gets started, the effects are confounded by other members of the quartet, making the situation worse. Even with the benefit of hindsight, unraveling the processes responsible for extinctions becomes difficult or impossible. Diamond and Pimm suggested that once one of the Evil Quartet comes into play, other members of the quartet can compound the effects of the first, take over as the principal cause of extinction, or enable other members of the quartet to amplify the extinction spasm. Diamond's Evil Quartet is discussed in Chapters 7 to 11.

Diamond's Evil Quartet is comprised of four mechanisms that are responsible for most modern extinctions. These mechanisms are

1. Overexploitation
2. Habitat fragmentation and destruction
3. Impacts of introduced species
4. Chains of extinctions

Most modern wildlife issues occur as a direct result of these mechanisms. Humans cause these mechanisms to occur, directly or indirectly, through their actions.

Today, the Endangered Species Act of 1973 (Chapter 2) is supposed to prevent the extinction of species. When bison were nearly exterminated in the United States before the turn of the 20th century, men armed with rifles actually tried to hunt down and kill the last few remaining bison in Yellowstone National Park. Although laws were in place to protect wildlife in the national parks, the dubious honor of killing the last bison was a coveted prize. Fortunately, the prize was denied, but laws were not an effective deterrent.

We must face the fact that no number of laws is enough to protect species from extinction. In 1996, the political agenda of the Republican party included not permitting any further additions to the federal endangered species list. The idea, apparently, was that no species could become endangered and thus be entitled to special protection. Species could then pass from being unprotected to extinct without the aid of protection programs. But can laws really protect the monarch butterfly that winters in Mexico and travels through the United States and into Canada? Simple human laws are not enough to protect wildlife. More often than not, the passage of laws is a knee-jerk reaction that serves to call attention to more fundamental problems. This is not to say that laws are bad; however, laws most often deal with symptoms rather than causes of the Evil Quartet.

Sometimes species are too abundant and thus become a wildlife issue. Surely, commonness precludes extinction for some length of time at least. However, when one species becomes too common, other species can decline as a result. One must look beyond the obvious to understand the profound consequences of species mismanagement. Too many deer, too many raccoons, too many blue jays, and too many brown-headed cowbirds mean too few songbirds. Thus, one issue connects with another in Diamond's last mechanism, chains of extinction. While some people in New Jersey think there are too many deer, a more serious result of this is that songbirds once found in the understory of forests are disappearing because of habitat loss caused by an overabundance of deer. Deer have increased in abundance because humans have caused fragmented habitat, putting into play the second member of the Evil Quartet. Thus, while some people focus on the obvious—that there are too many deer in certain regions—another more insidious quartet mechanism is acting to destroy songbird populations. This is why looking beyond the issue at hand is profoundly important.

Throughout this book, Diamond's Evil Quartet will be used as the common theme that ties together all wildlife issues. We will try to identify the mechanism that gives rise to the wildlife issue. Other mechanisms might come into play as well, acting perhaps to sustain the issue. Rather than deal with the wildlife issue directly (i.e., the symptom), we seek to identify the cause or causes of the issue, that is, the mechanism or mechanisms (one or more of the Evil Quartet) responsible for the issue in the first place. By addressing causes, our goal is to prevent wildlife from becoming an issue. We will take a closer look at examples of Diamond's Evil Quartet in Chapter 7.

USING THE WORLD WIDE WEB

Exploring for information on the World Wide Web is both interesting and informative—as well as addictive. Many people are familiar with electronic mail, or e-mail. The days of using surface mail to send information are rapidly coming to an end. Certainly solid objects must be sent from one point to another using a medium other than electrons. We cannot yet beam anything anywhere. However, when only information needs to be communicated, e-mail, and not so-called snail mail, is the method of choice. Sending information and receiving information sent from someone else via the Internet have become commonplace. The next logical step on the Information Superhighway is information proliferation. Information made available by users of the Internet for users of the Internet is as accessible as your patience will allow.

Many users of the Internet have what is called a World Wide Web address, or Web address for short. Using one of the popular Internet browsing tools such as Excite, Lycos, Explorer, or Netscape, a user can access anyone's Web pages—worldwide and at the speed that information can travel the Internet. Thus, an online user anywhere on earth can access information from anywhere else on earth nearly instantly. Information available on the Web is generally put there by people who want the information to be disseminated. What is impressive is the amount of information on the Web. Simply put, the amount of freely accessible information on the Web is staggering, and information continues to be added daily. Web sites get updated and new Web sites are added by the minute across the world. There is almost no limit to what kind of and how much information can exist on the Web. Each and every computer user could potentially have hundreds of megabytes of information on his or her personal Web site just waiting to be explored by an unknown user on another continent. No one really knows where all this will lead us, except that the changes that are occurring are occurring more frequently and more rapidly than ever before. Today is the age of information proliferation. Free information via the Internet is brought to you virtually anywhere you are and customized to make it easy to use. The only problem is finding the information you want. Information mining is an acquired skill. The more time you spend, the more skilled you will become.

The Internet browsing tools display both text and graphics with ease. Many Web site owners have customized their Web pages to use graphics, text, and sound, and some are interactive. A Web page is created or written in HyperText Markup Language (HTML) and has a Uniform Resource Locator (URL) address. HyperText Transfer Protocol (HTTP) is the set of rules that computers use to send HTML documents over the Internet. Thus, the University of Florida has a home page located at the Internet address

http://www.ufl.edu

When a home page is displayed, there are usually places where the user can point to select new addresses to access. These addresses are linked via the Internet and are not usually on the same storage device. Thus, one can think of the Web quite literally as an electronic, interconnected information network. The usual way of presenting choices is in a menu format that lists what is available and allows the user to select from various choices. The choices can be other addresses at other sites. Thus, the Web is highly distributed geographically.

The World Wide Web is the fruition of a project begun in 1990 at CERN, the European Particle Physics Laboratory in Switzerland. HTML and HTTP protocols, conventions, and standards are defined and supported there. CERN's home page can be found at

http://info.cern.ch/hypertext/WWW/TheProject.html

Access to information on the World Wide Web is gained by using one of the popular Web browsers. Mosaic was developed at the National Center for

Supercomputing Applications at the University of Illinois. Netscape Communications Corporation, founded by several members of the original Mosaic team, also produces a popular Web browser called Netscape. Presuming you have access to the Web through one of the tools, typing Netscape brings up a home page. Netscape offers an array of search engines to help you locate and access information on the Web. For instance, there are at least a dozen search engines that do keyword searches. Alta Vista is a search engine produced by Digital Equipment Corporation that is particularly effective, but others are also useful and generally find different Web addresses for the same keyword.

Search engines such as Alta Vista create an index using a software "spider" to visit all Web sites. They run night and day. The Alta Vista spider visits 2.5 million pages a day. The index in April 1996 was 30 gigabytes. There were an estimated 21 million pages and 10 billion words. The Alta Vista search engine was handling two million search requests a day back in December 1995. In February 1996, it was processing four million a day.

The amount and quality of information on the World Wide Web vary enormously. Most information on the Web is put forth in good faith. However, misinformation can also be found on the Web. For instance, in the summer of 1996, one of the authors (JGS) planned a visit to eastern Democratic Republic of the Congo (DRC, formerly Zaire) to investigate an isolated group of gorillas living on a remote mountaintop. Using the Internet browser Netscape and the search engine Alta Vista, he typed in the name of the mountain and did a keyword search worldwide. In a few minutes, up came the name of the mountain in the DRC that was to be explored. An Australian had written an article about the region, including the mountain, and put his article on the Web for all to see. He reported that the forest surrounding the mountain had been completely destroyed, and all wildlife on the mountain had been eaten by Rwandan refugees spilling into the DRC from their war-torn country. This was blatant misinformation. Thus, suffice it to say that all information should be cross-checked for authenticity and accuracy.

One way to do a search is to bring up a Web browser such as Netscape, invoke a search engine such as Alta Vista, and do a keyword search. Then you briefly explore the list of the top ten keyword matches. At the bottom of the page are usually many more listings that match some or all of your keywords. For instance, suppose we type in the two words *harp seal* and see what comes up. Information on the musical instrument, the harp, is interesting. Also, the state seal of Alabama might be of interest, but it is not really a wildlife issue. Further exploration reveals information about harp seals. We now begin exploring the information network, descending down fruitful paths and blind alleys. The *back-up* feature allows you to drop "bread crumbs" along the path so you can back out gracefully if necessary. Another useful tool is the *bookmark* feature, which allows storage of important addresses that survive after you log off.

Printing and copying whole Web pages are also easy. Indeed, one way to create a Web page is to find one you like on the World Wide Web, copy the entire page, and then modify it to fit your needs. Pictures can be scanned in and stored in a number of different formats. Your images will then be accessible by any popular Web browser when you link them through your page. Thus, creating your own Web page is relatively easy. Some Web pages contain copyrighted information, and these copyrights should be respected. By protecting the rights of others to own information, your own right to own will be protected.

The popularity of the Web is astounding, and no one can say where it will lead. One thing we can say is that there is a universe of information available on the Web. Combined with a local library, the problem we must deal with today is not a lack of information but a veritable flood of information. Managing the information, making sense of it, and summarizing it from disparate sources require persistence and patience.

You are never alone on the Web. According to a 1995 survey by the Georgia Institute of Technology, there were an estimated 20 to 40 million users of the Web. The number of users was growing at the rate of 10% per month. The average age of all users was about 35. The previous survey showed that the average user was 28. Of these users, 15.5% were female and 82% were male; the rest did not report their sex. Of U.S. users, 36% had a college degree, 18.8% had a master's degree, and 4.06% had a doctoral degree. The average user's income was $69,000 a year. The survey found that 50.3% of the users were married and 45.7% were single; the rest did not report. Fully 31.3% of the users on the Internet had used the Internet for less than 6 months, and 19.0% had used the Internet for 6 to 12 months. An amazing 70% browse the Web once a day, and 50% spend two to six hours a week on the Web.

What all of this means is that the popularity of the Web is growing dramatically. New users are older, which means that most young people are already users. The average user is well paid and highly educated. Users are obviously gaining new information from the Web or they would not spend so much time browsing. In short, this means that if you are not "virtual," you are missing out on an adventure in learning. We will use the Web to explore information space just as we would use a library to explore printed information. Hopefully, you will realize that you, too, should add to the information on the Web. Not only will we synthesize information, but we will create new information. New information should find its place on the Web. One night, a spider will crawl your Web site and explore your information and assimilate it. Across the globe, someone, somewhere, for whatever reason, will run a search and your site may well show up. No one knows where information is taking us. Changes are occurring faster and across a broader horizon. All we can do is surf high on the information wave and ride it for what it is worth. The alternative is to be left behind in an information desert.

EXERCISES

1.1 On April 5, 1996, the *Tampa Tribune* reported that a tourist attraction in Florida had offered "to adopt a huge rogue sea lion and four of his buddies who have been condemned to die for decimating the salmon population in the state of Washington while the fish try to make it to spawning grounds." Comment on this statement and suggest how we, as scientists, can contribute to understanding the issues at stake.

1.2 Practice using several search engines on the World Wide Web to locate some interesting Web sites. Start by typing in "crested caracara" and see what comes up. Next try "gorilla." Explain, in your own words, how to use the Web.

1.3 Using the World Wide Web, search for the phrase "Wildlife Services" and compile a list of interesting sites. What animals are becoming a nuisance? What makes an animal a nuisance?

1.4 Stamp collectors often assemble interesting stamps from across the world. One of the authors collects bird stamps, for instance. Using the World Wide Web, locate a source for wildlife stamps.

1.5 Repeat the previous exercise for coins. Make a list of coins that have wildlife on them.

LITERATURE CITED

Dayton, P.K. 1998. Reversal of the burden of proof in fisheries management. *Science* **279**:821–822.

Diamond, J.M. 1984. "Normal" extinctions of isolated populations. pp. 191–246 *in* Nitecki, M.H. (Ed.). *Extinctions.* University of Chicago Press, Chicago.

Lakatos, Imre. 1978. Falsification and the methodology of scientific research programmes. *in* Worrall, J. and G. Currie (Eds.). *The Methodology of Scientific Research Programmes,* Volume I. Cambridge University Press, Cambridge.

Pimm, S.L. 1996. Lessons from the kill. *Biodiversity and Conservation* **5**:1059–1067.

WILDLIFE LAWS   2

Wildlife issues affect all of us nearly every day. Not a day passes without some wildlife issue confronting us. Perhaps it is a road-killed squirrel that we drive past and hardly notice or the endangered bald eagle we see circling above in the afternoon. The small stream we cross during our evening walk may have a locally endemic fish in it. The nest box we put out for birds attracts raccoons. The hamburger we have at a fast-food restaurant may have come from beef cattle raised in the American West on public land. The fuel we use in our vehicle may have come from oil pumped from a Papua New Guinea rain forest. The fact is, wildlife issues surround us. Sometimes we simply ignore them. Other times we are gripped by them, such as when three whales were stranded in ice-bound waters of the Arctic a few years ago.

CHRONOLOGY OF IMPORTANT LAWS

To understand wildlife laws, it is necessary to understand the recent history of how and why those laws came into being. A chronology of historic developments can be found on the Web site of the U.S. Fish and Wildlife Service. Beginning in 1900 with the Lacey Act, the chronology is a fascinating look at the history of wildlife law because these laws were passed by Congress in response to critical wildlife needs. For instance, in 1935 it was recognized that foreign commerce in illegally taken wildlife was becoming a serious threat to wildlife populations in the United States. The Lacey Act was then expanded to include a prohibition against such commerce. A shortened chronology of the more important wildlife laws is included here.

1900—The Lacey Act was enacted as the first federal law protecting game. It prohibited the interstate shipment of illegally taken wildlife as well as the importation of injurious species (i.e., those species that might cause humans harm).

1913—The Federal Migratory Bird Law became effective and the first migratory bird hunting regulations were adopted on October 1.

1913—The United States signed the Migratory Bird Treaty with Great Britain (for Canada), recognizing migratory birds as an international resource.

1918—The Migratory Bird Treaty Act became law, making it unlawful to take, possess, buy, sell, purchase, or barter any migratory bird, including feathers, parts, nests, or eggs.

1926—The Black Bass Act became law, making it illegal to transport across state boundaries for the purpose of commerce black bass taken, purchased, or sold in violation of state law.

1929—The Migratory Bird Conservation Act of 1929 authorized the National Wildlife Refuge System.

1934—The Migratory Bird Hunting Stamp Act became law, requiring all waterfowl hunters age 16 and over to possess a "Duck Stamp."

1935—The Lacey Act was expanded to prohibit foreign commerce in illegally taken wildlife.

1936—The United States signed the Migratory Bird Treaty with Mexico.

1937—The Pittman–Robertson Act established a means to restore wildlife populations to the United States.

1940—The Bald Eagle Protection Act became law. The act prohibited a variety of activities involving the species, making it illegal to import, export, take, sell, purchase, or barter.

1962—The Bald Eagle Protection Act became the Bald and Golden Eagle Protection Act and extended protection to golden eagles.

1970—The Endangered Species Conservation Act of 1969 became effective, prohibiting the importation of into the United States species "threatened with extinction worldwide," except as specifically allowed for zoological and scientific purposes and propagation in captivity. The act amended the Black Bass Act to prohibit interstate and foreign commerce in fish taken in violation of foreign law, a provision that the Lacey Act had made in 1935 for wildlife. It also amended the Lacey Act so that its prohibition on interstate and foreign commerce applied not only to wild birds and mammals but to reptiles, mollusks, amphibians, and crustaceans. This amendment was made in an effort aimed primarily at protecting the American alligator. (There is now an alligator hunting season in Florida.)

1971—The Airborne Hunting Act was signed into law, prohibiting the use of aircraft to hunt or harass wildlife.

1972—The United States signed the Migratory Bird Treaty with Japan. The Migratory Bird Treaty with Mexico was amended to protect additional species, including birds of prey.

The Marine Mammal Protection Act of 1972 became law, establishing a moratorium on the taking and importing of marine mammals, such as polar bears, sea otters, dugongs, walrus, manatees, whales, porpoise, seals, and sea lions.

The Eagle Protection Act was amended to increase penalties from $500 or six months imprisonment to $5000 or one year and to add the provision that a second conviction was punishable by a $10,000 fine or two years imprisonment, or both. In addition, the amendment allowed for informants to be rewarded one-half of the fine, not to exceed $2500.

1973—The Endangered Species Act of 1973 became law, recognizing that "endangered species of wildlife and plants are of aesthetic, ecological, educational, historical, recreational, and scientific value to the Nation and its people." The act expanded the scope of prohibited activities to include not only importation but exportation, taking, possession, and other activities involving illegally acquired species, as well as interstate or foreign commercial activities. It implemented protection for a new "threatened" category—species likely to become in danger of extinction.

1975—The Convention on International Trade in Endangered Species of Wild Fauna and Flora (CITES) entered into force, regulating the importation, exportation, and reexportation of species listed in its three appendices (Hemley 1994).

1976—The United States signed the Migratory Bird Treaty with the Union of Soviet Socialist Republics.

1981—The Black Bass and Lacey Acts were repealed and replaced by the Lacey Act Amendments of 1981. A comprehensive statute, the Lacey Act

Amendments, restored protection for migratory birds, removed from the act in 1969, and initiated protection for plants. The Lacey Act Amendments increased penalties and included a felony punishment scheme to target commercial violators and international traffickers, with fines of up to $20,000 or five years imprisonment, or both.

1982—The Endangered Species Act was amended to include a plant-taking prohibition on federal lands and a new exception allowing the inadvertent noncommercial transshipment through the United States of endangered fish or wildlife.

1988—The African Elephant Conservation Act became law, providing additional protection for the species, whose numbers had declined by 50% in the last decade. The Lacey Act was amended to include, among other things, felony provisions for commercial guiding violations.

1992—The Wild Bird Conservation Act of 1992 was signed into law to address problems with international trade of wild-caught birds, which contributes both to the decline of the species and to unacceptably high mortality rates.

According to Lund (1980), American wildlife policy from the colonial period to the turn of the 20th century was based on "taking" as its goal. Steep declines in wildlife from the Northeast through Florida had occurred. When the Pilgrims arrived, they were amazed at the abundance of deer and turkey. Yet by 1900, deer had disappeared from the East. Turkey and beaver were hunted out. Those who traveled by rail thousands of miles west were "never out of sight of a dead buffalo, and never in sight of a live one" (Matthiessen 1964). The passenger pigeon and the Eskimo curlew, whose numbers staggered the imagination, were hunted down relentlessly and wiped out forever.

Commercial exploitation 100 years ago decimated many species. Millions of buffalo were shot just so their tongues could be shipped back east in new refrigerated railroad cars. Elk were common across the Great Plains of America, although they had been driven west. The last wild elk in New York was killed in 1845 (Day 1949). Fifteen years later, the last moose was gone from New York as well. They were soon killed out, and what was left retreated to the mountains. Today, we think of elk as occupying the Rocky Mountains and other similar regions. Humans drove them there. Men's hats were made of beaver skins, and ladies' fashions demanded plumes from millions of Florida's wading birds. In fact, wading birds were killed during the breeding season for a few feathers from the backs of the birds. The carcasses were left to rot. Spring shooting of waterfowl in Rhode Island was outlawed in 1846. Even as early as

1864, Idaho had a closed season on buffalo. Enforcement was impossible, however. Waterfowl were especially hard hit. Bag limits were nonexistent. At the same time, the U.S. population was growing—from 17 million in 1840 to 23 million in 1850 and 32 million in 1860. Imagine the view seen by Colonel R.I. Dodge along the Arkansas River in Colorado back in 1871, when his patrol, over the course of six days and nights, was among a herd of buffalo he estimated "contained no fewer than 4 million head" (Trefethen 1975). "At times they pressed before us in such numbers as to delay the progress of our column, and often a belligerent bull would lower and shake his craggy head at us when we passed him a few feet distant."

In 1865, the national buffalo kill was about one million head. By 1875, Kansas and Colorado closed buffalo hunting. There were none left anyway. In 1881, buffalo hunters had, quite literally, killed themselves out of business. By 1890, a small herd of mountain buffalo were the sole survivors of the tens of millions of buffalo that had roamed the great grasslands of the United States. Perhaps the saddest chapter in U.S. history was closed—a carnage Americans must accept as part of their history.

Market hunters used cannons to kill thousands of ducks—at night. Sporting games were held to see who could kill the most ducks with a single shot—81 dead, 46 crippled—from a double-barreled, four-gauge shotgun—a mighty stroke for humans (Wheeler 1947). Shooting 50 or more at once with this gun was easy. The carnage was massive and largely out of sight. No species was immune. The speed of the pronghorn antelope was no match for a bullet at 400 yards.

The belief was that wildlife resources were unlimited and the harvests could continue forever. They did not. Wildlife populations fell and species became extinct. Toward the turn of the century, sportsmen began to question commercial harvesting of wildlife resources. Indeed, it was sport hunting that pushed through the founding of wildlife agencies and specified how these agencies would be funded. Laws were passed to improve wildlife populations. The demand for sport hunting rescued wildlife across America from total destruction. Predators, however, would have to wait for their saviors. Predator control programs were designed to wipe out predators completely, and they did. Full-time government hunters were assigned the job of hunting down and shooting predators or poisoning them.

In summary, early America wildlife laws permitted the cruelest form of complete exploitation. State laws were formulated by sportsmen to protect certain species considered game animals. At the turn of the century, American wildlife stocks were at their lowest level. Federal wildlife laws were first established to work cooperatively with the states to support the sport laws. The aesthetic value of wildlife was still not recognized. Today, we realize that many wildlife themes interact in the United States as well as other parts of the world. Butterflies attract not hunters but rather collectors. Global commerce is accepted as commonplace. Wildlife laws must be agreed upon among nations. U.S. trade sanctions against China for illegally importing tiger parts from India could be used as a corrective measure. Today, wildlife issues take place on the global stage for all to see. While residents of northern Florida reject as outlandish the reintroduction of the secretive panther where it once was common, donors from across the United States support research into why lion populations in Africa are crashing. Predators are okay, but only in someone else's backyard it seems.

DEFINITIONS

Today, there are many laws that protect all kinds of wildlife. A summary is provided later in this chapter. Indeed, there are so many laws that protect wildlife that it is surprising there are any wildlife issues at all. And yet, wildlife issues are surfacing more and more rapidly. There are laws that govern our air and water quality, laws that govern policy in our national forests, laws that protect elephants and migratory birds, and laws that protect threatened and endangered species. There is a law protecting Antarctica and one that prevents shooting animals from an airplane. To understand these laws and where they come from, several terms need to be defined.

- **Misdemeanor**—An offense of lesser gravity than a felony, for which punishment may be a fine or imprisonment in a local rather than a state institution.
- **Felony**—Any of several crimes such as murder, rape, or burglary considered more serious than a misdemeanor and punishable by a more stringent sentence.
- **Civil law**—The body of law dealing with the rights of private citizens in a particular state or nation as distinguished from criminal law, military law, and international law.
- **Criminal law**—Pertaining to the administration of penal law as distinguished from civil law.
- **Standing**—To have standing means that one is a "proper party to request adjudication" of a particular issue (Littell 1992).

Wildlife law has four main goals according to Lund (1980). First and foremost to our predecessors, laws facilitated the sustained harvest of wildlife.

Wildlife management became a bean-counter's paradise. Second, wildlife laws can be formulated to control human behavior. For instance, the use of weapons could be controlled through wildlife laws. Third, special groups can be favored by wildlife laws. Wildlife could be treated as a form of wealth so that special groups would be favored for protecting or providing habitat for wildlife. Lastly, wildlife laws protect wildlife.

Bounties have always played an interesting role in wildlife laws because a bounty is a reward for the hunter but, on the other hand, a cost to the government in most cases. The government would have preferred to pay nothing to the hunter for services rendered. Thus, the bounty hunter's wits were matched against the government as lawmaker and bounty payer. Early on in U.S. history, bounty hunters aided farmers and settlers.

By placing a bounty on an animal's head, government was encouraging hunters to eliminate the animal. Hunters, however, would have preferred a more sustained yield approach to collecting bounties. Thus, when a hunter could ensure exclusivity to an area containing wolves, he would take only young wolves and male wolves, fully realizing that the females could provide a sustained bounty income. To ensure the killing of breeding females, bounties had to be increased.

Different counties sometimes paid different bounties. Hunters then shot wolves in one place and transported them to the county that paid the most. The local legislature moved to require that animals be killed in the county. Hunters acted swiftly by catching animals, transporting them to the appropriate county, and killing them locally. Long ago, New York City developed a thriving trade in wolves' ears when the government required proof of a wolf kill but did not want the body delivered. For panthers and foxes, however, the body had to be delivered. Stale wolf heads were always suspect because hunters would ship them to the highest payoff areas; thus, local paymasters were required to make sure the heads were "green."

At first, the highest bounties were paid on panthers. Wolves were more valuable than bears and wildcats, such as bobcats. In Virginia, the bounty for wolves and panthers was a license to kill a hog. Thus, very few dead panthers and wolves were recorded in Virginia; however, neighboring states had seemingly enormous populations. Ethical or aesthetic wildlife laws had not even been dreamed of 200 years ago. No one would have been able to foresee that in a little under 150 years, North America would require laws to protect wildlife.

U.S. customs law prohibits the importation of most wild bird feathers, mounted birds, and skins. Most migratory birds are protected by international treaties. Just because something is for sale in another country does not make it legal for a U.S. citizen to purchase it. Many live birds and animals can be imported but must be quarantined in U.S. Department of Agriculture Animal Import Centers or quarantine stations for 30 days after entry.

Taking is a specific term used in laws regarding plants and animals. Taking refers to harassing, harming, pursuing, hunting, shooting, wounding, trapping, killing, capturing, and collecting species of plants or animals. So-called "take" permits are available from the federal government. For instance, a large wind-mill electricity generation project in Wyoming was required to apply for a "take" permit when researchers discovered that the location of the windmills was also a flyway for migratory birds, especially birds of prey such as eagles. Thus, the project was required to file for a "take" permit, requesting permission to "take" no more than a certain number of bald and golden eagles.

Injurious wildlife is wildlife that can cause harm. Animals such as fruit bats, mongooses, walking catfish, and Java sparrows cannot be imported into the United States because they have been declared harmful to people, animals, plants, or the environment and are thus injurious. These creatures need not be rare or endangered. Some states even have laws preventing the introduction of nonnative species.

Wildlife trophies other than marine mammals or threatened, endangered, or CITES Appendix I species lawfully taken by U.S. residents in the United States, Canada, or Mexico can be imported or exported for personal use.

UNITED STATES CODE

The United States Code (U.S.C.) is comprised of 50 titles that contain the laws by which the United States functions. Some of these titles are as follows:

> **Title 1 General Provisions**
> **Title 2 The Congress**
> **Title 3 The President**
> **Title 4 Flag, Seal, Seat of Government, and the States**
> ...
> **Title 16 Conservation**
> ...
> **Title 18 Crimes and Criminal Procedure**
> ...
> **Title 27 Intoxicating Liquors**
> ...
> **Title 50 War and National Defense**

Title 16 contains laws referring to conservation:

INTERNATIONAL CONVENTIONS AND AGREEMENTS

In addition to the U.S. Code, there are also many treaties and international agreements. Many of the above laws refer to these international agreements and specify that the U.S. government will provide "assistance in the development and management of programs in that country which the Secretary [of the Interior] determines to be necessary or useful for the conservation of any endangered or threatened species." In other words, the U.S. government will actively seek partnerships in encouraging the protection of certain species beyond the borders of the United States. Some of these treaties and agreements as are follows:

- Agreement on the Conservation of Polar Bears of 1973
- Antarctic Treaty of 1959
- Convention Concerning the Conservation of Migratory Birds and Their Environment of 1976
- Convention Concerning the Protection of the World Cultural and Natural Heritage of 1972
- Convention for the Conservation of Antarctic Seals of 1972
- Convention for the Establishment of an Inter-American Tropical Tuna Commission of 1949
- Convention for the Preservation of the Halibut Fishery of the Northwest Pacific Ocean and Bering Sea of 1953
- Convention for the Protection of Migratory Birds of 1916
- Convention for the Protection of Migratory Birds and Birds in Danger of Extinction, and Their Environment of 1972
- Convention for the Protection of Migratory Birds and Game Mammals of 1936
- Convention for the Protection, Preservation, and Extension of the Sockeye Salmon of the Fraser River System of 1930
- Convention for the Regulation of Whaling of 1931
- Convention on Fishing and Conservation of the Living Resources of the High Seas of 1958
- Convention on International Trade in Endangered Species of Wild Fauna and Flora of 1973 (CITES)
- Convention on Nature Protection and Wildlife Preservation in the Western Hemisphere of 1940
- Convention on the Conservation of Antarctic Marine Living Resources of 1980
- Convention on the Territorial Sea and Contiguous Zone of 1958

- Declaration of the United Nations Conference on the Human Environment of 1972
- Interim Convention on Conservation of North Pacific Fur Seals of 1957
- International Convention for the High Seas Fisheries of the North Pacific Ocean of 1952
- International Convention for the Northwest Atlantic Fisheries of 1949
- International Convention for the Regulation of Whaling of 1946
- Treaty for the Preservation and Protection of Fur Seals of 1911
- Treaty with the Yakimas of 1855

The Convention on International Trade in Endangered Species of Wild Fauna and Flora of 1973 (CITES) is one of the most famous international conventions. Many U.S. laws refer to CITES and pledge the support of the United States to support CITES-listed species through international cooperation.

Convention on International Trade in Endangered Species of Wild Fauna and Flora

A conference in Washington, D.C. in early 1973 led to the signing of the Convention on International Trade in Endangered Species of Wild Fauna and Flora on March 3, 1973. CITES is a multinational agreement among 132 nations (as of January 1998) that attempts to regulate trade in specific plants and animals to prevent their excessive exploitation. Scientific authorities in each country are designated to issue permits to control the taking of listed species. National laws also protect CITES-listed species. In the United States, the African Elephant Conservation Act, the Eagle Protection Act, the Endangered Species Act, the Lacey Act, the Marine Mammal Protection Act, and the Migratory Bird Treaty Act are domestic laws that cover many species, including, but not limited to, CITES-listed species.

CITES listings include some 30,000 plants and over 2500 animal species. Listed in Appendix I of CITES are species threatened with extinction that are or might one day be affected by trade. Export and import permits are required for CITES-listed species. Appendix II of CITES includes species that are not threatened with extinction yet, but soon might be if trade in those species is not controlled. For instance, animal parts are often used as medicinal treatments or as aphrodisiacs. Ground-up antlers are used to treat certain illnesses. American black bears are listed in Appendix II because their gallbladders look like any other bear gallbladder and Asians use bear gallbladders for medicinal purposes.

Appendix III contains a list of species that individual nations wish to protect within their borders. To prevent international trade in those species, a certificate of origin is required.

CITES is an agreement among nations (a list of signatories to CITES is provided in Appendix 2.1) and is not a law. Therefore, no fines or jail sentences apply. However, with proper certification, animals and plants can be confiscated and repatriated. Section 1532 Definitions in paragraph (4) refers specifically to the term "Convention," which means the Convention on International Trade in Endangered Species of Wild Fauna and Flora. Section 1537 of the Endangered Species Act commits the United States to support CITES financially and encourages the secretary of the interior to assist other countries in support of their species protection programs.

A CITES permit (Figure 2.1, right) allows the holder to take specimens of CITES-listed species across country borders. The permit is granted by the appropriate government authorities and carries a special stamp. Often an export permit is also required to remove the species from the country of origin (Figure 2.1, left). The original export and CITES permits are taken by the U.S. Fish and Wildlife Service upon entry into the United States after the specimen is examined at customs. Failure to produce a legal CITES permit results in confiscation of the specimen and a possible fine and jail sentence.

EXAMPLES OF U.S. CODES

To better understand U.S. wildlife codes, let's take a closer look at some of the acts. For instance, Congress decided to do something about protecting coral reefs and enacted 16 U.S.C., Chapter 25A—Crown of Thorns Starfish. This will help us understand what Congress does when the American people decide to pass a law to protect a species or to protect a species from other species. Unless otherwise noted, all acts are 16 U.S.C.

Chapter 25A. Crown of Thorns Starfish

§ 1211. Congressional Statement of Purpose
§ 1212. Investigation and Control of Crown of Thorns Starfish
§ 1213. Authorization of Appropriations

§ 1211. Congressional Statement of Purpose

For the purpose of conserving and protecting coral reef resources of the tropical islands of interest and concern to the United States in

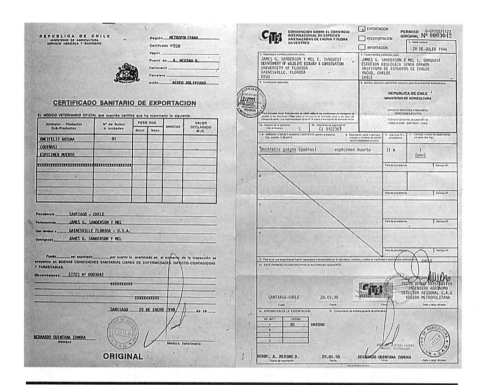

FIGURE 2.1 An export permit (left) and a CITES permit (right) permitting one dead *Oncifelis guigna* to be reexported (exported and then imported back) from Chile to the United States and then to Chile. The guigna, a small spotted cat, is listed in Appendix II of CITES, as are all spotted cats not listed in Appendix I. Note that Dr. Mel Sunquist's last name has been inadvertently omitted from the reexport permit.

the Pacific and safeguarding critical island areas from possible erosion and to safeguard future recreational and esthetic uses of Pacific coral reefs, the Secretary of Commerce and the Secretary of the Smithsonian Institution are authorized to cooperate with and provide assistance to the governments of the State of Hawaii, the territories and possessions of the United States, including Guam and American Samoa, the Trust Territory of the Pacific Islands, and other island possessions of the United States, in the study and control of the seastar "Crown of Thorns" (*Acanthaster planci*).

§ 1212. Investigation and Control of Crown of Thorns Starfish

In carrying out the purposes of this chapter, the Secretary of Commerce and the Secretary of the Smithsonian Institution are authorized—

(1) to conduct such studies, research, and investigations, as they deem desirable to determine the causes of the population increase of the Crown of Thorns, their effects on corals and coral reefs, and the stability and regeneration of reefs following predation;

(2) to monitor areas where the Crown of Thorns may be increasing in numbers and to determine future needs for control;

(3) to develop improved methods of control and to carry out programs of control in areas where these are deemed necessary; and

(4) to take such other actions as deemed desirable to gain an understanding of the ecology and control of the seastar "Crown of Thorns."

§ 1213. Authorization of Appropriations

For the purpose of carrying out the provisions of this chapter, there is authorized to be appropriated for the period commencing on September 26, 1970, and ending June 30, 1975, an amount not to exceed $4,500,000.

Unlike the crown of thorns, which Congress was trying to control, African elephant populations were being decimated all across Africa in the 1980s when Congress decided to act. The African Elephant Conservation Act of 1988 was passed to help protect and conserve African elephants. The act makes it illegal to import raw ivory such as elephant tusks from any country other than an African country that contains any part of the range of African elephants. If an African country prohibits exportation of raw or worked ivory, then importation of that ivory into the United States is prohibited. Exporting African ivory from the United States is also prohibited. If the ivory is derived from illegal sources, importation of the ivory is illegal, even if the country allows exportation. However, on June 9, 1989, the United States established a moratorium on all ivory imports from all countries, including Asian countries, except where elephant trophy hunting is permitted and also for certain ivory antiques.

Chapter 62. African Elephant Conservation Act of 1988

§ 4201. Statement of Purpose
§ 4202. Findings
§ 4203. Statement of Policy

§ 4201. Statement of Purpose

The purpose of this chapter is to perpetuate healthy populations of African elephants. The Congress finds the following:

(1) Elephant populations in Africa have declined at an alarming rate since the mid-1970's.

(2) The large illegal trade in African elephant ivory is the major cause of this decline and threatens the continued existence of the African elephant.

(3) The African elephant is listed as threatened under the Endangered Species Act of 1973 (16 U.S.C. 1531 et seq.) and its continued existence will be further jeopardized if this decline is not reversed.

(4) Because African elephant ivory is indistinguishable from Asian elephant ivory, there is a need to ensure that the trade in African elephant ivory does not further endanger the Asian elephant, which is listed as endangered under section 4 of the Endangered Species Act of 1973 (16 U.S.C. 1533) and under Appendix I of CITES.

(5) In response to the significant illegal trade in African elephant ivory, the parties to CITES established the CITES Ivory Control System to curtail the illegal trade and to encourage African countries to manage, conserve, and protect their African elephant populations.

(6) The CITES Ivory Control System entered into force recently and should be allowed to continue in force for a reasonable period of time to assess its effectiveness in curtailing the illegal trade in African elephant ivory.

(7) Although some African countries have effective African elephant conservation programs, many do not have sufficient resources to properly manage, conserve, and protect their elephant populations.

(8) The United States, as a party to CITES and a large market for worked ivory, shares responsibility for supporting and implementing measures to stop the illegal trade in African elephant ivory and to provide for the conservation of the African elephant.

(9) There is no evidence that sport hunting is part of the poaching that contributes to the illegal trade in African elephant ivory, and there is evidence that the proper utilization of well-managed elephant populations provides an important source of funding for African elephant conservation programs.

§ 4203. Statement of Policy

It is the policy of the United States—

(1) to assist in the conservation and protection of the African elephant by supporting the conservation programs of African countries and the CITES Secretariat; and

(2) to provide financial resources for those programs.

§ 4211. Provision of Assistance

(a) In general—The Secretary may provide financial assistance under this subchapter from the African Elephant Conservation Fund for approved projects for research, conservation, management, or protection of African elephants.

(b) Project proposal—Any African government agency responsible for African elephant conservation and protection, the CITES Secretariat, and any organization or individual with experience in African elephant conservation may submit to the Secretary a project proposal under this section. Each such proposal shall contain—

(1) the name of the person responsible for conducting the project;

(2) a succinct statement of the need for and purposes of the project;

(3) a description of the qualifications of the individuals who will be conducting the project;

(4) an estimate of the funds and time required to complete the project;

(5) evidence of support of the project by governmental entities of countries within which the project will be conducted, if such support may be important for the success of the project; and

(6) any other information the Secretary considers to be necessary or appropriate for evaluating the eligibility of the project for funding under this chapter.

(c) Project review and approval—The Secretary shall review each project proposal to determine if it meets the criteria set forth in subsection (d) of this section and otherwise merits assistance under this chapter. Not later than six months after receiving a project proposal, and subject to the availability of funds, the Secretary shall approve or disapprove the proposal and provide written notification to the person who submitted the proposal and to each country within which the project is proposed to be conducted.

(d) Criteria for approval—The Secretary may approve a project under this section if the project will enhance programs for African elephant research, conservation, management, or protection by—

(1) developing in a usable form sound scientific information on African elephant habitat condition and carrying capacity, total elephant numbers and population trends, or annual reproduction and mortality; or

(2) assisting efforts—

(A) to ensure that any taking of African elephants in the country is effectively controlled and monitored;

(B) to implement conservation programs to provide for healthy, sustainable African elephant populations; or

(C) to enhance compliance with the CITES Ivory Control System.

(e) Project reporting—Each entity that receives assistance under this section shall provide such periodic reports to the Director of the United States Fish and Wildlife Service as the Director considers relevant and appropriate. Each report shall include all information requested by the Director for evaluating the progress and success of the project.

§ 4223. Prohibited Acts

Except as provided in section 4222(e) of this title, it is unlawful for any person—

(1) to import raw ivory from any country other than an ivory producing country;

(2) to export raw ivory from the United States;

(3) to import raw or worked ivory that was exported from an ivory producing country in violation of that country's laws or of the CITES Ivory Control System;

(4) to import worked ivory, other than personal effects, from any country unless that country has certified that such ivory was derived from legal sources; or

(5) to import raw or worked ivory from a country for which a moratorium is in effect under section 4222 of this title.

§ 4224. Penalties and Enforcement

(a) Criminal violations—Whoever knowingly violates section 4223 of this title shall, upon conviction, be fined under title 18 or imprisoned for not more than one year, or both.

(b) Civil violations—Whoever violates section 4223 of this title may be assessed a civil penalty by the Secretary of not more than $5,000 for each such violation.

(c) Procedures for assessment of civil penalty—Proceedings for the assessment of a civil penalty under this section shall be conducted in accordance with the procedures provided for in section 1540(a) of this title.

(d) Use of penalties—Subject to appropriations, penalties collected under this section may be used by the Secretary of the Treasury to pay rewards under section 4225 of this title and, to the extent not used to pay such rewards, shall be deposited by the Secretary of the Treasury into the Fund.

(e) Enforcement—The Secretary, the Secretary of the Treasury, and the Secretary of the department in which the Coast Guard is operating shall enforce this subchapter in the same manner such Secretaries carry out enforcement activities under section 1540(e) of this title. Section 1540(c) of this title shall apply to actions arising under this subchapter.

§ 4225. Rewards

(a) In general—Upon the recommendation of the Secretary, the Secretary of the Treasury may pay a reward to any person who furnishes information which leads to a civil penalty or a criminal conviction under this chapter.

(b) Amount—The amount of a reward under this section shall be equal to not more than one-half of any criminal or civil penalty or fine with respect to which the reward is paid, or $25,000, whichever is less.

(c) Limitation on eligibility—An officer or employee of the United States or of any State or local government who furnishes information or renders service in the performance of his or her official duties shall not be eligible for a reward under this section.

Example—While visiting Tanzania, you notice an attractive wastepaper basket made from an elephant's foot. Your offer to purchase the basket for five dollars is accepted. So generous is your offer that the astounded merchant throws in a tiny hand-carved elephant of poor quality made, he says, from bone. He also throws in an old document that says something about CITES. You pack up your purchases and return home. At the airport, you are one of only a handful of people to have their luggage inspected. The customs official reads the yellowed document and ushers you aside as he points out that the carving is not bone but ivory. You are charged with violating the African Elephant Conservation Act.

Penalties—A misdemeanor carries a $5000 fine for an individual. Felony convictions carry a $100,000 fine and up to one year in jail for individuals. Organizations can be fined up to $200,000.

7 U.S.C. Chapter 54. Transporting, Sale, and Handling of Certain Animals

This is the so-called Animal Welfare Act of 1966 (1970, 1976) which deals with interstate transportation and commercial transactions involving animal species. The act outlaws animal fights, although gamebird fights are outlawed only where state law has precedence.

Chapter 30. Wild Free-Roaming Horses and Burros Act of 1971

President Eisenhower signed the first federal law written to protect wild horses and burros. The Wild Free-Roaming Horses and Burros Act of 1971 and as amended in 1978 provides for the protection from various forms of cruelty and the capture and killing of wild horses which live on federally owned public land (Cooper 1987). Specifically, unbranded and unclaimed horses and burros on public lands of the United States were protected as "living symbols of the historic and pioneer spirit of the West." The act applies to all lands administered by the Forest Service and the Bureau of Land Management.

The act is not limitless, however. It describes the range of the animals as "the amount of land necessary to sustain an existing herd or herds of wild free-roaming horses and burros, which does not exceed their known territorial limits." The act also denies any authority "to relocate wild free-roaming horses or burros to areas of public lands where they do not presently exist." If an overpopulation exists, under the 1978 amendment, the secretary of the interior can "immediately remove excess animals." However, the act also states that protected horses and burros are to "be considered...an integral part of the natural system of the public lands." Since horses and burros are both nonnative introduced species, one has to wonder how a "natural" ecological balance is to be preserved and maintained. Furthermore, it seems equally impossible to maintain a "thriving natural ecological balance" and a "multiple-use relationship," as the act suggests (Bean 1983).

Example—On your vacation to Nevada, you attend a wild horse auction. The horses are up for adoption, and you choose five of them. As you are loading them into your trailer, a sheriff walks up and asks to see your letter from the secretary of the interior that permits you to adopt five horses. You are arrested when you cannot produce the letter. You can, under the act, only adopt four horses.

Penalties—Violators are subject to a $2000 fine and one year in prison.

Chapter 9. Fish and Wildlife Service

In 1971, Congress amended the Fish and Wildlife Act of 1956 by adding the so-called Airborne Hunting Act. Section 742j-1 makes it unlawful to harass or take wildlife from aircraft except when protecting wildlife, livestock, and human health and safety as authorized by a federal- or state-issued license or permit. Permits have been issued in Alaska to allow the shooting of wolves from aircraft to protect other wildlife such as moose and caribou.

Example—You are a passenger in your friend's airplane and are out for a day of leisure flying. The pilot decides to get a closer look at some elk. He circles and flies directly over their heads, scattering them. Unknowingly, your friend has violated the Airborne Hunting Act. It is unlawful for anyone to knowingly participate in using an aircraft to "harass any bird, fish, or animal."

Penalties—Violators are subject to a $5000 fine, one year in prison, or both. In addition, hunters can forfeit their guns and aircraft, as well as the birds, fish, or animals they shoot.

Chapter 44. Antarctic Conservation Act

To protect the unique fauna and flora of Antarctica, this act makes it illegal for any U.S. citizen to carry, deliver, export, import, possess, receive, sell, take, or transport any native wildlife from Antarctica or any plant from specially protected areas in Antarctica. The act also expressly prohibits introduction of any animal or plant not indigenous to Antarctica.

Penalties—The amount of the civil penalty shall not exceed $5000 for each violation unless the prohibited act was knowingly committed, in which case the amount of the civil penalty shall not exceed $10,000 for each violation. A criminal offense is punishable by a fine of $10,000 or imprisonment for not more than one year, or both.

Chapter 5A. Protection and Conservation of Wildlife

Subchapter I. Game, Fur-Bearing Animals, and Fish

This act, which is often referred to as the Lacey Act of 1900 (1948, 1949, 1960, 1969, 1981), ensures that fish and wildlife shipped to the United States receive humane and healthful treatment. It also designates wildlife that is potentially harmful to humans and prevents its importation. Under the Lacey Act, importing, exporting, transporting, selling, receiving, acquiring, or buying any fish, wildlife, or plant that was taken, transported, or possessed illegally under state, federal, tribal, or foreign law is a federal offense. The United States can and does prohibit the export of certain species and can also ban the importation of certain species.

The Lacey Act is often used to prosecute smugglers of illegally taken fish and wildlife, including birds and reptiles. In 1981, amendments were passed that strengthen federal laws and broaden federal participation to states and foreign governments with respect to enforcement.

Example—In an effort to stock your home aquarium, you visit a local pet store that sells tropical fish. Many of the fish are new and different to you, and you purchase several of the more colorful ones. The fish are loaded into a plastic bag and put into a cardboard box for transport. Your receipt says you purchased "tropical guppies." On the way to your car, a U.S. Fish and Wildlife Service agent notices the box and asks you to show off your purchase. The agent quickly identifies the fish as rare and endangered tropical species from Papua New Guinea coastal waters. You are charged with violating the Lacey Act because you have purchased fish that are prohibited from export from Papua New Guinea.

Penalties—Individuals convicted of violating the Lacey Act can be sentenced to up to $100,000 and one year in jail for misdemeanors. Fines of up to $250,000 and five years in jail can be imposed for felony convictions. Businesses can be fined up to $250,000 and $500,000 for misdemeanors and felonies, respectively. Illegal wildlife, fish, and plants as well as vehicles, airplanes, boats, and other equipment used during the crime can be confiscated by the government.

Subchapter II. Protection of Bald and Golden Eagles Act of 1940 (1959, 1962, 1972)

This act was amended to include golden eagles as well and is now known as the Bald and Golden Eagle Protection Act. To import, export, or take bald or golden eagles or to sell, purchase, or barter their parts or products made from them, including their nests or eggs, is illegal. Furthermore, pursuing, shooting, poisoning, wounding, killing, capturing, trapping, collecting, molesting, or disturbing eagles is also illegal.

Exceptions—Permits can be granted for scientific or exhibition use or for traditional and cultural use by Native Americans. For instance, the Hopi Indians of Arizona use golden eagles in ceremonies. No permits can be issued for commercial or barter activities.

Example—In Anchorage, Alaska, bald eagles are ubiquitous. During a visit to a local park, you find a bald eagle tail feather. Apparently, the feather has molted (i.e., it has been replaced naturally by another feather). You stick the feather in your hatband, where it is prominently displayed. A local man offers to purchase the feather for five dollars, and you sell it. The man turns out to be a police officer working undercover and places you under arrest for selling a bald eagle part.

Penalties—Fines of up to $100,000 for individuals and $200,000 for organizations and one year in jail are possible for misdemeanors. Fines of up to

$250,000 and $500,000 for individuals and organizations, respectively, can result from felony convictions.

Pittman–Robertson Act of 1937

The so-called Pittman–Robertson Act of 1937 must be regarded as one of the most important wildlife acts ever enacted. Prior to the 1930s, wildlife was mercilessly exploited. By the turn of the 1900s, there were an estimated 500,000 deer in the entire nation, few bison remained, and waterfowl populations had been reduced to a fraction of their former numbers. Wildlife research was almost nonexistent. The new era began when the U.S. Congress extended the life of an existing 10% tax on ammunition and firearms used for sport hunting and earmarked the proceeds to be distributed to the states for wildlife restoration. The Federal Aid in Wildlife Rehabilitation Act (popularly referred to as the Pittman–Robertson Act) empowered states with a continued source of funding to carry out wildlife research programs. The act was signed into law by President Franklin D. Roosevelt on September 2, 1937. Senator Key Pittman of Nevada and Representative A. Willis Robertson of Virginia were the principal sponsors of the bill in the Senate and House, respectively.

The Pittman–Robertson Act was one of three acts passed and signed into law that impacted the nation's wildlife at no charge to the common taxpayer. The Migratory Bird Conservation Act of 1934 (Duck Stamp Act) and the Federal Aid in Sport Fish Restoration Act of 1950 (Dingell–Johnson Act) protected waterfowl and aquatic species, respectively. The former was funded by a special annual fee paid by hunters and friends of waterfowl. The latter was principally funded by an excise tax on fishing equipment. The Pittman–Robertson Act had two principal purposes: (1) reestablishment of wildlife populations to natural habitats and (2) wildlife research was to become the basis for a science-based management program. The impact of this act on wildlife in the United States was profound, and the wildlife we see today, in many ways, is the living legacy of the vision of Pittman and Robertson. We will learn more about the profound and long-lasting impact of this act in Chapters 13 and 14.

Note in particular in this act how funding is obtained to support wildlife programs. Be careful to understand how the funds are allocated. Most states did not even have fish and game departments in 1937.

Chapter 5B. Wildlife Restoration Act (The Pittman–Robertson Act of 1937)

§ 669.   Cooperation of Secretary of the Interior with States; Conditions
§ 669a.   Definitions

§ 669b.   Authorization of Appropriations; Disposition of Unexpended Funds

§ 669b-1. Authorization of Appropriation of Accumulated Unappropriated Receipts

§ 669c.   Apportionment of Funds; Expenses of Secretary

§ 669d.   Apportionment; Certification to States and Secretary of Treasury; Acceptance by States; Disposition of Funds Not Accepted

§ 669e.   Submission and Approval of Plans and Projects

§ 669f.   Payment of Funds to States; Laws Governing Construction and Labor

§ 669g.   Maintenance of Projects; Expenditures for Management of Wildlife Areas and Resources

§ 669g-1. Payment of Funds to and Cooperation with Puerto Rico, Guam, American Samoa, Commonwealth of the Northern Mariana Islands, and Virgin Islands

§ 669h.   Employment of Personnel; Equipment, etc.

§ 669i.   Rules and Regulations

§ 669j.   Repealed

§ 669. Cooperation of Secretary of the Interior with States; Conditions

The Secretary of the Interior is authorized to cooperate with the States, through their respective State fish and game departments, in wildlife-restoration projects as hereinafter in this chapter set forth; but no money apportioned under this chapter to any State shall be expended therein until its legislature, or other State agency authorized by the State constitution to make laws governing the conservation of wildlife, shall have assented to the provision of this chapter and shall have passed laws for the conservation of wildlife which shall include a prohibition against the diversion of license fees paid by hunters for any other purpose than the administration of said State fish and game department, except that, until the final adjournment of the first regular session of the legislature held after September 2, 1937, the assent of the Governor of the State shall be sufficient. The Secretary of the Interior and the State fish and game department of each State accepting the benefits of this chapter shall agree upon the wildlife-restoration projects to be aided in such State under the terms of this chapter and all projects shall conform to the standards fixed by the Secretary of the Interior.

§ 669a. Definitions

For the purposes of this chapter the term "wildlife-restoration project" shall be construed to mean and include the selection, restoration, rehabilitation, and improvement of areas of land or water adaptable as feeding, resting, or breeding places for wildlife, including acquisition by purchase, condemnation, lease, or gift of such areas or estates or interests therein as are suitable or capable of being made suitable therefor, and the construction thereon or therein of such works as may be necessary to make them available for such purposes and also including such research into problems of wildlife management as may be necessary to efficient administration affecting wildlife resources, and such preliminary or incidental costs and expenses as may be incurred in and about such projects; the term "State fish and game department" shall be construed to mean and include any department or division of department of another name, or commission, or official or officials, of a State empowered under its laws to exercise the functions ordinarily exercised by a State fish and game department.

§ 669c. Apportionment of Funds; Expenses of Secretary

(a) So much, not to exceed 8 per centum, of the revenues (excluding interest accruing under section 669b(b) of this title) covered into said fund in each fiscal year as the Secretary of the Interior may estimate to be necessary for his expenses in the administration and execution of this chapter and the Migratory Bird Conservation Act (16 U.S.C. 715 et seq.) shall be deducted for that purpose, and such sum is authorized to be made available therefor until the expiration of the next succeeding fiscal year, and within sixty days after the close of such fiscal year the Secretary of the Interior shall apportion such part thereof as remains unexpended by him, if any, and make certificate thereof to the Secretary of the Treasury and to the State fish and game departments on the same basis and in the same manner as is provided as to other amounts authorized by this chapter to be apportioned among the States for such current fiscal year. The Secretary of the Interior, after making the aforesaid deduction, shall apportion, except as provided in subsection (b) of this section, the remainder of the revenue in said fund for each fiscal year among the several States in the following manner: One-half in the ratio which the area of each State bears to the total area of all the States, and one-half in the ratio which the number of paid hunting-license hold-ers of each State in the second fiscal year preceding the fiscal year for which such apportionment is made, as certified to said Secretary by the State fish and game departments, bears to the total number of paid hunting-license holders of all the States. Such apportion-ments shall be adjusted equitably so that no State shall receive less than one-half of 1 per centum nor more than 5 per centum of the total amount apportioned. The term fiscal year as used in this chap-ter shall be a period of twelve consecutive months from October 1 through the succeeding September 30, except that the period for enumeration of paid hunting-license holders shall be a State's fiscal or license year.

(b) One-half of the revenues accruing to the fund under this chapter each fiscal year (beginning with the fiscal year 1937) from any tax imposed on pistols, revolvers, bows, and arrows shall be appor-tioned among the States in proportion to the ratio that the population of each State bears to the population of all the States: Provided, that each State shall be apportioned not more than 3 per centum and not less than 1 per centum of such revenues and Guam, the Virgin Islands, American Samoa, and the Northern Mariana Islands shall each be apportioned one-sixth of 1 per centum of such revenues. For the purpose of this subsection, population shall be determined on the basis of the latest decennial census for which figures are available, as certified by the Secretary of Commerce.

§ 669e. Submission and Approval of Plans and Projects

(a) Setting aside funds—Any State desiring to avail itself of the benefits of this chapter shall, by its State fish and game department, submit programs or projects for wildlife restoration in either of the following two ways:

(1) The State shall prepare and submit to the Secretary of the Interior a comprehensive fish and wildlife resource management plan which shall insure the perpetuation of these resources for the economic, scientific, and recreational enrichment of the people. Such plan shall be for a period of not less than five years and be based on projections of desires and needs of the people for a period of not less than fifteen years. It shall include provisions for updating at intervals of not more than three years and be provided in a format as may be required by the Secretary of the Interior. If the Secretary

of the Interior finds that such plans conform to standards established by him and approves such plans, he may finance up to 75 per centum of the cost of implementing segments of those plans meeting the purposes of this chapter from funds apportioned under this chapter upon his approval of an annual agreement submitted to him.

(2) A State may elect to avail itself of the benefits of this chapter by its State fish and game department submitting to the Secretary of the Interior full and detailed statements of any wildlife-restoration project proposed for that State. If the Secretary of the Interior finds that such project meets with the standards set by him and approves said project, the State fish and game department shall furnish to him such surveys, plans, specifications, and estimates therefor as he may require. If the Secretary of the Interior approves the plans, specifications, and estimates for the project, he shall notify the State fish and game department and immediately set aside so much of said fund as represents the share of the United States payable under this chapter on account of such project, which sum so set aside shall not exceed 75 per centum of the total estimated cost thereof.

The Secretary of the Interior shall approve only such comprehensive plans or projects as may be substantial in character and design and the expenditure of funds hereby authorized shall be applied only to such approved comprehensive wildlife plans or projects and if otherwise applied they shall be replaced by the State before it may participate in any further apportionment under this chapter. No payment of any money apportioned under this chapter shall be made on any comprehensive wildlife plan or project until an agreement to participate therein shall have been submitted to and approved by the Secretary of the Interior.

(b) "Project" defined—If the State elects to avail itself of the benefits of this chapter by preparing a comprehensive fish and wildlife plan under option (1) of subsection (a) of this section, then the term "project" may be defined for the purposes of this chapter as a wildlife program, all other definitions notwithstanding.

(c) Costs—Administrative costs in the form of overhead or indirect costs for services provided by State central service activities outside of the State agency having primary jurisdiction over the wildlife resources of the State which may be charged against programs or projects supported by the fund established by section 669b of this title shall not exceed in any one fiscal year 3 per centum of the annual apportionment to the State.

§ 669g. Maintenance of Projects; Expenditures for Management of Wildlife Areas and Resources

(a) Maintenance of wildlife-restoration projects established under the provisions of this chapter shall be the duty of the States in accordance with their respective laws. Beginning July 1, 1945, the term "wildlife-restoration project", as defined in section 669a of this title, shall include maintenance of completed projects. Notwithstanding any other provisions of this chapter, funds apportioned to a State under this chapter may be expended by the State for management (exclusive of law enforcement and public relations) of wildlife areas and resources.

(b) Each State may use the funds apportioned to it under section 669c(b) of this title to pay up to 75 per centum of the costs of a hunter safety program and the construction, operation, and maintenance of public target ranges, as a part of such program. The non-Federal share of such costs may be derived from license fees paid by hunters, but not from other Federal grant programs. The Secretary shall issue not later than the 120th day after the effective date of this subsection such regulations as he deems advisable relative to the criteria for the establishment of hunter safety programs and public target ranges under this subsection.

§ 669g-1. Payment of Funds to and Cooperation with Puerto Rico, Guam, American Samoa, Commonwealth of the Northern Mariana Islands, and Virgin Islands

The Secretary of the Interior is authorized to cooperate with the Secretary of Agriculture of Puerto Rico, the Governor of Guam, the Governor of American Samoa, the Governor of the Commonwealth of the Northern Mariana Islands, and the Governor of the Virgin Islands, in the conduct of wildlife-restoration projects, as defined in section 669a of this title, and hunter safety programs as provided by section 669g(b) of this title, upon such terms and conditions as he shall deem fair, just, and equitable, and is authorized to apportion to Puerto Rico, Guam, American Samoa, the Commonwealth of the Northern Mariana Islands, and the Virgin Islands, out of the money available for apportionment under this chapter, such sums as he shall determine, not exceeding for Puerto Rico one-half of 1 per centum, for Guam one-sixth of 1 per centum, for American Samoa one-sixth of one per centum, for the Commonwealth of the Northern Mariana Islands one-sixth of 1 per centum, and for the Virgin Islands one-sixth of 1 per centum of the total amount apportioned, in any one year, but the Secretary shall in no event require any of said cooperating agencies to pay an amount which will exceed 25 per centum of the cost of any project. Any unexpended or unobligated balance of any apportionment made pursuant to this section shall be available for expenditure in Puerto Rico, Guam, American Samoa, the Commonwealth of the Northern Mariana Islands, or the Virgin Islands, as the case may be, in the succeeding year, on any approved project, and if unexpended or unobligated at the end of such year is authorized to be made available for expenditure by the Secretary of the Interior in carrying out the provisions of the Migratory Bird Conservation Act (16 U.S.C. 715 et seq.).

Chapter 35. Endangered Species Act of 1966, 1969, 1973, and 1988

The Endangered Species Act (ESA) is probably the most widely recognized law protecting plants and animals in the United States. The act was amended and signed into law on December 28, 1973 by President Richard M. Nixon. The ESA is the law that is used to designate threatened and endangered species of plants and wildlife. The term *endangered species* means that a plant or animal is listed by regulation as being "in danger of extinction." A *threatened species* is a plant or animal that is likely to become extinct within the foreseeable future. More than 1000 plants and animals are listed as endangered or threatened. The courts have consistently upheld the ESA despite legislative and administrative attempts to weaken it (Easter-Pilcher 1996).

The ESA prohibits importing, exporting, and taking endangered species in the United States. Taking is prohibited in U.S. territorial and international waters. In addition, carrying, delivering, possessing, selling, shipping, and transporting endangered species unlawfully taken in the United States, its territorial waters, and international waters is prohibited. Interstate and international commerce in endangered species is also punishable by law. Despite the vague definitions of endangered and threatened, the ESA continues to be a powerful law (Bean 1983, Wilcove et al. 1993).

Example—During a trip to Arizona, you notice an interesting cactus. Rather than dig up the cactus for later identification, you decide instead to take a flower. You insert the flower into your plant press book and return to the university museum to identify the curious cactus. The curator of the museum notices the flower and immediately calls state law enforcement officials. You are arrested for violating the ESA, having harmed an endangered species of cactus. If you are convicted, your car, shovel, notebooks, boots, and plant press with specimens can be confiscated.

Penalties—The ESA is considered to be a law "with teeth" (Kohm 1991). Violators can be fined up to $100,000 and can serve up to one year in jail. Organizations can be fined up to $250,000. Items used during the crime can be confiscated.

Chapter 31. Marine Mammal Protection Act of 1972 (1973, 1976)

In 1972, the United States halted the harvesting and importation of marine mammals such as dugongs and manatees, polar bears, porpoises, all seals, sea lions, sea otters, walruses, and whales and any of their parts such as tusks, bones, and hides. The law specifically prohibits individuals or groups of individuals from importing any marine mammal or marine mammal product into the United States and from taking any marine mammal in international waters or waters or lands under U.S. jurisdiction.

Additionally, the Marine Mammal Protection Act makes it illegal to use any U.S. port or harbor for any purpose connected with the taking or importation of marine mammals. It is also unlawful for any individual to possess parts or products of marine mammals. Thus, offering for sale, purchasing, selling, or transporting any marine mammal, including parts or products, is illegal.

Exceptions—Alaskan Native Americans such as Aleuts, Eskimos, and Indians residing in Alaska are permitted subsistence use of marine mammals. Subsistence uses include the taking of marine mammals such as walruses and whales for human consumption and the making of native handicrafts and products. The secretaries of the interior and commerce can grant permits for purposes that support scientific research or public displays such as zoo exhibits. With a permit, exportation of handicrafts and products for scientific or display purposes is not prohibited provided a "Certificate of Origin" accompanies the product.

Example—You are casually walking along a northern California beach when you come across a young sea otter with a fish line wrapped around its body. The twine has cut into its skin in places. Realizing the small sea otter can be saved, you decide to bring the animal to a nearby veterinarian. By putting the sea otter in your car, you have violated the Marine Mammal Protection Act and are subject to the following penalties.

Penalties—A fine of up to $100,000 and one year in jail are possible for individuals in violation of the act. Organizations can be fined up to $200,000. Additionally, all cargo can be seized from transport vessels such as airplanes, ships, or other means of transport.

Chapter 7. Protection of Migratory Game and Insectivorous Birds

This act provides protection for migratory birds. It is unlawful to pursue, hunt, shoot, poison, wound, kill, capture, possess, buy, sell, trap, collect, or barter any migratory bird, including feathers or parts, nests, eggs, or migratory bird products. The act is based on treaties with Great Britain, Mexico, and Japan and puts restrictions on specific birds covered under the act.

Exceptions—The U.S. Fish and Wildlife Service establishes bird-hunting regulations and during hunting season allows for the taking of ducks, geese, doves, rails, woodcocks, and other species. Other permits may be granted for scientific research.

Example—As an avid bird-watcher, you pride yourself on knowing the local birds, their calls, and their habits. One day while birding, you find on the forest floor the nest of a Wilson's warbler. You correctly identify the nest for your companions and then stuff it into a plastic bag for later examination at home. By taking the nest, you have violated the Migratory Bird Treaty Act and are subject to the full penalties under the law. Suppose, for instance, that on the way home, you see a road-killed hummingbird and put it in your car to give to a museum. Again, you are in violation of the Migratory Bird Treaty Act unless you have a permit that allows you to collect.

Penalties—Individuals and organizations can be fined up to $5000 and $10,000, respectively, and can receive six months in prison for misdemeanor violations. For felony violations, individuals and organizations can be fined up to $250,000 and $500,000, respectively, and receive up to two years in jail.

Chapter 69. Wild Exotic Bird Conservation Act of 1992

This act prohibits the import of all CITES-listed birds into the United States except for those species included on an approved list either by country of origin or approved captive breeding facilities or for wild-caught birds. For wild-caught approved birds, a plan that provides for the conservation of the species and its habitat is required. The act also established the Exotic Bird Conservation Fund, which is funded by fines and penalties and is to be used to assist exotic bird conservation projects in their native countries.

Penalties—Civil violators can be fined up to $25,000. Criminal violations call for a fine specified in Title 18 and a jail term of not more than two years.

The international regulation of the trade in plant and animal products is an enormous undertaking. Certainly, finding a six-foot elephant's tusk is easy compared to examining a bottle of pills that contain ground-up tiger bones. Crates of snakes are imported every day into the United States. Discovering which are listed species, and therefore protected, and which are common is a formidable task for customs agents. Unfortunately, the profit from one successful theft can easily make up for ten intercepted shipments. The rich nations are by far the largest importers of rare and exotic plants and animals. Whales are still consumed at sushi bars in Asia. In the final analysis, each individual must decide what is right and what is wrong. Just because something is not prohibited by law does not make it right. Even when a "Certificate of Authenticity" is available, it is difficult to tell the difference between an authentic certificate and a fraud. Would you be prepared to discuss the situation with a customs agent representing the U.S. government?

Sometimes the situation in foreign countries is horrific, yet almost nothing can be done. While in Namibia, one of the authors (JGS) was offered two cheetah cubs for $2.50 each. The farmer (a rancher in our terms) said that if the cubs were not purchased, they would be drowned. Cheetahs are, of course, an endangered species under the ESA. Their international trade is strictly prohibited. Healthy cheetah populations still exist in Namibia, and the cat is considered vermin to farmers.

A trip down the Congo River through the Democratic Republic of the Congo in Central Africa is the experience of a lifetime. The trip is, at best, arduous and brings to mind the phrase "survival test." Along the way, local people canoe out to the steamship and sell food such as bananas and various other fruits, bread, fish, and animals. Often the animals are alive and bound with twine. During one trip, a canoeist offered two live young chimpanzees for sale as food. Of course, chimpanzees are endangered. Needless to say, cash is an even rarer commodity in remote parts of Africa. Although the author refused to purchase the chimpanzees, another local passenger bought both. During the trip, the chimps were fed fruit and kept in a grass basket typical of the region. The small male was eaten first, while his sister whimpered inside her basket. Later that week, the basket contained a small forest antelope. Local enforcement of international law in certain parts of the world is impossible, and sometimes the situation seems hopeless.

Other U.S. Codes also specify many useful wildlife laws, including the following:

- Classification and Multiple Use Act of 1964, 1976, 43 U.S.C.
- Clean Air Act of 1970, 1980, 42 U.S.C.
- Clean Water Act of 1981, 33 U.S.C.
- Comprehensive Environmental Response, Compensation, and Liability Act of 1980, 42 U.S.C.
- Federal Environmental Pesticide Control Act of 1981, 7 U.S.C.
- Federal Fruit Fly and Tick Act of 1938 and 1942, 7 U.S.C.
- Federal Land Policy and Management Act of 1976, 1981, 43 U.S.C.
- Federal Noxious Weed Act of 1974, 1976, 7 U.S.C.
- Fishermen's Protection Act of 1967, 1981, 22 U.S.C.
- Intervention of the High Seas Act of 1976, 33 U.S.C.
- Mineral Leasing Act of 1976, 1981, 30 U.S.C.
- Mineral Leasing Act for Acquired Lands of 1976, 30 U.S.C.
- National Environmental Policy Act of 1976, 42 U.S.C.
- Outer Continental Shelf Lands Act of 1953, 1976, 1981, 43 U.S.C.
- Plant Pest Act of 1976, 7 U.S.C.
- Resource Conservation and Recovery Act of 1976, 1981, 42 U.S.C.
- River and Harbor Act of 1899, 1976, 33 U.S.C.
- Submerged Lands Act of 1976, 43 U.S.C.
- Tariff Act of 1930, 1976, 19 U.S.C.
- Tariff Act of 1962, 1976, 19 U.S.C.
- Tariff Classification Act of 1976, 19 U.S.C.
- Taylor Grazing Act of 1976, 1981, 43 U.S.C.
- Walrus Protection Act (omitted from 16 U.S.C. as obsolete)
- Wilson Original Package Act of 1976, 1980, 27 U.S.C.

U.S. FISH AND WILDLIFE SERVICE

The U.S. Fish and Wildlife Service is the principal federal agency responsible for conserving, protecting, and enhancing fish and wildlife and their habitats for the continuing benefit of the American people. The service manages 511 national wildlife refuges covering 92 million acres, as well as 72 national fish hatcheries. The agency enforces federal wildlife laws, manages migratory bird populations, stocks recreational fisheries, conserves and restores wildlife habitat such as wetlands, administers the Endangered Species Act, and helps foreign governments with their conservation efforts. It also oversees the federal aid program that funnels federal excise taxes on fishing and hunting equipment to state wildlife agencies. This program is a cornerstone of the nation's wildlife management efforts, funding fish and wildlife restoration, boating access, hunter education, shooting ranges, and related projects across America.

Division of Law Enforcement

The U.S. Fish and Wildlife Service (USFWS) Division of Law Enforcement is charged with upholding the laws and acts of the United States and making certain that the laws are obeyed by U.S. citizens and their entities, such as corporations, and any foreign citizens in the United States. The USFWS protects fish, wildlife, and plants through a wide variety of law enforcement techniques, including:

- Surveillance of areas where priority wildlife and fishery resources are concentrated. To protect the dwindling numbers of ducks and geese, the USFWS routinely creates task forces during the waterfowl season in many states.
- Inspection of shipments entering or leaving the United States at designated ports, border ports, and special ports; investigation of known and suspected violations; and distribution of information about federal wildlife regulations and their enforcement.
- Submission of evidence of violations to regional solicitors of the Department of the Interior or U.S. attorneys for consideration of civil penalties or criminal prosecution.

The objectives of the USFWS Division of Law Enforcement are to enforce the Lacey Act, Migratory Bird Treaty Act, Migratory Bird Hunting and Conservation Stamp Act, Eagle Protection Act, Endangered Species Act, Airborne Hunting Act, Marine Mammal Protection Act, Convention on International Trade in Endangered Species of Wild Fauna and Flora, National Wildlife Refuge System Administration Act, Antarctic Conservation Act, Archaeological Resources Protection Act, Wild Bird Conservation Act, and African Elephant Conservation Act. The Division of Law Enforcement is also charged with uncovering major commercial activity involving illegal trade of protected wildlife and wildlife products. When appropriate, the division also issues permits.

The USFWS is also charged with enhancing legitimate use and enjoyment of migratory birds and other wildlife and with keeping U.S. citizens informed of various federal laws and regulations relating to the protection of fish, wildlife, and plants.

The USFWS maintains a Branch of Special Operations comprised of eight special agents who are assigned full-time undercover duties and an Intelligence Section that is composed of one intelligence research specialist. The special agents employ highly sophisticated and innovative investigative techniques to uncover and document large-scale violations involving the illegal taking, importing, and/or commercialization of wildlife. We will see examples of their successes in subsequent chapters. The intelligence research specialist provides analytical and strategic intelligence to personnel within the Branch of Special Operations and to field agents located throughout the country.

The duties and responsibilities of the branch have taken on heightened importance due to the ever-increasing difficulty of detecting wildlife crime through more conventional methods and enforcement activities. Today, wildlife violations are often clandestine in nature, committed by organized and sophisticated persons and groups, and are complex in terms of the conspiracies and fraud associated with them and the worldwide arena in which they take place. While such undercover operations are time consuming and potentially dangerous, they nevertheless are very effective in combating significant and ongoing exploitation of wildlife resources and in deterring such activities.

In addition, wildlife inspectors monitor and assist in halting illegal trade. The USFWS employs approximately 85 uniformed wildlife inspectors. These inspectors work with U.S. Customs and Department of Agriculture inspectors to provide expertise in wildlife law and species identification. In addition to scrutinizing the legality of accompanying permits, wildlife inspectors conduct physical inspections, targeting repeat offenders and checking shipments on a random basis. Inspectors give priority to processing importation of live wildlife.

The Division of Law Enforcement of the USFWS has been particularly effective (or has chosen to prosecute only the most egregious cases).

	1992	*1993*	*1994*	*Total*
Cases brought to trial	75	64	123	262
Indictments	103	88	150	341
Sentences	97	82	138	317
Not guilty verdicts	6	6	11	23

U.S. Fish and Wildlife Service Agents Stop Illegal Market

On November 21, 1996, the USFWS announced "Operation 4-Corners Feather Sales." Special agents had served search and arrest warrants to 35 individuals and businesses in Arizona, Colorado, and New Mexico suspected of taking part in the killing and selling of birds and bird parts protected by the Migratory Bird Treaty Act, the Endangered Species Act, and the Bald and Golden Eagle Protection Act. The investigation was terminated prematurely to prevent the loss of more birds.

Undercover agents posing as traders of Indian artifacts infiltrated a commercial trapping ring. The agents were sold dead eagles caught with baited steel-jawed leg-hold traps located on Indian pueblos. Eagles are considered sacred in many Native American cultures, and take permits are available for a small number of eagles that can be used in religious ceremonies. Commercial uses are prohibited however.

In addition to eagles and eagle parts, commercial products of owls, hawks, kestrels, magpies, flickers, scissor-tailed flycatchers, and anhingas were also uncovered. Feathers are popular in making Native American gift items sold at so-called trading posts and other popular tourist stops. One golden eagle feather can bring $100. A golden eagle fan requires a complete golden eagle tail of 12 feathers and can sell for $1000.

The Lacey Act carries a maximum penalty of five years imprisonment and a $250,000 fine. The Bald and Golden Eagle Protection Act has a one-year prison term and a $100,000 fine. The Migratory Bird Treaty Act carries a two-year prison term and a $250,000 fine. The Department of Justice's U.S. attorney's offices in Albuquerque, New Mexico, and Phoenix, Arizona, will prosecute the case. "The United States government has a strong interest in the preservation of these magnificent animals. This kind of large-scale commercial trapping must end before the resource is depleted entirely," said John J. Kelly, U.S. attorney for the District of New Mexico.

U.S. Fish and Wildlife Agents Stop Illegal Alaska Hunting

A 50-year-old man from Mount Washington, Massachusetts, who hunted illegally in Alaska, was fined nearly $30,000 and will be prevented from hunting during his two-year probation as punishment for violating federal wildlife protection laws. Because of the felony conviction, he will never again be able to own a firearm. In a week-long trial in 1996 in a Springfield, Massachusetts, courtroom filled with life-size mounted game animals, a federal jury convicted the man of six felony charges for hunting without a valid license in Alaska and then transporting the illegally taken game across state lines, clearly in violation of the Lacey Act.

During the March 27, 1997 sentencing, U.S. District Court Judge Michael A. Ponsor fined the hunter $20,000. In addition, Judge Ponsor ordered him to pay the state of Alaska $9994 in restitution for lost hunting license revenue and placed him on supervised probation for two years. As a condition of his probation, he will not be allowed to hunt or be in the company of people engaged in hunting anywhere in the world during his probation. The judge also ordered that firearms would not be allowed in his residence during the probationary period and advised the hunter that, as a convicted felon, he would not be authorized to possess a firearm for the rest of his life.

Judge Ponsor also approved a forfeiture order for six big game mounts that wildlife agents had seized from the hunter's home in February 1995. These mounts included Dall sheep, moose, and caribou that had been killed illegally in Alaska. A life-size brown bear mount previously seized from the residence was forfeited to the government as part of a civil action.

During the criminal trial, the prosecution presented evidence that the man illegally hunted in Alaska from 1990 through 1994 without a valid Alaska hunting license and then transported the illegally killed animals to Massachusetts. The prosecution provided documentation showing that he falsely claimed to be an Alaska resident on his application for an Alaska hunting license, thereby saving thousands of dollars in hunting and licensing fees. Because he purchased the services of Alaska guides during the course of his illegal hunting activities and then transported the unlawfully taken wildlife across state lines, his conduct violated the felony section of the Lacey Act.

While searching the hunter's residence, state and federal wildlife agents discovered a live black bear in a cage behind the house and a live copperhead snake that is an endangered species in Massachusetts. Possession of these wildlife is a violation of state law, and they were seized by the Massachusetts Environmental Police, with the assistance of officers of the Animal Rescue League of Boston. The hunter paid a $5000 fine in state court stemming from these violations.

This case was investigated by special agents of the USFWS from Boston, New York, and Anchorage; the Massachusetts Environmental Police; the Massachusetts State Police; and investigators from the Alaska State Troopers Fish and Wildlife Protection. The case was prosecuted by Assistant U.S. Attorney Nadine Pellegrini of the Major Crimes Unit and trial attorney Charles W. Brooks of the Justice Department's Environmental Division.

U.S. Fish and Wildlife Service Agents Bust Snake Dealer

A 31-year-old man from Marion, Illinois, admitted his role in an international wildlife trafficking scheme that included smuggling rare and protected reptiles from Spain, as well as shipping nearly 70 poisonous snakes through the U.S. mail in unmarked packages to avoid detection by authorities. The smuggler pleaded guilty to one felony count of conspiracy to smuggle wildlife into the

United States and to trade in protected species in interstate commerce. A reptile dealer known for his captive breeding success with small lizards called geckos, the man entered his guilty plea before Judge J. Phil Gilbert in U.S. District Court in Benton, Illinois, and now faces five years of incarceration and/or a $250,000 fine.

The investigation into his activities began in 1994 at Kennedy Airport in New York, where USFWS inspectors discovered a mail parcel from Spain addressed to him. Hidden within the parcel were 13 Lilford's wall lizards, small blue lizards that inhabit the Balearic Islands off the coast of Spain. These lizards are protected by CITES, of which both the United States and Spain are signatory countries.

Following the package to its destination in southern Illinois, a USFWS special agent worked with U.S. postal inspectors, Illinois Conservation Police officers, and other USFWS law enforcement officers to carry out a federal search warrant at the smuggler's residence. They found records and documents chronicling ten years of smuggling reptiles to and from Spain, France, and South Africa. Among the reptiles seized at his home were the 13 Lilford's wall lizards, European ladder rat snakes also smuggled from Spain, box turtles illegally collected from a national wildlife refuge, venomous massasauga rattlesnakes mailed illegally from Florida, a timber rattlesnake and Great Plains rat snakes listed as threatened species in Illinois, and two desert tortoises, a species considered threatened under the U.S. Endangered Species Act of 1973.

The smuggler actively solicited and traded reptiles through the mail with a reptile supplier in Barcelona, Spain. Each would ship parcels containing live reptiles in plastic containers, using fictitious names and addresses. Packages were unmarked and declared as "books" to avoid detection. Search warrants were also served on the supplier by authorities in Barcelona. Portions of the investigation are ongoing in Spain and several U.S. states, and additional people may be charged.

Investigators found that the smuggler frequently traded venomous snakes, collecting from the wild and subsequently mailing copperheads, timber rattlers, massasaugas, and speckled and diamondback rattlesnakes, in violation of U.S. postal laws. He was also found to have collected turtles and snakes from national wildlife refuges and national forests. These reptiles were then traded or sold to reptile collectors around the country.

Among the wildlife laws the smuggler violated are the U.S. Endangered Species Act of 1973 and CITES. In addition, his trading activities violated the Lacey Act, a federal statute which prohibits interstate commercialization of wildlife in violation of state laws. Some of the species traded were protected by Illinois state law, including the Dangerous Animals Act, which prohibits the possession of dangerous wildlife, including venomous snakes.

This investigation was prosecuted by Assistant U.S. Attorney William E. Coonan, Southern District of Illinois, and Jonathon Blackmer, U.S. Department of Justice, Wildlife and Marine Resources Section, Washington, D.C.

In a related smuggling investigation, a man from St. Charles, Missouri, pleaded guilty in April 1996 to violations of the Lacey Act. He was fined $10,000 for unlawfully importing 18 live Hermann's tortoises through the mail. These protected tortoises were sent by the same supplier in Barcelona in the same manner that the previously mentioned smuggler from Illinois had smuggled reptiles.

U.S. Fish and Wildlife Agents Bust Another Trafficker

On January 10, 1997, one of the most severe sentences ever handed down in a reptile-smuggling case was imposed against a German national for his involvement in an international smuggling ring. In Orlando, Florida, Federal Judge Ann Conway sentenced the 33-year-old smuggler from Rauenberg, Germany, to serve 46 months in jail for his role in a reptile-smuggling scheme. He was also fined $10,000. His partner in the smuggling conspiracy, from Blairgowrie, South Africa, received three years probation and six months in a community corrections facility for his role in the conspiracy.

Both men and four others were indicted by a federal grand jury in August 1996 for participating in an international wildlife-smuggling conspiracy. They moved hundreds of protected reptiles from Madagascar through Europe and Canada into the United States. In October 1996, the South African smuggler pleaded guilty to charges of smuggling, conspiracy, Lacey Act violations, money laundering, and attempted escape.

Their most recent smuggling attempt was intercepted at Orlando International Airport on August 14, 1996, when officials found 61 Madagascar tree boas and 4 spider tortoises concealed in the South African smuggler's personal baggage. He had arrived on a commercial flight from Frankfurt, Germany, to attend a large commercial reptile trade show. He cooperated with the investigators and identified the German national as a partner in the conspiracy and the intended recipient of the smuggled reptiles. The German national was arrested two days later.

In this case alone, the wildlife had an estimated commercial value of more than $250,000. The United States is the world's largest importer of wildlife, and the demand for live reptiles has increased rapidly in the past few years. During a two-year period, the individuals involved in this conspiracy smuggled at least 107 Madagascar tree boas, 25 spider tortoises, 51 radiated tortoises, and 2 Madagascar ground boas into the United States, where they are prized by collectors of exotic reptiles and commercial reptile breeders.

These species occur naturally only in Madagascar. Each is protected under CITES, an international treaty signed by more than 130 nations designed to regulate and monitor the trade of rare plants and animals throughout the world. The radiated tortoise is also classified as endangered on the U.S. endangered species list. The radiated tortoise is considered one of the most brilliant species of tortoises, with a bright yellow head and high-domed black shell with yellow starburst designs.

Four additional defendants in the case remain outside the United States, which has begun formal extradition procedures against one defendant from Windsor, Ontario, Canada. The three other defendants, all from Germany, have not yet been arraigned.

This case was investigated by special agents from the USFWS. The prosecution was led by the U.S. attorney's office in Orlando, Florida, assisted by the Wildlife and Marine Resources Section of the Environment and Natural Resources Division of the U.S. Department of Justice.

U.S. Fish and Wildlife Service Web Page

The USFWS maintains a Web page (http://www.fws.gov) that is updated regularly. You can also subscribe to receive free via e-mail the latest news on the service's activities. Mail is sent out periodically, and subscribers are not bombarded with junk mail. To subscribe, send the following e-mail message:

send mail to	mail majordomo@www.fws.gov
leave the subject blank	subject:
make the first line in the body	subscribe fws-news
send the message	

EXAMPLE OF A "TAKE" PERMIT UNDER THE ENDANGERED SPECIES ACT OF 1973

Karen Bouma's article in the *Orlando Sentinel* on October 16, 1996 described the use of the USFWS's federal "take" permit process. In Osceola County, Florida, a small developer planned to build 30 single-family homes on 12 acres. A pair of bald eagles had previously built one of their characteristically large nests in a tree on the property. The bald eagle, the national bird of the United States, is protected by the Endangered Species Act of 1972. USFWS officials granted the developer the first-ever "take" permit for bald eagles, but this does not mean the developer can harm the eagles.

Since the birds were nesting, the nest and tree had to be protected. The developer paid about $8000 to replan the development to leave the nest tree standing. In addition, $15,000 was paid in mitigation fees, and the "take" permit cost the developer $25,000 that is refundable if the newborn eagles fledge and leave the nest. If the nesting pair of eagles abandon the nest, the developer forfeits the $25,000 deposit. Also recall that the Bald and Golden Eagle Protection Act carries a one-year prison term and a $100,000 fine Thus, the developer has a strong incentive to allow the eagles to successfully raise their young. Because the eagles were already sitting on eggs, USFWS scientists felt that the eagles would not abandon their nest if limited disturbance took place away from the nest tree; thus the developer was able to proceed with the project.

ANIMAL RIGHTS GROUPS

Animals do not have standing; therefore, they must be represented by a group that does have standing. Sometimes the Sierra Club or the Wilderness Society represents forests or species. Dr. Albert Schweitzer, perhaps the greatest humanitarian to ever live, held a belief that life, all life, was sacred. In his world-famous hospital in Gabon, Dr. Schweitzer treated people and animals, and he won a Nobel Prize for his outstanding work. Animal rights groups also believe in the sanctity of all life. European animal rights groups are especially active. Animal testing labs, fox hunts, chicken factories, and kennel clubs have been their targets. Of course, not all animal rights groups are forthright about their concerns. Wildlife issues have a way of exciting people into actions they would not otherwise take.

A report issued by Direct Action for Animal Liberation summarizes recent actions taken against those perceived to have violated the sanctity of life. Many actions occur each month. Some highlights are given here. In January 1994, the Animal Liberation Front (ALF) smashed 17 animal abusers' windows. A horse was rescued from its owner, who was taken to court for abuse. The owners of a hunt club in the United Kingdom received a fake parcel bomb. The Hunt Retribution Squad planted three land mines at a horse racing club. A U.K. group called the Justice Department (JD) sent a hoax parcel bomb to a hunt club. In February 1994, an incendiary device was dropped through a mailbox at a leather boot factory. The ALF claimed to have freed, in separate incidents, 3 sheep, 158 hens, and 10 dogs. The JD sent a boot manufacturer an incendiary device. The ALF sabotaged a meat factory in April 1994. At Northeast Surrey College, the ALF freed more than 200 animals, including 15 rabbits, 52 guinea pigs, 58 hamsters, and more than 90 rats. The cages were smashed and the rooms were flooded. A bomb exploded at a pig company, injuring one man. No group claimed responsibility. Two incendiary devices caused severe damage to a boot factory in Cambridge, U.K. The Animal Rights Militia took credit.

In Finland, the ALF smashed the windows and glued the locks at a fur store. In Germany, Autonome Tierschutzerinnen set off stink bombs in nine fur shops. In March 1994, 2500 worms and maggots were freed from anglers. In Frankfurt, Germany, the toilets were blocked at the local McDonald's, Burger King, and Wienerwald. In Northern Ireland, the ALF smashed the windows at the local McDonald's. The ALF also sabotaged a fur shop in Norway. A Norwegian-registered outlaw whaling ship was scuttled and the engine room flooded. The Sea Shepard Conservation Society took credit for sinking a second outlaw whaler. A chicken-breeding company in Scotland was targeted with a letter bomb by the JD. In Sweden, Animal Avengers sabotaged more than a dozen fur shops.

Abandoned Sheep

On September 6, 1996, the Panamanian livestock carrier *Uniceb* was adrift and burning in the Indian Ocean 700 km east of the Seychelles Islands. The *Uniceb* was en route from Australia to the Jordanian port of Aquaba with 67,448 live Australian sheep aboard. The fire was described as "horrific," and a salvage tug was on the way from Djibouti. The crew had already abandoned the *Uniceb*. The worst disaster in the history of the livestock shipping business was unfolding, but even this debacle would not slow the live sheep trade that Australia developed during the 1950s.

The previous record disaster occurred in 1980, when 40,605 sheep died from a fire aboard the ship *Farid Fares* en route from Tasmania to Iran. In 1989, Saudi Arabia refused delivery of 37,000 sheep, and the ship carrying them was turned away from port with its cargo, resulting in their complete loss. Jordan alone bought 1.5 million sheep in 1995. In 1996, Australia earned more than US$500 million, so the live sheep trade is unlikely to falter. Indeed, Australian statistics show that 100,000 sheep (2%) die each year from stress or disease during the long voyage aboard ship to the Middle East, Australia's primary importer of sheep. Another 3% die while in feedlots in the Middle East, but these are counted as delivered and paid for. Muslim beliefs also require that animals be slaughtered according to halal procedures, without the preslaughter stunning that Australian's use.

Customers in North Africa and the Middle East prefer fresh meat over frozen, and the most effective means of transportation is by ship. The *Uniceb*'s safety record was clean before it left Australia. A fire reportedly broke out in the engine room, and eventually the ship had to be abandoned. One member of the crew died and 54 others were rescued, a death rate of 2%. The Royal Society for the Prevention of Cruelty to Animals suggested that an Australian stockman experienced with sheep accompany every vessel. When the Australian public discovered that live horses were being shipped to Japan for meat, there was a public outcry and the trade was halted. British animal rights activists routinely protest the shipment of veal calves to France, and the sheep trade from Britain has also been under recent attack.

Although the treatment of livestock is not technically a wildlife issue, people are concerned abut the ethical treatment of animals. Once educated about an issue, people can force it onto the ballot during an election. In this way, such trade can be debated in public and decided upon for the common good. If the sheep trade is thought to be morally unacceptable, then the voters of Australia can make the trade illegal or at least more acceptable.

APPENDIX 2.1: CITES PARTICIPATING NATIONS AS OF JANUARY 1998

Nation	Signature date	Nation	Signature date
Afghanistan	January 28, 1986	Canada	July 9, 1975
Algeria	February 21, 1984	Central African	November 25, 1980
Argentina	April 8, 1981	Republic	
Australia	October 27, 1976	Chad	May 3, 1989
Austria	April 27, 1982	Chile	July 1, 1975
Bahamas	September 18, 1979	China	April 8, 1981
Bangladesh	February 18, 1982	Colombia	November 29, 1981
Barbados	March 9, 1993	Comoros	February 21, 1995
Belarus	November 8, 1995	Congo	May 1, 1983
Belgium	January 1, 1984	Costa Rica	September 28, 1975
Belize	September 21, 1981	Cote d'Ivoire	February 19, 1995
Benin	May 28, 1984	Cuba	July 19, 1990
Bolivia	October 4, 1979	Cyprus	July 1, 1975
Botswana	February 12, 1978	Czech Republic	January 1, 1993
Brazil	November 4, 1975	Denmark	October 24, 1977
Brunei	August 20, 1990	Democratic	October 18, 1976
Bulgaria	April 16, 1991	Republic of	
Burkina Faso	January 15, 1990	the Congo	
Burundi	November 6, 1988	(Zaire)	
Cameroon	September 3, 1981	Djibouti	May 7, 1992

Nation	Signature date	Nation	Signature date
Dominica	November 2, 1995	Malta	July 16, 1989
Dominican Republic	March 17, 1987	Mauritius	July 27, 1975
		Mexico	September 30, 1991
Ecuador	July 1, 1975	Monaco	July 1978
Egypt	April 4, 1978	Mongolia	January 5, 1996
El Salvador	July 29, 1987	Morocco	January 14, 1976
Equatorial Guinea	June 8, 1992	Mozambique	June 23, 1981
		Namibia	March 18, 1991
Eritrea	January 22, 1995	Nepal	September 16, 1975
Estonia	October 20, 1992	Netherlands	July 18, 1984
Ethiopia	July 4, 1989	New Zealand	August 8, 1989
Fiji	January 1, 1998	Nicaragua	November 4, 1977
Finland	August 8, 1976	Niger	December 7, 1975
France	August 9, 1978	Nigeria	July 1, 1975
Gabon	May 15, 1989	Norway	October 25, 1976
Gambia	November 24, 1977	Pakistan	July 19, 1976
Germany	June 20, 1976	Panama	November 15, 1978
Ghana	February 12, 1976	Papua New Guinea	March 11, 1976
Greece	January 6, 1993		
Guatemala	February 5, 1980	Paraguay	February 13, 1977
Guinea	December 20, 1981	Peru	September 25, 1975
Guinea-Bissau	August 14, 1990	Philippines	November 16, 1981
Guyana	August 25, 1977	Poland	March 12, 1990
Honduras	June 13, 1985	Portugal	March 11, 1981
Hungary	August 29, 1985	Romania	November 16, 1994
India	October 18, 1976	Russian Federation	December 8, 1976
Indonesia	March 28, 1979		
Iran	November 1, 1976	Rwanda	January 18, 1981
Israel	March 17, 1980	Saint Kitts and Nevis	May 15, 1994
Italy	December 31, 1979		
Japan	November 4, 1980	Saint Lucia	March 15, 1983
Jordan	March 14, 1979	Saint Vincent and the Grenadines	February 28, 1989
Kenya	March 13, 1979		
Korea, Republic of	October 7, 1993		
		Senegal	November 3, 1977
Liberia	June 9, 1981	Seychelles	May 9, 1977
Liechtenstein	February 28, 1980	Sierra Leone	January 26, 1995
Luxembourg	March 12, 1984	Singapore	February 28, 1987
Madagascar	November 18, 1975	Slovak Republic	January 1, 1993
Malawi	May 6, 1982	Somalia	March 2, 1986
Malaysia	January 18, 1978	South Africa	October 13, 1975
Mali	October 16, 1994	Spain	August 28, 1986

Nation	Signature date	Nation	Signature date
Sri Lanka	August 2, 1979	Uganda	October 16, 1991
Sudan	January 24, 1983	United Arab	May 12, 1990
Suriname	February 15, 1981	Emirates	
Sweden	July 1, 1975	United Kingdom	October 31, 1976
Switzerland	July 1, 1975	United States	July 1, 1975
Tanzania	February 27, 1980	Uruguay	July 1, 1975
Thailand	April 21, 1983	Vanuatu	October 15, 1989
Togo	January 21, 1979	Venezuela	January 22, 1978
Trinidad and	April 18, 1984	Vietnam	April 20, 1994
Tobago		Zambia	February 22, 1981
Tunisia	July 1, 1975	Zimbabwe	August 17, 1981

EXERCISES

2.1 Find examples of wildlife crimes and cite the laws that were violated during these crimes. How do the sentences meted out compare with the possible sentences the laws provide? Consider looking at http://www.fws.gov/~r9dle/div_le.html on the Internet.

2.2 Inspect http://www.law.cornell.edu/uscode. What exactly is the Migratory Bird Act?

2.3 The United States has many laws that protect wildlife. Find examples of laws from other countries that protect wildlife.

2.4 Despite all these laws, the list of threatened and endangered species continues to grow. Explain why this is so.

2.5 Use your favorite search engine to surf the World Wide Web for international agreements.

2.6 Are there laws preventing harmful bacteria or viruses from entering the United States?

2.7 Find a list of all CITES Appendix I, II, and III species. Compare the most recent lists with older lists. Are trends evident?

2.8 Make a list of threatened and endangered species included in the Endangered Species Act. Examine historical lists for trends. Based on a comparison, make predictions about possible additions to the list.

2.9 The Endangered Species Act of 1973 was reauthorized and strengthened in 1988. The updated version of the Endangered Species Act required all federal agencies to undertake programs for the conservation of endangered and threatened species and prohibited them from authorizing, funding, or carrying out any action that would jeopardize a listed species or destroy or modify its "critical habitat." For the first time, plants and all classes of invertebrates were eligible for protection, as they were under CITES. Assemble a list of threatened and endangered species of mammals, birds, reptiles, amphibians, fishes, snails, clams, crustaceans, insects, arachnids, and plants. How has this list changed since 1973?

2.10 How is "critical habitat" defined in the Endangered Species Act? What does the law say about habitats that are used only rarely or seasonally, such as ephemeral wetlands in South Dakota that are used by migratory waterfowl?

2.11 Update the actions by animal rights groups. What is happening in the United States? Try doing a search on "Animal Liberation Front."

2.12 Trace the history of animal rights groups. Are there any clear trends in the issues these groups raise? Start your search at http://www.envirolink.org/adn/alt.old or do a search on "Direct Action for Animal Liberation."

2.13 Investigate more thoroughly which animals were heavily exploited during the early history of the United States.

2.14 Is there a wildlife issue with the American bison anymore?

2.15 Will we protect more wildlife more effectively by enacting laws, arresting more violators, and levying more and heavier fines? What is the most effective way to protect wildlife? Explain your position using examples.

2.16 Since 1939, the Pittman–Robertson Act has provided more than $3.7 billion for wildlife restoration and hunter education. The total for each state is listed below. What can be done to increase your state's share of this funding?

State Apportionments of Pittman–Robertson Funds from 1939 to 1997

State	Total	State	Total
Texas	152,087,376	Florida	58,274,361
Alaska	135,233,411	Nevada	57,372,271
Pennsylvania	131,313,817	Iowa	57,065,806

State	Total	State	Total
California	125,666,293	Arkansas	54,664,379
Michigan	123,605,212	Nebraska	53,251,341
New York	99,705,091	Kentucky	53,030,315
Wisconsin	97,247,591	South Dakota	52,068,324
Minnesota	90,908,879	Mississippi	51,431,695
Montana	88,401,343	North Dakota	43,610,786
Missouri	84,036,603	West Virginia	39,586,390
Colorado	80,110,723	South Carolina	38,583,830
Oregon	78,092,652	Maine	36,679,945
Ohio	77,437,164	Maryland	31,526,819
Tennessee	76,373,542	New Jersey	30,103,763
Arizona	72,339,073	Massachusetts	28,373,232
Illinois	70,833,217	Connecticut	23,078,388
Georgia	69,341,125	Vermont	19,210,762
New Mexico	68,118,962	New Hampshire	18,593,899
Washington	68,067,503	Delaware	18,504,523
North Carolina	67,339,133	Rhode Island	18,490,899
Virginia	66,659,464	Hawaii	18,126,117
Oklahoma	60,844,220	Puerto Rico	11,943,636
Wyoming	60,606,589	Virgin Islands	4,688,194
Idaho	60,324,963	Guam	4,575,194
Utah	59,770,910	Mariana Islands	3,497,782
Louisiana	59,299,439	American Samoa	3,062,416
Alabama	58,492,218		

LITERATURE CITED

Bean, M. 1983. *The Evolution of National Wildlife Law.* Praeger Publishers, New York.

Cooper, M.E. 1987. *An Introduction to Animal Law.* Harcourt Brace Jovanovich, New York.

Day, A. 1949. *North American Waterfowl.* Stackpole and Heck, New York.

Easter-Pilcher, A. 1996. Implementing the Endangered Species Act. *BioScience* **46**(5): 355–363.

Hemley, G. 1994. *International Wildlife Trade: A CITES Sourcebook.* Island Press, Washington, D.C.

Kohm, K.A. 1991. *Balancing on the Brink of Extinction.* Island Press, Washington, D.C.

Littell, R. 1992. *Endangered and Other Protected Species: Federal Law and Regulation.* The Bureau of National Affairs, Washington, D.C.

Lund, T.A. 1980. *American Wildlife Law.* University of California Press, Berkeley.

Matthiessen, P. 1964. *Wildlife in America.* Viking Press, New York.

Trefethen, J. 1975. *An American Crusade for Wildlife.* Winchester Press and the Boone and Crockett Club, New York.

Wheeler, C.E. 1947. *Duck Shooting Along the Atlantic Tidewater.* William Morrow, New York.

Wilcove, D.S., M. McMillan, and K.C. Winston. 1993. What exactly is an endangered species? An analysis of the U.S. endangered species list: 1985–1991. *Conservation Biology* 7(1):87–93.

EXAMPLES OF
WILDLIFE ISSUES

3

Why, with so many laws in place, are there any wildlife issues left to resolve? How is it that new wildlife issues arise nearly every day? There is no law that prevents habitat destruction (except where threatened or endangered species exist), no act that prevents birds from escaping from aviaries, no decree against constructing new highways that fragment habitat, and almost no universally accepted laws that apply globally. In fact, the laws are difficult to enforce, and when enforced, many judges apply only a so-called slap on the wrist for most criminal violations. As long as there is wildlife, there will be wildlife issues to resolve.

Few articles grab the attention of newspaper and magazine readers as stories about wildlife issues do. How many of us have seen a picture of a wolf on the cover of a newspaper or magazine? Not a day passes without someone mentioning one wildlife issue or another. This chapter is a sampler of issues. Remember, an issue is created by at least two people with a difference of opinion. Few people would suggest, for instance, that we set aside a preserve to harbor the Ebola virus, but when it comes to wolves, the two sides are often far apart. From insects and parrots to the consumption of gorillas as food, wildlife issues demand our attention and often our emotions.

The United States is by far the world's largest importer of wildlife. Demand for exotic pets in the United States is astounding, and where there is demand, there will be supply. According to TRAFFIC–USA (1996), 10,000 to 12,000 live primates enter the United States each year and are used primarily for research. A staggering 100,000 to 200,000 live birds are imported each year; an estimated 50,000 are smuggled in. Even more lizards are imported, 1 to 2 million each year, in addition to some 3 to 4 million whole skins and 25 to 30 million manufactured goods. Who buys this stuff?

The United States imports 200 to 250 million ornamental fish each year. The majority come from fish farms in Southeast Asia. Though little is known of the mollusk trade, some 10 to 15 million raw shells make it to the United States each year. Few mollusks are protected by any laws, national or international, so why should we care? Some 200,000 to 300,000 live corals make it to the United States as well. World trade in the above species is enormous. For instance, some 50 million manufactured reptile products are imported worldwide each year. Many species are poached for international trade. Certain special lizards and snakes are traded in the hundreds of thousands, presumably to end up as pets, many of which get flushed down toilets.

The Endangered Species Act of 1973, the Convention on International Trade in Endangered Species of Wild Fauna and Flora (CITES), and the Lacey Act of 1900 and its subsequent amendments have all been used recently to stop international trade in rare insects. Recently, the U.S. Justice Department successfully prosecuted professional butterfly collectors running international businesses. The U.S. Fish and Wildlife Service (USFWS) Division of Law Enforcement has put on notice all potential collectors of living organisms, from plants to animals. Collecting listed species, even insects, is against the law; violators will be prosecuted under the law and, if found guilty, can be fined and sent to prison.

SAMPLE WILDLIFE ISSUES

Butterflies

Williams (1996) reported that Stanford University's Center for Conservation Biology and its climate-controlled laboratory were used by a criminal to raise butterflies and their food plants. Oddly, no one reported the fact that some of the butterflies were threatened or endangered species. When one of the guilty men was apprehended in his home, federal agents seized 2375 specimens worth more than $300,000. Some 14 butterfly species were listed as threatened and endangered under the Endangered Species Act. Another poacher connected to the case was found guilty of smuggling more than 30,000 Mexican butterflies, moths, and beetles. Another two poachers were also found guilty. The investigating team believed that with sufficient resources, hundreds of other collectors could have been indicted. Instead, as Williams reported, the U.S. District Court served 57 grand jury subpoenas and offered poachers the option of forfeiting the specimens in lieu of further criminal action. The U.S. government's conviction rate for wildlife violations is a whopping 94%, which would indicate that only the most flagrant cases are prosecuted. Most potential cases are not prosecuted.

Sometimes wildlife issues never get decided in a court of law. The migration of the monarch butterfly (*Danaus plexippus*) is one of nature's most spectacular events. According to Brower (1995), the monarch's overwintering colonies rank as one of the great biological wonders of the world. Unfortunately, there are only 12 such colonies. Human population growth and deforestation in the Oyamel fir forest enclaves threaten the colonies. Also, increasing use of herbicides across North America kills both larval and adult food resources. Monarchs co-evolved with milkweed (*Asclepias syriaca*), a toxic plant. The reduced area of the fir forest (40,000 to 50,000 ha) and patchy distribution make the forest more vulnerable to deforestation pressures than any other type of forest in Mexico. The monarchs survive from September to March in these forests.

The migratory and overwintering biology of the eastern monarch has become, in Brower's words, "an endangered biological phenomenon." Overwintering occurs in Mexico. The monarchs then migrate northward into the United States at the end of March and early April. They lay their eggs on southern milkweeds and then die. The eggs hatch, and eventually the monarchs fly farther north to southern Canada, laying eggs and then dying along the migration route. After the first spring and two or three subsequent summer generations, the monarchs enter reproductive diapause and begin migrating to their respective overwintering sites.

During December 1995, heavy snow fell in five of the monarch butterfly sanctuaries in the rugged mountains of Michoacan State, west of Mexico City. Millions of butterflies overwinter in the mountains, where cold weather is rare. However, snowfall in 1992 killed an estimated 80% of the wintering butterflies, and the monarchs had not recovered from that episode before the most recent snows hit. What is the wildlife issue here? Surely, something can be done to protect the butterflies.

Perhaps the issue is climate change in the form of global warming. Global warming does not imply that the earth will uniformly warm. One of the critical hypotheses of global warming is that there will be increasing variability in the earth's climate. That is, while global average temperatures rise, climate variability will increase. Whereas snow in the monarch refuges might have occurred once every 25 years, snow will likely occur more frequently while generally higher temperatures will occur throughout the year, particularly during summer. If the monarch population is unable to recover in favorable years, perhaps these mountain populations are doomed. Thus, the issue might be larger in scope than turtles versus tires or ranchers versus wolves. The issue at stake addresses a more fundamental question: Will humans be able to accommodate other species' requirements or will they continue to garner more of the earth's finite resources for their own benefit and leave less and less for all other species?

In Papua New Guinea, native villagers sell tropical insects for cash. Endemic birdwing butterflies are among the most beautiful and prized collectibles in the butterfly world. Some birdwing butterflies being sold have been on the CITES list since 1977. Insects such as giant walking sticks are collected from the wild and sold to exporters. At a time when habitat loss through forest clearing and agricultural schemes is occurring elsewhere in the world, this seems like a ludicrous strategy.

Certainly insects have enormous reproductive capacity. Conservation requirements for vertebrates are much different than requirements for insects and other invertebrates. Parks and reserves are established mainly to protect vertebrates. Papua New Guinea is rich in unique birds, insects, and plants. Several large reserves may not be the answer in Papua New Guinea. Perhaps many smaller reserves would serve the needs of conservation better. Also, insect ranching might best serve the needs of butterfly conservation.

Parrots

On January 4, 1996, the *Chicago Tribune* and *USA Today* reported that Tony Silva intended to plead guilty to federal charges of conspiring to smuggle protected parrots into the United States. On January 29, he pleaded guilty to violating wildlife protection and customs laws and filing a false income tax return. A co-defendant in the case against Silva turned government witness and was sentenced to 3 years probation, 6 months home detention, and 300 hours of community service, in addition to a $10,000 fine and $200 in court fees. One conspirator remains a fugitive.

Tony Silva and his co-conspirators were indicted for conspiracy to smuggle or attempt to smuggle into the United States some of the world's most rare and endangered wild birds, such as parrots and macaws. The value of the birds was estimated to be $1,386,900. Some shipments included extremely rare hyacinth macaws, whose wild population is between 2000 and 5000. Hyacinth macaws, valued at between $5000 and $12,000, are included in Appendix I of CITES. Silva also confessed to filing false income tax returns for failing to claim the income from his crimes.

"Operation Renegade" was a three-year international effort by the USFWS Division of Law Enforcement's Branch of Special Operations to investigate the illegal trade in wild birds. Tony Silva was just one of many egregious violators who were apprehended.

On November 19, 1996, U.S. District Court Judge Elaine Bucklo in Chicago sentenced Tony Silva to nearly seven years in prison without parole for leading an international parrot-smuggling ring and for income tax violations. In addition to imprisonment, Silva was fined $100,000 and ordered to perform 200 hours of community service during a three-year supervised release program that will begin when he is released.

Silva's mother, Gila Daoud, was sentenced to 27 months in prison, a one-year supervised release program, and 200 hours of community service for her part in the bird-smuggling operation.

For the world-renowned parrot expert and his mother, such charges were unthinkable a decade ago. But that was before USFWS investigators began a probe code-named Operation Renegade six years ago that culminated in 35 convictions. The guilty party was a world expert on rare birds, especially rare and endangered parrots. At 35, he was already at the top of his field as the curator of birds at Loro Parque in the Canary Islands, which has one of the finest collections of parrots in the world. Today, there is but one male Spix macaw in the wild left on earth. The curator had convinced private owners of Spix macaws to establish a captive breeding program. As a result of his efforts, the captive population has tripled to 37, and these will be the founders of the next generation that will hopefully be released in the wild.

The author of two popular coffee table books on parrots, Silva was in demand as a speaker and was an outspoken advocate for bird protection. However, an avalanche of criminal evidence against him would have been released had his case gone to trial. He and his mother were charged with smuggling into the United States $1.3 million worth of rare birds, including blue-throated conures, Lilacine Amazons, African grey parrots, and red-vented cockatoos. Some 180 hyacinth macaws were also among the booty recovered. There are only a few thousand hyacinths left in the wilds of Bolivia, Brazil, and Paraguay. He had shipped some 30 hyacinths to a private collector and all died in transit. The live birds could have been worth $9000 apiece. Assistant Attorney General Lois Schiffer said, "These crimes threaten our endangered species and global biodiversity." His crimes carry penalties of up to $2.5 million in fines and 45 years in prison. His mother is also charged with possession of elephant tusks and with aiding her son in filing false income tax returns. She faces up to 50 years in prison.

Operation Renegade also netted an animal keeper at the Playboy Mansion. She was sentenced to 37 months in prison for illegally smuggling eggs of threatened species from Australia to the United States. As of April 1996, 15 individuals received 186 months in prison (TRAFFIC–USA 1996) and the animal keeper was free pending an appeal of her sentence. The smuggling ringleader received a $10,000 fine and a five-year prison sentence, and another defendant received 41 months in prison and three years probation for his participation in the smuggling ring. More than 800 Australian cockatoo eggs worth more than

$1.5 million were illegally taken from the wild and smuggled into the United States. All cockatoos are protected by CITES.

In Central and South America, 16 species of macaws, large members of the parrot family, are found. Nine species of macaws, including the Spix macaw mentioned above, are endangered. Macaws are a poacher's favorite bird. They are intelligent, large, and colorful, and demand for them in the pet trade is therefore high. Guatemala's scarlet macaw is on the verge of extinction (Colclough 1996). Some 500 are estimated to remain in the wild. Habitat destruction, hunting, poaching, and taking of nestlings have contributed to their demise. While a chick can bring $300, an adult macaw, depending on the species, can be sold for $4000 outside the country. Smugglers, of course, pocket most of the money. A captive breeding program might be able to repopulate the forest, but unless habitat is saved, there will be no room for the scarlet macaw.

In 1990, some one-third of the 450,000 live birds imported into the United States were parrots. Japan, Belgium, the Netherlands, and Germany are also major importers of parrots. The vast majority of parrots sold on the import market are caught in the wild. In the 1980s, at least 50,000 parrots were smuggled into the United States each year. No one knows how many parrots died in transit and were never counted. Asian species come mainly from Indonesia, although Central and South America supply the bulk of the parrot trade.

On behalf of all citizens of the United States, the prosecution was a cooperative effort between the U.S. attorney's office in Chicago, the U.S. Internal Revenue Service, and the Wildlife and Marine Resources section of the Environment and Natural Resources Division of the U.S. Department of Justice. "The severity of the sentence in this case sends a clear signal that the United States will absolutely not tolerate the depletion of irreplaceable natural resources for personal gain," said USFWS Acting Director John Rogers.

Wolves

A picture of a wolf (*Canis lupus*) showing its teeth in a newspaper or magazine article is a real attention-grabber. This is no small feat because wolves tend to growl and growling does not fully expose the canines. Yawning and howling expose the canines, but a yawning wolf is not very threatening. On March 14, the *Tampa Tribune* printed an article on wolves in Yellowstone National Park. The headline in bold letters read "a pack of trouble." The Montana Stockgrowers Association had apparently printed a pamphlet showing a wolf dressed as a Washington bureaucrat. It seems that the Montana ranchers did not fully appreciate that the Yellowstone wolves are part of the native wildlife. Of course, visitors and tourists have shown overwhelming support for the reintroduction of wolves. The thrill of seeing a wolf in the wild can be the highlight of a trip to Yellowstone National Park. Anyone can just walk up to Old Faithful and sit down on a bench to see it vent, but seeing a wolf is extremely rare and thus exciting.

Wolves roam widely and will kill rodents, deer, and elk, as well as sheep and cattle. When a local resident shot and killed one of the Yellowstone wolves that had left the park, he was fined $10,000 and sentenced to six months in prison. Local ranchers fear the worst for their livelihood. Perhaps as many as three million tourists visit Yellowstone each year, and most of them are on the lookout for wolves. The issue is clear. Should the U.S. government reintroduce wolves into Yellowstone National Park?

An editorial in the *Seattle Times* reported that "fears of ranchers outside the park...have not been realized." No livestock were killed in 1995 and only nine sheep were killed in the first half of 1996. Polls of visitors to Yellowstone National Park show that the public, by a significant margin, supports the reintroduction of the wolf.

Corals

Corals are marine animals of the invertebrate phylum Cnidaria. They are related to jellyfish and also sea anemones. Millions of individual coral, living in colonies, make up coral reefs. Coral reefs are known throughout the world for the variety of fish and other organisms that live in and around them. Reef-building corals are only found in warm tropical waters. Other corals do not form reefs. Although no species of coral is endangered, local reefs around the world have been severely damaged. The World Conservation Union (IUCN) considers all coral to be commercially threatened however, because if these reefs are commercially harvested, corals will become endangered. All reef-building corals are listed in Appendix II of CITES.

Reef-building corals are used commercially for a large variety of industrial uses, such as road construction, building material, bricks, and tile (TRAFFIC–USA 1996). Living corals are collected and sold for aquariums. Often coral is blasted loose with dynamite, and large sections of reef are destroyed. From 1989 to 1993, 12 million pieces of raw and 1.6 million pieces of live reef-building coral were reported in world trade. This figure is probably low, as domestic use of coral is not reported. The United States is the largest consumer of coral and imports over 90% of all commercial coral.

The origin of each piece of coral is difficult to determine and is generally not recorded. TRAFFIC–USA (1996) recommends that "it is best to refrain from buying *any* coral, whether for decoration, for your aquarium, or as jewelry." No one knows if yields are sustainable, and very few countries have any management plans for reefs.

Spiders

Many people dislike spiders (class Arachnid), and some even find them offensive. Others, such as the authors, are fascinated by spiders and appreciate them for what they are. One spider family, tarantulas (Theraphosidae), is in demand by collectors (Fitzgerald 1989). Pet connoisseurs in the United States prize a Mexican tarantula, the red-kneed tarantula (*Brachypelma smithi*), because it is docile and beautiful. Although *B. smithi* is listed in CITIES Appendix II, many illegally make it into the United States each year. Almost all of them are wild-caught. They sold for about $12 in 1989. The species has a low reproductive rate. Attempts to monitor trade in spiders are almost impossible and the ban is not always enforced. Who would want a pet spider? Apparently, 30,000 people did in 1989.

Manatees

The West Indian manatee (*Trichehus manatus*) is an aquatic herbivore that seems to permit more children than adults to pet it. Indeed, the shy manatees of Silver Springs, Florida, seem attracted to children frolicking in the clear water of the springs and often approach children to be stroked. When pursued, the manatee can swim faster than the best of swimmers. It is not unusual to hear residents of Florida claim that "there were more manatees 100 years ago than there are today" or "there are fewer manatees now than there ever were." Most manatees in Florida have motorboat scars on their backs. Researchers studying manatees recognize individuals by their unique scars. Recreational boaters have been successful in blocking attempts to establish speed limits in rivers with high manatee populations. Many see the manatees as putting a crimp in their recreational style.

During the first three months of 1996, 258 manatees died in Florida, 57 more than the total for all of 1995. Some 2639 manatees remain in Florida. News of the deaths spread around the world. As one newspaper suggested, "If the death rates continue, the population of manatees in this region [the western part of the state] could be wiped out by fall." Disease is a suggested killer, but pathologists were investigating each death. The manatee is listed as endangered. Although many residents were concerned, no large preserve has been set aside strictly for the manatees. The wildlife issue here is clear. If manatees are endangered, why is there no large preserve for them where they will be protected from motorboats? Can a large reserve be set aside for this wide-ranging mammal? In 1995, a manatee nicknamed Chelsea showed up off Rhode Island and was airlifted back to Florida. During 1996, manatees have been sighted in Texas coastal waters. Are these dispersing individuals seeking out new territories or are they fleeing an overabundant Florida population?

Cockatoos

In August 1994, six people were indicted for smuggling into the United States cockatoo eggs obtained from the wilds of Australia. One suspect had 49 eggs hidden on his body, as reported in the *Daily News* on May 18, 1994. An adult cockatoo can sell for more than $12,000. Once the eggs pass customs, they are incubated, and the young are raised and then sold. Smuggling eggs has become a popular way of transporting species. The eggs can be kept in a vest close to the body and thus are very difficult to detect. The United States is the largest market for wild-caught birds. Seventeen people in Australia, New Zealand, and the United States were convicted, and more convictions were expected. Smuggled species include the rose-breasted, Major Mitchell, red-tailed black, white-tailed black, greater sulphur crested, and slender-billed cockatoos. Of course, Australian cockatoos are protected by CITES.

Wildebeest

The European Economic Community (EEC) signed an agreement to import beef from Botswana, a landlocked semi-arid and desert country just north of South Africa. Botswana contains part of the Kalahari Desert, and water is scarce throughout the country except for the Okavango Delta in northern Botswana. During epidemics of hoof-and-mouth disease, the EEC refused to purchase beef from Botswana. The Department of Animal Health, a government agency, was given the task of devising a way to prevent hoof-and-mouth disease from contaminating the domestic stock. Since the rich and powerful owned the largest cattle ranches, something had to be done by the government. Hoof-and-mouth disease is carried by African Cape buffalo and some species of antelope, and these carriers could presumably infect beef cattle, although it is not known whether this has actually happened. The government of Botswana decided that fences were the most effective means of segregating wild animal populations from cattle. There are, of course, two ways to consider the situation: either the fences serve to keep beef cattle in and wild animals out or, conversely, they serve to keep beef cattle out and wild animals in. The result was ultimately the same: a massive die-off among the wild herds occurred as a direct result of the fences (Owens and Owens 1984).

FIGURE 3.1   Two hyenas capture and eat alive a wildebeest in Ngorongoro Crater National Park, Tanzania.

The cycle of rain and drought in the Kalahari, the great desert of southern Africa, ebbs and flows over time. In some years, the rains are good and the great herds of herbivores are nowhere to be found. Instead, the animals are scattered across the desert, following the rain clouds and feeding on the grasses. From 1974 to 1979 the rains had been generous, and great numbers of wildebeest (Figure 3.1), oryx (Figure 3.2), springbok, and eland penetrated deep into the desert. In 1979, the rains failed to come and the vegetation withered. The wildebeest knew, somehow, that they must head north or east to the only permanent water sources.

Slowly the wildebeest funneled north and east. Groups of tens became thousands, and thousands became tens of thousands. Some 90,000 strong plodded north, their heads hung low as the sun beat down on them. Dust rose from their hooves and turned the sky brown. They trekked for more than 300 miles, only to meet with the fences. They could smell the waters of Lake Xau, but they could not break through the fences. More than 100,000 wildebeest were forced to stumble along the fences, driven by instinct and the smell of water. Forage and shade were nonexistent. Now they could see the water, but their route was blocked by villages of herdsmen and their dogs. The last great die-off of Botswana's great herds had begun.

FIGURE 3.2 A water hole at Etosha National Park, Namibia. A giraffe, zebras, kudus, and oryx find water in the desert.

Some 87% of all wildebeest died. The herd of more than 100,000 animals was reduced to 13,000. Emaciated animals piled up along the fences. Some walked along, stumbled, rose, and fell again. Their chests heaving, they were unable to run when the dogs set upon them. Villagers stoned wildebeests to death. An almost surrealistic horror unfolded day by day, and the stench of death filled the air. The government of Botswana continues to erect fences today in compliance with the EEC's wishes to enjoy healthier beef cattle. One of the most powerful wildlife issues of our time was settled when the second largest migration of herbivores was wiped off the face of the planet within six months.

Sea Turtles

On December 31, 1995, the headline "beach battle pits turtles vs. tires" appeared in the *Palm Beach Post.* The article reported that "Endangered sea turtles may push cars off Daytona's beach. But tourism would take a hit, the county fears." For thousands, perhaps tens of thousands of years, sea turtles have been nesting on Daytona Beach. Daytona is one of a few towns that still permits motorists and their vehicles on the beach, for a five-dollar fee. The fee generates millions of dollars for the city, and local businesses thrive from the tourism. Businesspeople fear that a lawsuit filed on behalf of the sea turtles, if successful, would cause an economic disaster. An environmentalist claimed the county was "living in the Dark Ages."

As is typical of a wildlife issue, the different interest groups take sides and try to create an either/or issue. Either you are for business or against business. Either you are for the turtles or against the turtles. Maybe a different breed of tourist would show up during the sea turtle nesting and hatching season and would be willing to pay even more money to see young hatchling sea turtles make their way to the sea. Certainly vehicles would not be restricted from the entire beach. Perhaps the solution lies somewhere between the interests of the vested parties. In either case, the wildlife issue is clear: the Endangered Species Act of 1973 protects the sea turtle. By March 1996, county officials had submitted to the USFWS a conservation plan that specifically addressed the turtles' requirements. A federal judge postponed the trial, issued a temporary restraining order against nighttime driving and parking on county beaches, and imposed a protection area from the sea turtle nesting dunes to the beach. Is this an either/or issue?

On January 8, 1996, the *New York Times* reported that on December 29, 1995, the U.S. Court of International Trade issued an order against the Department of Commerce for "failure to bar shrimp imports into the U.S. from nations that are not implementing procedures to reduce sea turtle deaths." The judge cited the Department of Commerce for failing to carry out the Endangered Species Act, which requires the department to ban shrimp exports from all countries that do not have so-called "turtle excluder devices" on their nets.

On February 10, 1996, Reuters News Service reported that three people were caught stealing 372 endangered loggerhead sea turtle (*Caretta caretta*) eggs from their nests. The eggs are delicacies in local ethnic communities in southern Florida. An extraordinary Palm Beach police officer noticed a large cloth bag containing the eggs in the back seat of a car and was credited with the bust. One man received 2 years in jail and another 15 months for violating the Endangered Species Act.

Sharks

A recent advertisement in *American Health* magazine quoted an executive medical director: "Research on shark cartilage has been headline news, so more of my patients have been asking me about it. I've been impressed with the research, and Lane Labs, the maker of BeneFin (a trademark of Lane Labs–USA, Inc.), does more than anyone." The advertisement also quoted Dr. Lane, author of the book *Sharks Don't Get Cancer* and an authority on shark cartilage, who said, "BeneFin is the most effective shark cartilage I have studied."

Researchers agree that sharks do not get cancer. The claim is that if you eat their cartilage (i.e., if you take BeneFin), you won't get cancer either. Although clinical trials have only just begun, some 25,000 to 100,000 people have purchased and presumably taken shark cartilage. There is one certainty, however. Sharks are being killed indiscriminately. One small factory in Costa Rica harvests more than 111,000 large coastal sharks annually. The U.S. commercial shark fishery annual quota is 150,000 large coastal sharks.

The numbers from Costa Rica indicate intense fishing pressure (Dold 1996). Researchers suggest that the shark fishery will be decimated in a few years as some species require 30 years or more to reach sexual maturity and some have only one young every other year. In U.S. coastal waters, shark populations have declined 70 to 80% in the last two decades. Factory ships lay so-called "longlines" of hooks for a hundred miles and harvest anything that bites, including sea turtles and billed fish such as sailfish. There are virtually no controls beyond U.S. borders. Indeed, few people believe sharks should even be protected. Nevertheless, environmentalists convinced the CITES signatory nations to investigate and document the international trade in sharks. While scientific clinical tests of the efficacy of shark cartilage in cancer treatment will take years to complete, sharks will continue to be harvested for meat, fins, jaws, and cartilage. The fins remain the most valuable part of the shark. Asians in Hong Kong are willing to pay upwards of $20 per pound for the fins to turn them into shark fin soup.

Jellyfish

"Aggressive sea lice ambush swimmers," read the *Palm Beach Post* on April 17, 1996. Palm Beach County's famous beaches were the scene of reports from swimmers who were "covered with sea lice" and endured three to five days of itching, rashes, fever, hives, and general discomfort. Outbreaks are possible from March to August. The greatest infestation occurs in May, when the Gulf Stream is closest to the shore. Researchers believe that a microscopic juvenile jellyfish floating in the Gulf Stream causes the welts. As more people crowd the beaches, more cases are expected.

Alligators

There are many alligator (*Alligator mississippiensis*) farms where animals are raised for food and their skins sold. The wild population of alligators in Florida is estimated to be between one and two million animals. Alligators are legally hunted in Florida. So successful is protection of the American alligator that

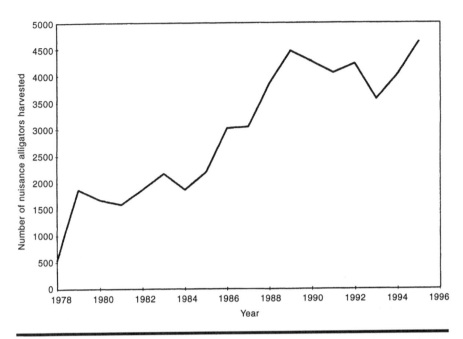

FIGURE 3.3 The number of nuisance alligators harvested in Florida since 1978.

some alligators have become a nuisance. Residents who decide that an alligator has become a nuisance need only pick up the phone and call their local state game and freshwater fish representative. The nuisance alligator will be dispatched without charge. An alligator simply sunning itself on someone's lawn can be considered a nuisance. The definition of "nuisance" is left to the individual. One in three calls results in a removal. Figure 3.3 shows the number of nuisance alligators killed in Florida over the last 15 years. More nuisance alligators are killed each year than are killed during alligator hunting season.

Harp Seals

On December 18, 1995, Canadian Fisheries Minister Brian Tobin announced that seals were to blame for the 99% decline in cod in the North Atlantic. To stem the decline in one of the world's most productive fisheries, the Canadian government announced that the seal harvest would be expanded from 186,000 permits to 250,000 in 1996. From 1983 to 1994, approximately 57,000 seals were harvested every year. Primarily so-called young of the year (from ten days to several months old) are harvested by shooting or clubbing them.

The *Detroit News* wrote that Tobin estimated the seal population to have more than doubled since the 1970s, to 4.8 million seals, and could reach 6 million by the year 2000 if no seal hunt is conducted. Tobin suggested the seals eat 6.9 million tons of fish and other prey in a year, including a billion cod, weighing 142,000 tons. The International Wildlife Coalition and the International Fund for Animal Welfare are among some 40 organizations that oppose the seal hunt. Since most Atlantic fishermen are out of work, the seal hunt provides them the opportunity to earn cash from the harvested seals. Seal penises sell for over $200 in oriental markets, for instance.

What is the wildlife issue? The Atlantic fishery has been decimated, and many fisherman, especially those from Newfoundland, are out of work. Is the seal harvest the real issue? Certainly not. The heart of this issue lies in the Atlantic fishery. How can the scientific method be used to address this question?

Reptiles

According to a February 1, 1996 CITES Update, USFWS special agents served search warrants in Florida, New York, North Carolina, New Mexico, Indonesia, and the Netherlands against suspected smugglers of reptiles. As is so typical, USFWS agents were selected for their "expertise in wildlife import, export, smuggling, and illegal commercialization offenses." The international investigation involved suspected reptile smugglers operating in Australia, Indonesia, and the Netherlands. The United States is the world's largest importer of wildlife, and live exotic pets are in demand. In addition to frilled lizards (*Chlamydosaurus kingii*) from Indonesia, the Fly River turtle (*Carettochelys insculpta*), the green tree python (*Morelia viridis*), and two species of blue-tongued skinks (*Tiliqua gigas* and *T. multifasciata*) were among the species smuggled. All these reptiles are protected by law in their countries of origin, and hence their trade is strictly forbidden. However, the high demand for new and exotic specimens encourages smugglers to pursue wild-caught reptiles.

The investigation began in September 1994, when Netherlands police gathered information from Indonesia. Through an agreement known as the Mutual Legal Assistance Treaty, the United States and the Netherlands cooperate with each other in criminal matters. Indonesia is becoming a popular target for reptile smugglers. For instance, a firm owned by an Indonesian made over 1700 shipments totaling approximately 250,000 animals, mostly reptiles, to the United States. Shipping records were obtained by the International Primate Protection League using the Freedom of Information Act. An analysis of the firm's customers revealed many nationally known reptile dealers across the United States. Unfortunately, shipping records are not specific enough when it comes to reptiles, and generic names are often used for species that even an expert would have trouble identifying. The declared value of the reptiles was $2 million. Further analysis is under way.

Iguanas

Imagine standing before a federal judge and entering a felony guilty plea for "knowingly and unlawfully receiving, concealing, and facilitating the transportation of endangered Fiji banded iguanas (*Brachylophus fasciatus*)." That is exactly what one man from Bushnell, Florida, did. Since the Fiji banded iguana was protected by CITES, the Lacey Act, and the Endangered Species Act, the case was investigated by the USFWS and prosecuted by the U.S. attorney's office and the Environment and Natural Resources Division of the U.S. Department of Justice (U.S. Attorney's Office, Middle District of Florida, press release April 19, 1995).

On April 19, 1995, the defendant received five months in prison, five months home detention, two years supervised release, and a $3000 fine.

Rio Grande Silvery Minnow

The New Mexico Department of Game and Fish lifted the bag limit on most species of fish in the Rio Grande over a particular stretch of river in April 1996, the *Albuquerque Tribune* reported. The fish would likely become stranded when the spring irrigation diversion began. Agricultural lands along the river are irrigated in the spring, and stretches of the river are nearly dry. The Rio Grande silvery minnow remains a protected species, however. After Navajo Dam was built on the Rio Grande to control flooding and support spring irrigation, much changed along the Rio Grande, a unique riparian habitat that stretches longitudinally through New Mexico. Deep water in the dam is cold, and trout were introduced. Native squawfish soon disappeared, and snails were eaten by the trout and displaced by the cold water near the dam. New Mexico's beautiful cottonwood trees also suffered, as they require spring floods to set seed. Habitat alteration occurred, and nonnative species such as tamerisk invaded, destroying more wildlife habitat.

On April 19, 1996, The Associated Press reported that the endangered Rio Grande silvery minnow is threatened because the flow of the Rio Grande near Socorro, New Mexico, has been diverted for irrigation projects. The USFWS, the press release claims, had informed the Middle Rio Grande Conservancy

District that water diversion for agricultural crops, such as New Mexico's famous green chili peppers, would completely drain the Rio Grande for a stretch of 30 to 40 miles all the way to Elephant Butte Lake. The Rio Grande below the San Acacia diversion dam near Socorro supports 70% of the known populations of the Rio Grande silvery minnow.

The Middle Rio Grande Conservancy District has, in the past, steadfastly refused the USFWS's request to divert less water to support the endangered minnow. This is, of course, a violation of the Endangered Species Act because the act prohibits habitat destruction. The district claims it has a legal obligation to provide 461 farms along the river with adequate irrigation water. Yet no legal agreement between the state of New Mexico and local farmers can usurp federal law, including the Endangered Species Act.

A temporary solution recently agreed upon will be to release some water owned by the USFWS to support the minnow instead of directing the water to Bosque del Apache National Wildlife Refuge, where water is also needed. This will keep the minnows in water for 30 days while a long-term solution is negotiated.

Farther north, the city of Albuquerque was considering whether to release "unused" water from its 1995 allotment from one of the many water diversion projects. Albuquerque would lose the water anyway. Albuquerque's water could then be redirected into a dry portion of the river about 12 miles north of Socorro. The water would help preserve the minnow population. The city has to pay for the water even if it is not used. At least for the month of May 1996, the minnows were safe. But how long can solutions such as these go on saving the minnows?

Semi-Tropical Insects

On April 30, 1996, the *Gainesville Sun* reported that millions of cockroaches, locusts, and grasshoppers had invaded the southern Iranian town of Baft, according to an Iranian official news agency report. After record rains fell in the preceding months, the insects swarmed the streets. Pesticides were being sprayed to combat the outbreak. The residual effects of the pesticides were not mentioned.

Why is this a wildlife issue? Normally, when large concentrations of insects "descend" on an area, large flocks of insectivorous birds "descend" upon the insects. No mention of birds appeared in the article. The work performed by the birds (that is, the consumption of insects) must now be done at great cost by humans using mechanical or chemical means. Are the insect outbreaks the problem or merely a symptom of a greater problem?

Australian Reptiles

The investigation by USFWS and Australian wildlife authorities started in Sydney and ended in a Wal-Mart store in Minnesota. On January 18, 1996, 40 reptiles, including rare and beautiful mountain dragons (*Amphibolurus diemensis*), geckos (*Gekkonidae* spp.), skinks (*Scinidae* spp.), and black-headed pythons (*Aspidites melanocephalus*), were smuggled via a courier flight to the United States. The lizards were kept in socks and pillowcases marked "fragile electronic equipment." Many of the reptiles perished in transit. Three days later in Tennessee, police dogs picked up the scent. The USFWS was notified, and the package was forwarded intact to its destination in Florida. Police then followed the package and its carrier to a private residence. The resident was confronted and quickly agreed to cooperate with federal officials. Another package was sent to a Wal-Mart store in Minnesota. When it arrived, store employees notified the USFWS, and $10,000 worth of dead reptiles were found. The investigation is still under way, and no arrests have been made yet. On February 6, 1996, the Australian Associated Press reported that a widely advertised "alternative pet" supplier in Minnesota and the role of courier service employees were being investigated.

Frog Gigging

About two tons of frogs were being taken from Big Cypress National Preserve in Florida every month, according to the page one headline of the *Miami Herald* on April 29, 1996. What the froggers, or giggers, as they are known, were doing was illegal according to the chief ranger of the preserve. One environmentalist referred to the taking of frogs as "amphibian clear-cutting." Moreover, two federally endangered species of frogs live in the preserve.

Giggers come from as far away as Georgia and Louisiana to get in on the bounty. During the month of March, a park service ranger counted 216 airboats being used to hunt frogs. On a routine check, one airboat operator caught 235 pounds, or about 1200 frogs, in one night. No one knows how many giggers there are or how many frogs are taken each season, let alone if listed species are taken. The frogs are systematically hunted at night with an airboat and a three-spiked jabbing stick called a gig. The airboat is driven slowly over the reeds to flatten them, and then the gigger circles back for the harvest. Using a flashlight, the gigger carefully scans the reeds for two eyes. When the eyes are seen, the gig is thrust forward, and the frog is impaled, lifted out of the water, and scraped into a plastic bucket strategically placed a gig's length away. The giggers are well practiced and deadly. Rarely does a frog escape.

European Birds

Studies in the Netherlands showed that acidic pollution is harming wild bird populations. Acid rain and other forms of acidic pollution leach calcium from the soil, which has led to a decrease in snail populations. With fewer snails to eat, birds have been getting less calcium in their diets. With less calcium, the birds' egg shells are thinner, and hence more porous, and many fail to hatch. Experiments performed by researchers confirmed that birds that were fed snails produced fewer defective eggs than birds that received fewer snails.

Kangaroo Rats

In March 1995, a California resident was introduced to the Endangered Species Act. A farmer wanted to plant part of his 700 acres of land with vegetables and market them in nearby Los Angeles. The land, however, was occupied by Tipton's kangaroo rats, a federally listed endangered species. When the farmer began plowing his fields, he was charged by the USFWS with not only killing an endangered species but also destroying its habitat. Under the Endangered Species Act, both killing and habitat destruction of endangered species are illegal.

Snail Darters

In 1978, the U.S. Supreme Court in *Tennessee Valley Authority v. Hill*, 437 U.S. 153, 184, rendered "the plain intent of Congress in enacting this [the Endangered Species Act] was to halt and reverse the trend towards species extinction, whatever the cost." The operation of the $100 million Tellico Dam across the Little Tennessee River was blocked by the snail darter, a small fish thought to be an endemic species.

The Tellico Dam was begun in 1967 by the Tennessee Valley Authority (TVA). When the project was 80% complete in 1973, a University of Tennessee ichthyologist discovered the snail darter, an unknown fish species. The snail darter was thought to exist only in the Little Tennessee River and was declared an endangered species under the newly enacted Endangered Species Act of 1973 (Littell 1992). A long court battle ensued, and in 1978 the Supreme Court affirmed an injunction against Tellico, thus halting the project as the dam was nearing completion. Congress was quick to act. It decided the snail darter was not worth saving and allowed the TVA to complete the project in 1979, creating a 33-mile reservoir and a 129-foot dam. Shortly thereafter, the snail darter was found in other rivers in Alabama, Georgia, and Tennessee. The same snail darter is no longer endangered but remains threatened.

In the *Hill* decision, the U.S. Supreme Court made clear that extinction of a species for any reason is not an acceptable alternative no matter how worthy the cause. Only Congress could circumvent the law, as it did in *Hill*. The intent of Congress in enacting the Endangered Species Act was to place the highest priority on endangered species, and Congress viewed the value of an endangered species as "incalculable." In *Hill,* the Court affirmed that Congress's command was "to halt and reverse the trend toward species extinction, whatever the cost." After the snail darter, time and again the Supreme Court has invoked precedence to block actions that threaten a species' survival. The snail darter case is one of the most famous in Supreme Court history, and it set the stage for the 1996 battles over the Endangered Species Act.

Pandas

In April 1991, *Time* magazine (Sachs 1991) reported on a visit to Quanzhou, China, under the sponsorship of TRAFFIC–USA (Trade Records Analysis of Flora and Fauna in Commerce), a joint program of the World Wildlife Fund (WWF) and the World Conservation Union (IUCN). The panda (*Ailuropoda melanoleuca*) is one of the most widely recognized animals on earth. It is the symbol of the WWF. In 1991, an estimated 1000 remained in the wilds of China, their populations decimated by the disappearance of their bamboo hillside habitat. Without a $10,000 deposit, a panda skin that was for sale would not even be available for inspection. Perhaps a better deal was in the offing.

Soon the smugglers made their pitch—two live young pandas, chained together and ready to go, for only $112,000. The coast between Taiwan and mainland China was crowded with shops carrying on a multitude of businesses, including wildlife smuggling. Live Amur leopards (approximately 40 remained in the wild in 1996), gibbons, eagles, and golden monkeys were for sale. There were more Amur leopard skins available at $380 apiece than there were live animals in the wild, a tragic fact. Taxidermy shops were crowded with stuffed animals and birds. Although two panda smugglers were executed, the wildlife trade was brisk. As far as the WWF could tell, there was little or no enforcement of wildlife laws or international agreements.

Apparently, exotic pets are still status symbols, and pelts of rare animals are often hung in homes. People believe that eating elaborately prepared dishes featuring endangered animals carries mystical connotations of power. For example, if you eat a tiger's eye, your eyes will see like that of the tiger. Perhaps this explains why tiger penises are in demand (Figure 3.4). One has to wonder what you are supposed to do with it—sleep with it under your pillow, carry it around in your pocket, eat it, or just fondle it? Actually, the tiger penis is put into a bottle of table wine that is sipped during dinner. The sad fact is that

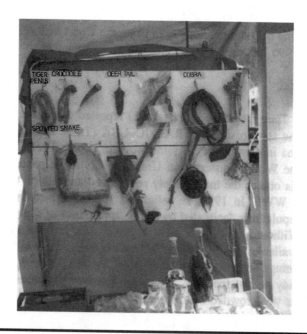

FIGURE 3.4   Display in a local shop in China.

China's wildlife and its unique biodiversity are on their way out. The panda will likely go extinct in the next 20 years. Is this a wildlife issue, and if so, what can we as individuals do about it? Are we willing to live in a world without pandas? Will we deny future generations the sight of living, wild pandas? Concern for species other than humans has a long way to go in Asia and especially China, where more than a billion people reside.

Whales

A White House press release on February 9, 1996 confirmed that President Clinton had delivered a message to Congress aimed at curbing Japan's insatiable desire for whale meat under the Pelly Amendment. Japan continues to whale in the newly created Southern Ocean Whale Sanctuary in the Antarctic. This reserve circles the globe in the only part of the world where this is possible, the Antarctic. The sanctuary is 30 million square miles and was established in 1994 to protect whale populations. The Japanese government was the sole objector to the establishment of the sanctuary, and the Japanese continue to violate its boundary.

Japan's whaling is conducted under the guise of "taking whales for scientific purposes," which is a loophole in CITES. Japan announced its intention to kill 440 minke whales (*Balaenoptera acutorostrata*) in 1996 for "study." The catch ends up as high-priced sushi that sells for $200 a pound.

Kangaroos

Certain restaurants in Australia serve kangaroo. At Denekas Cafe Restaurant in Leederville, Western Australia, the chef's special recipe is "seared kangaroo." A 100-g kangaroo steak is brushed with olive oil and seared on an open grill until medium rare. The steak is served with beetroots and a sauce of parsnip and garlic. If the grilled dish is not to your liking, you can try the Asian stir-fried kangaroo.

Harvesting some species of kangaroos has become a way for farmers in Australia to earn extra cash and also protect their farms. The market for kangaroo meat and skin has developed in Australia and elsewhere. Kangaroo carcasses are inspected by authorities representing the state and territory and then sold to a processor. There are only a small number of so-called shooters and processors licensed to deal in kangaroos. In addition, there are penalties for killing kangaroos illegally.

Kangaroo harvesting does not occur in national parks or reserves (Caughley et al. 1987). A reported 95% of the harvested kangaroos are the eastern grey kangaroo (*Macropus giganteus*), the red kangaroo (*M. rufus*), and the western grey kangaroo (*M. fuliginosus*). Wallaroos (*M. robustus*) and whiptail wallabies (*M. parryi*) make up the bulk of the remaining 5% of the harvest.

Certainly there are alternatives to shooting kangaroos. Deterrents such as electric fences, noise cannons, and other methods can exclude kangaroos from selected areas. Contraceptives work, but only on kangaroos that can be treated. Quotas set by the Australian government ensure that the populations of the harvested kangaroos do not exceed established limits based on scientific recommendations. The Wildlife Protection Act of 1982 requires that management plans be in place and approved before harvesting is permitted.

Tropical Fish

On February 2, 1996, U.S. customs agents busted a smuggler and confiscated his 1980 Chevy pickup. He won't need it for perhaps five years. At a border crossing from Mexico to the United States, customs agents discovered 36 tropical fish hidden in five plastic bags inside a converted gas tank on the pickup (TRAFFIC–USA 1996). The bags included garibaldi (*Hypsypops runicundus*), California's state marine fish. Although not listed as endangered or threatened, there is a three-year commercial moratorium in effect in Mexico and California.

The smuggler faces a penalty of five years in prison and a $100,000 fine. "Only 15 of the 36 fish were still alive when agents removed them from the leaking gas tank. Four more died from lack of oxygen and 11 garibaldi were taken to Sea World to recover" (TRAFFIC–USA 1996).

Bears

The demand for bear bile from South Korea, Japan, and Taiwan has had a detrimental effect on bear populations worldwide. Indeed, bear bile is more valuable than its weight in gold. In October 1991, a black bear was found dead in Riding Mountain National Park in Canada with only its gallbladder removed. International trade in animals and animal parts is regulated by CITES, but some countries such as Taiwan and South Korea are not signatories of the agreement. Moreover, CITES does not address bile taken from live bears.

Traditional Asian medicine claims that patients suffering from deafness, baldness, rheumatism, and numerous other ailments can be cured with a small dose of bear fat or meat, blood, bones, or bile, depending on the illness. Menus in restaurants often contain something from bears. In fact, bear gallbladders rank with rhino horn, ginseng, and deer musk as having a most powerful curative effect. Wildlife farms are a solution to the dilemma of taking animals from the wild, and in 1992 the Chinese were reported to have between 5000 and 8000 bears on bile-milking farms (Mills 1992). Asian bears are endangered.

On the last day of 1995, the *San Francisco Examiner* announced that on January 1, 1996 California would increase the penalties for illegal killing and selling of fish and wildlife to a maximum sentence of one year in jail and/or a $30,000 fine. Penalties were increased because a hunt club in northern California was killing bears and selling the parts to Korean counterparts.

The North American Bear Society reported that human garbage is one of the biggest threats to American black bears. Bears are opportunistic feeders and will eat garbage if given the opportunity. As bears lose their fear of humans, the result will be more nuisance bears that must be destroyed. Each spring in New Mexico, bears forage for food and water in the mountain communities east of Albuquerque. Dry winters and summers are hard on bears because their food supplies dwindle. Lack of quality food forces bears to search over wider areas, and this brings them into contact with humans. About 45 black bears live on the eastern slopes of the Sandia Mountains east of Albuquerque, New Mexico's largest city. The bears overturn barbecues, rummage through picnic sites, and search in open dumpsters for food. In 1989, 22 bears were captured in Albuquerque and returned to the mountains. Bear attacks on humans are rare in New Mexico; however, human attacks on bears are usually fatal. In 1995, two cubs were killed in a botched attempt to relocate them along with the sow as they were searching for food.

Bears seem to attract attention wherever they are found. The *Okeechobee News* reported on July 3, 1996 that a black bear from the Florida panhandle was captured in Louisiana and relocated to the Apalachicola National Forest southwest of Tallahassee. The bear, outfitted with a radio-collar transmitter, had apparently wandered nearly 350 miles west. Wildlife biologists are unsure, the article reported, why the bear roamed so far away from Florida. This bear previously had been captured and relocated three times in Florida. Why might this bear migrate? If habitat conditions provide enough food and breeding opportunities are available, would a bear migrate? How might we investigate these concerns scientifically? Perhaps there were other male bears protecting and maintaining territories, and this particular bear was unable to compete. By leaving the area, he might have been able to breed elsewhere. Indeed, one Louisiana resident was quoted in the article as saying, "If that bear came 400 miles from Florida, he needs to stay here. He's a good breeder." Perhaps the opposite is really true. He may have been a poor breeder forced out by better breeders. Perhaps he should have been left in Louisiana. How can we test these ideas using the scientific method?

Elephants

On January 15, 1996, Chinese customs officials seized 1600 pounds of Asian ivory valued at about $840,000. Reuters News Service made the announcement on January 21, 1996. Finding the ivory was no easy task; 72 tusks had been cut up into 133 pieces and hidden in 22 bags of teak intended for a crafts company. China has expanded the investigation because of obvious smuggling. These 72 tusks from 36 elephants represent a substantial portion of the remaining population of some 150 to 300 wild Asian elephants. Considerably more elephants remain in Africa (Figure 3.5), but their populations have been dramatically reduced.

Saiga

Most people have probably never heard of the odd-looking saiga antelope (*Saiga tatarica*), but medicinal manufacturers in China and Hong Kong have. TRAFFIC–USA (1995) reported that trade in saiga horns has reached unsustainable

FIGURE 3.5   A herd of elephants stop for a drink in a sand river in Ruaha National Park, Tanzania.

levels. The saiga is a tiny, horned antelope with a strange humped snout. The snout, however, is inconsequential compared to the animal's tiny pointed horns. Saiga live in the steppes of Kazakhstan to Mongolia in three populations. The Mongolian herds once numbered in the thousands. Today, only an estimated 350 remain. The Kalmykia population is about 130,000, but that is about one-fifth the number in the 1970s, before the breakup of the Soviet Union. The number of saiga in Kazakhstan is estimated at one million.

Because only males have horns, poaching occurs in one sex only. However, commercial harvesting takes place in both sexes. In Kazakhstan, about 500,000 saiga were harvested in some years. Kalmykia outlawed hunting of saiga in 1991; however, poaching takes about 12,500 animals each year. In short, all three populations are in danger of being wiped out. This is an example of an animal similar to the America buffalo in numbers. Its supply was thought to be inexhaustible. At present harvest rates, the last ones will end up in a museum.

TRAFFIC–USA (1996) reported that 44 metric tons of saiga horn was smuggled mainly into China, Japan, and South Korea in 1994. One metric ton is equivalent to about 5000 horns. The horns sold for about $30 per kilogram. At one small stall in China, the owner stated that he could guarantee delivery of 6000 horns in a few days, which suggests that there are warehouses of horns in China that serve as distribution centers. While traditional Chinese shops were loaded with saiga horns, only 5 of 76 shops in Taipei had saiga horns for sale. Before the government ban in September 1994, saiga horns could be easily had. In Japan, only 6 of 110 shops sold saiga horns. In Malaysia, 35 of 39 Chinese shops sold saiga horns. Clearly, the Chinese believe that saiga horns cure something, and the Chinese make up fully one-fifth of the world's population. Can 1.1 billion people be wrong about the curative powers of this strange antelope?

Snakes and Frogs

On July 20, 1995, the U.S. Attorney's Office, Southern District, Miami, reported that a smuggler was sentenced to six months in jail, three years probation, and a $1000 fine for smuggling boa constrictors and poison-arrow frogs into the United States from Venezuela. Customs officers at Miami International Airport found 14 boas and 300 frogs, along with 200 bird-eating spiders and six sacks containing 500 to 600 eggs, inside the smuggler's luggage. No one knows how many successful trips he had made. All the smuggled species were CITES listed, and their street value was approximately $10,000.

Passenger Pigeons

Learning, understanding, and appreciating both sides of a wildlife issue are the first steps in resolving the issue. Resolution of the wildlife issue is the goal of all parties concerned. Back in the early 1800s, John James Audubon wrote about and painted the birds of North America. In his writings, he mentioned the passenger pigeon. Obviously, back when Audubon collected and painted birds, there were no wildlife laws. And why should there have been laws to protect wildlife? Wildlife was plentiful from the Atlantic to the Pacific. One day, Audubon attempted to count the number of passenger pigeons he observed migrating. "The air was literally filled with Pigeons; the light of noon-day was obscured as by an eclipse; the dung fell in spots, not unlike melting flakes of snow; and the continued buzz of wings had a tendency to lull my senses to repose." His estimate of the number of pigeons he saw that one day alone was "one billion, one hundred and fifteen million, one hundred and thirty-six thousand pigeons in one flock." Later, he commented that even though so many were killed, their numbers seemed not to be reduced. By 1914, the last passenger pigeon died in her cage, her species a victim of unmitigated greed and the vanity trade. The wildlife issue was resolved by default. One purpose of this book is to help resolve wildlife issues to prevent extinction by default.

Caimans

The following information was found on the USFWS World Wide Web site. On September 14, 1995, the USFWS's Albuquerque area office announced that a boot company in El Paso, Texas, forfeited caiman lizard skins and boots valued at more than $1 million.

An investigation into illegal trade in exotic reptile skins resulted in the forfeiture, which included 907 pairs of caiman lizard cowboy boots and 2554 pairs of boot vamps (pieces used in the lower part of a boot, between the sole and the upper portion). In addition, a 15-count felony indictment for smuggling and violations of the Lacey Act (see Chapter 2) was returned by the grand jury in the Western District of Texas against two people who sold the skins to the company using fraudulent export permits.

"This is one of the largest forfeitures of protected animal parts ever undertaken by the U.S. Fish and Wildlife Service and the U.S. Attorney's Office," said Nancy Kaufman, regional director of the USFWS southwest region. "We trust the forfeiture and our prosecution of smugglers involved in this trade will send a message to anyone contemplating the idea of making money from trading in protected wildlife species."

The caiman (*Dracaena guianensis*) is an olive-brown lizard that inhabits the Amazon Basin of South America; it can grow up to 4.5 feet in length. Its lustrous skin is prized for cowboy boots, which retail for $700 to $1000. Four lizards are used to make one pair of boots. More than 13,800 caiman lizards were sacrificed for the skins sold to the boot company.

Because its populations are threatened by commercial trade, the caiman lizard is protected under CITES (see Chapter 2). As of January 1996, 131 countries were parties to CITES, a treaty that obligates signatories to regulate international trade of plants and animals that could become extinct without this protection. Both the United States and Brazil are parties to CITES. Effective June 10, 1996, Saudi Arabia became a party.

Under CITES, member nations agree to control the import and export of plant and animal species listed by the convention. Species listed in Appendix II of CITES, including the caiman lizard, are not immediately threatened with extinction but may become so if trade in the species is not controlled. Countries may allow some commercial exportation of these species, but export permits are required.

The USFWS investigation into the illegal trade of exotic reptile skins used in the leather boot industry began in 1993. The illegal trade of caiman lizard skins was initially traced through a computer database that records wildlife imports and exports governed by CITES. Agents and wildlife inspectors discovered that U.S. CITES permits and Mexican export permits were being fraudulently acquired and illegally used for the worldwide movement of caiman lizard skins and products.

Fish Stocks

Looking out from a coastal beach, the ocean seems immense. The place we call "earth" should rightly be called "water" because the earth's surface is nearly 70% water. The average depth of the sea is about 3.8 km. If you were given the task of emptying the ocean of fish, you would probably conclude that this would be an impossible mission. Indeed, individual female fish lay thousands of eggs. How could you possibly catch all those fish? First, you would realize that you can only see about 50 feet into the water at most. If there were any waves at all, your vision would be completely obscured. The fish would be impossible to find. Even if you did locate fish, their evasive abilities are extraordinary. Your ability to move a net around underwater would be severely impaired. You would be better off dropping explosives into the water. Catching all the fish in a small lake seems impossible. Today, however, most of the world's fish stocks have been depleted to the point where they might never recover.

How could this have occurred? Coastal industrial development has led to increased pollution and runoff, destroying nearby coastal fisheries. The increasing use of technology to locate fish, the efficient use of many types of nets and longlines, and the increasing number of factory ships able to stay at sea longer have made year-round exploitation possible. According to Safina (1995), the worldwide catch of fish excluding aquaculture peaked in 1989 at 82 million metric tons. Since then, the catch trend has been negative, while the fishing effort has increased. Some of the world's best fishing areas have been depleted so badly that they have been closed to fishing altogether. Worldwide, only the Indian Ocean fisheries have yet to reach peak production, and this is because modern fishing methods have not been fully utilized yet. The Atlantic Ocean has been severely impacted. Sadly, the United Nations reported that an amazing 70% of the world's edible crustaceans, fish, and mollusks are in desperate need of managed conservation.

In every fishery, target species are not the only part of a catch. In the world's catch, one of every four creatures caught is "bycatch" or "bykill" and is "discarded" (Safina 1995). Discarded means that the remains of the creature are thrown overboard, since most likely it is dead by the time it is discovered. Discards are often fish that are too small to sell, sea turtles that are threatened or endangered, birds that are not protected, or sea mammals such as dolphins or otters. In 1990, high-seas drift nets alone accounted for 42 million animals that were discarded. Although the United Nations placed a global ban on the use of drift nets, Italy, France, and Ireland continue to use them.

Wild Horses: Introduced Species

The Bureau of Land Management (BLM) announced plans to round up 750 wild horses from Nellis Air Force Base in Nevada in January 1997. In 1996, the BLM rounded up 9365 wild horses and burros. More than half of those rounded up came from Nevada, where about 60% of the nation's wild horses and burros are found. The BLM estimated there were 1350 wild horses on the Nellis rangelands, many suffering from poor range conditions. The horses were offered to the public for adoption at $125. Any horses that were sick or lame and could not be transported were to be euthanized. However, none of the horses could be sent to slaughterhouses—dead or alive. A previous BLM investigation concluded that some wild horses did end up in slaughterhouses in the United States and Canada, and the horse meat had been subsequently sold in Europe. An Associated Press investigation in 1996 revealed that some BLM employees could not account for the horses they purchased at previous auctions, while others acknowledged that some were sent to slaughterhouses after the mandatory one-year holding period. Some 32,774 horses and burros are unaccounted for because the BLM's record-keeping procedures were faulty and, according to The Associated Press, some records were falsified to cover up bureaucratic mistakes.

In 1971, Congress passed the Wild Horses and Burros Protection Act (Chapter 2). The law called for excess animals to be put up for adoption at a modest price. In 1978, to better prevent their wholesale slaughter or sale, legal title was obtained by the adopter after a one-year holding period and health check. Until title was issued, the horse or burro was legally the property of the U.S. government. Once title was issued, however, the animal became private property and could be sold for whatever reason. The Associated Press reported that the adoption program was flawed and that some BLM employees were profiting from the slaughter of horses. The former state director of the New Mexico BLM reported that one in five animals is never titled. In April 1996, the U.S. Department of Justice indicated that the BLM failed to screen adopters and failed to follow up on adoptions. One Justice Department attorney wrote that the adopt-a-horse program was seriously flawed and the BLM had ignored the problem.

Recipe for Sea Turtles in *The Joy of Cooking*

Karen Volyes reported in the *Gainesville Sun* on January 4, 1997 that the popular cookbook *The Joy of Cooking* contained a recipe for sea turtle. Although the book recommended using canned turtle meat, all sea turtles are listed as endangered or threatened, and hence possession of any sea turtle products is a violation of the Endangered Species Act of 1973. Serving turtle soup can earn you a $25,000 fine and a prison sentence. New editions of this book do not contain the recipe for sea turtle.

Louisiana Oysters

At least 21 people became ill after eating tainted Louisiana oysters around Christmas 1996. As a result, shellfish harvesting areas were closed as health department officials tried to trace the cause of the illnesses. People reported symptoms such as vomiting and 24-hour diarrhea about a day after eating the oysters. The tainted oysters were taken off the coast of Louisiana between December 15 and December 22, 1996. The oysters were tainted with human feces when harvesters emptied their sewage over the oyster beds. The organism, called the Norwalk virus, originated from untreated human feces. Thus, we can assume that anyone who ate those raw oysters also got a dose of some ship crew's raw sewage.

Elimination of Cats from Australia

On October 17, 1996, the *Seattle Times* reported that the Australian Parliament was urged to draw up a program to eradicate cats. Richard Evans, a lawmaker in Australia, called for a feline-free Australia by 2020. By requiring cat registry, cat curfews, and neutering and by unleashing a deadly feline virus, cats would eventually die out. Cats are notoriously devastating to wildlife, especially small birds and marsupials. Naturally, Evans angered animal rights groups and pet lovers. It is a sad commentary that native wildlife apparently has no rights, while an introduced species such as the common house cat can kill millions of birds with impunity each year.

In Australia, an estimated tens of millions of wild cats roam freely, preying on native wildlife. Having evolved in Australia in the absence of cats, native Australian wildlife has proven to be easy prey. Many species have been pushed to the brink of extinction and some exist only in zoos. Cat are claimed to be responsible for 39 species becoming extinct, locally extinct, or near extinct in Australia. The National Parks and Wildlife Service claims that a wild cat can kill as many as 1000 animals and birds per year. Earlier in this century, rabbits were imported to Australia. They became a threat to crops when their numbers ballooned out of control. In October 1996, a virus was released into the wild rabbit population to reduce their numbers, now estimated at 150 million. The rabbits are widely blamed for widespread environmental damage. In 1910, the red fox was introduced for sport hunting in Australia. It was also thought that the fox would control rabbit numbers. Instead, foxes also found easy prey among the native species.

Because native species are at stake, a hatred for cats has swept across Australia, and cat owners are keeping their pets indoors. In areas of the tropical Northern Territory, flooding of wetlands forced cats into trees, where they were shot by park staff. Endangered lyrebirds in Victoria are now recovering because cat owners are keeping their pets inside. Conservationists have been urging Australians to adopt a native species as a pet.

Evans told Parliament that "the difference between the moggy [an English and Australian slang term for a pet cat] next door and the feral cat is only one meal, and a hungry moggy can and does kill native animals."

Cats can breed three to four times per year and can average more than three kittens per litter. Such output poses a serious biological threat to native wildlife. Ultimately, however, humans are responsible for wildlife losses in Australia. Humans have caused widespread habitat destruction. Humans also have pet cats and feed stray cats. In a survey by the Australian Museum, 61% of cat owners and 77% of noncat owners favored killing feral cats. Some have argued that capturing the cats, neutering them, and returning them to the wild is the most effective control program. Programs under discussion include killing all cats found outside city limits, large fees for owning unneutered cats, restrictions on the number of cats per household, fines of up to A$5000 for cat owners whose cats are caught roaming outside, a cat curfew from 8:00 P.M. to 6:00 A.M., keeping cats on leashes, allowing only neutered cats to be sold, and compulsory declawing and defanging. Does this problem exist elsewhere?

Most pet owners are responsible citizens. However, there is no doubt that cats pose a risk to native wildlife. In Australia, the labeling of cats as moggys is similar to the dehumanization of the enemy during war. A citizen is not shooting a neighbor's cat but rather ridding Australia of a dangerous moggy. One solution, of course, is to keep domestic house cats indoors at all times. All others could then be considered pests.

People for the Ethical Treatment of Animals (PETA, http://envirolink.org/arrs/peta/index.html) estimates that there are between 11 and 19 million feral cats in the United States. PETA acknowledges that because of a shortage of homes and the difficulty of training a feral cat, "it may be necessary to euthanize unwanted cats." It is a sad fact that in one of the wealthiest countries on earth, many millions of unwanted dogs and cats are destroyed each year as a result of human negligence.

Tropical Fish

According to the Florida Tropical Fish Farms Association, Florida has 205 aquaculture farmers, about two-thirds of whom raise tropical fish for consumers. About 10% of U.S. households, or roughly 9.4 million residences, keep freshwater fish. Some 850,00 homes have saltwater aquariums, and perhaps 200 million fish live in people's homes in the United States. Overharvesting of tropical saltwater fish has made obtaining a license to harvest wild fish more difficult. Collecting fish internationally might also become more difficult. As demand has pushed prices higher, breeding techniques have been perfected to allow farmers to raise their own tropical fish. Indeed, the tropical fish market is far ahead of the aquacultural attempts to raise catfish for food. Tropical fish exports into the United States totaled $54 million in 1995 because demand could not be met domestically.

Amerindians

On October 12, 1996, the *Gainesville Sun* reported that four Amazon Indian tribes laid claim to over four million acres of ancestral lands in the northern Amazon after "waiting in vain for the Brazilian government to draw the border of their reservations." The Ingariko, Macaxi, Taurepang, and Wapixana tribes marked off the boundaries of their reservation in Raposa Serra do Sol territory in northern Roraima State. The federal Indian Affairs Bureau of Brazil set aside the territory near the Guyana–Venezuela border in 1993. This entire region has been in dispute as both Guyana and Venezuela claimed ownership. Apparently, however, no country owns the land; Native Amerindians have simply claimed what is rightfully theirs. Who will own the wildlife?

Pollinators

Butterflies, bees, birds, and many other small forgotten organisms perform valuable ecosystem functions. Most of these functions are essential to human existence. Pollination of plants, especially food crops, is critical to human survival. Unfortunately, humans have waged war on the pollinators and now seem to be winning. Pollinators are disappearing across the United States.

The European honeybee is one of the most conspicuous pollinators. It is an introduced species in the United States, imported more than 375 years ago. Honeybees pollinate about 15% of U.S. crops. However, a severely cold winter in 1996 and a parasitic mite infestation have taken their toll on honeybees. Several pollination specialists have suggested that native pollinators could fill the role honeybees now occupy. Butterflies are excellent pollinators, but their numbers have dropped because of forest and native grassland fragmentation. Other native bees such as the alkali bee are used to pollinate alfalfa. Human use of pesticides has worked against increasing native pollinator populations. Has the time come to realize that native species provide ecosystem services for free and that given an opportunity they would do so again? Is it time to outlaw the use of toxic pesticides so that our dependence on chemicals decreases? Perhaps we should increase our use of pesticides and rid ourselves forever of all these pesky insects.

HISTORICAL PERSPECTIVES ON WILDLIFE ISSUES

Taking a historical perspective on wildlife issues is invaluable. With some knowledge of history, how we got to where we are today is much clearer. Often the best solution to a problem is one that takes history into consideration. For instance, most people believe acid rain and the greenhouse effect are recent phenomena. At one time, 95% of western and central Europe was covered with forests; now only 20% remains. The last bear in the United Kingdom was killed back in the 1600s. Once China was 70% forested; a mere 5% remains today. North Americans typically forget that Asian and European civilizations have been around a very long time. Since the late 1700s, 75% of North America's forests have been cleared. How can we understand soil erosion, loss of wildlife habitat, desertification, salinization, siltation, dams, or industrial pollution without knowing the history of both human and naturally occurring events? We must learn from history, synthesize historical facts, and apply these facts as best we can today. Wildlife issues are really the symptoms of more insidious, widespread problems. By resolving wildlife issues in a timely and satisfactory manner, we will really be dealing with the much broader issues of environmental degradation and the loss of quality of life.

Ponting (1992) examined many societies beset with chronic environmental problems largely brought on by their own practices. To give a complete perspective on human history in one chapter, we would spend all but the last paragraph on hunter/gatherer groups of nomadic people. The last paragraph would then consider the history of humans for about the last 10,000 years, roughly the time when agriculture was first developed.

The first settlements of people took place as a result of agriculture in southwest Asia (Rzoska 1980), China, and Mesoamerica (Culbert 1973). These societies shared several attributes (Hughes 1973). For instance, farmers grew enough food to support a burgeoning class of nonproducing consumers such as priests, the military, and bureaucrats. Of course, agriculture involves a complete disruption of a natural ecosystem to make way for crops and livestock. Natural ecosystem functions are replaced with processes necessary to support people. For the first time in the history of the earth, humans were bending and twisting the environment to meet their needs.

The total population of the world 7000 years ago was a mere five million people. Just 2500 years ago, the population was 95 million. Agricultural systems and, in turn, technological developments throughout recent human history have allowed the population to explode to more than 5.5 billion. Pressure on natural ecological systems has increased. Ecosystem functions have been degraded and massive desertification has taken place, for instance. The all too brief histories of Easter Island, Mesopotamia, and Mesoamerica serve to illustrate what is in store for us in the future. Humans cannot simply continue to consume an increasing share of global resources while ignoring the consequences.

Because Easter Island is indeed an island, environmental problems were particularly acute because there was no place left for the inhabitants to go. Easter Island therefore serves as an example for inhabitants of Island Earth. There is no place left to go. Furthermore, even the most casual observer can see that the moon and other planets are not better places to live. Wildlife issues of today are harbingers of greater problems later. How much later depends on the size of the human population and the resources necessary to sustain that population.

Easter Island

Easter Island is famous for its gigantic stone heads and its remoteness. The island is just 64 square miles in area, located 2000 miles west of South America and 1400 miles from the nearest habitable islands. One has to wonder how the ancient Polynesians found the island, much less colonized it, but they did. Located at 27 degrees south latitude, the climate is mild. Since the island is of volcanic origin, the soil is fertile.

Today, the giant stone heads have fallen and the island appears denuded. When first observed by Europeans, there was not a single plant higher than a shrub. Naturally, this raises an interesting question. How did people quarry, cut, and erect such large statues in so remote a place with few natural resources and little food? Did people go to the island just to erect stone heads, bring everything with them to stay for a while, and then leave when the task was finished? Hardly. Diamond (1995) related the human history of Easter Island. We cannot help but see the parallels to what is happening today.

On Easter Day, April 5, 1722, when a Dutch explorer sighted and named Easter Island, it appeared from a distance that the island was a desert. On closer inspection, it was a grassland with a few ferns and bushes. Because the grass was dried out, it appeared to be a desert. There were no birds, bats, land snails, or lizards. The inhabitants kept chickens for food. The islanders who rowed out to greet the visitors did so in flimsy canoes that filled with water and required constant bailing. Surely the direct descendents of the Polynesians, known worldwide for their seafaring abilities, could do better. Only a few canoes were found on the island. It would have been nearly impossible to do any serious offshore fishing in so pitiful a craft.

The islanders were apparently unaware of other human existence. The Dutch found some 200 stone statues that once stood on the hillside looking out to sea. Some 700 more statues were found in a quarry in various stages of development. However, it was clear that none had been worked on for a very long time. The quarry was about six miles from where the statues were erected, and many weighed more than 75 tons. Some in construction were four times as heavy and some 60 feet tall. How could the statues be moved such a distance? The resident islanders had not a clue. Without ropes, timber, or even wheels, the task would have been impossible. Moreover, resources were distributed around the island. The best soils were quite far from the quarries. Red capstone was found in a quarry different from that used for the statues. The best fishing was on yet a different coast. The Dutch explorers were astonished. Although they had traveled the world and made many new discoveries, nothing rivaled the mysteries of Easter Island.

Modern archeology is beginning to answer some of the mysteries. The inhabitants of Easter Island were definitely Polynesian, and their language places its roots at about 400 A.D. Radiocarbon dating has been used to establish the earliest settlement times to be around that date as well. Crops of taro, bananas, sweet potatoes, and sugar cane combined with chicken, fish, and marine mammal were standard table fare.

Pollen analysis has established that Easter Island was once covered with forest. Indeed, for at least 30,000 years before humans arrived, the island was a subtropical forest covered with trees, shrubs, herbs, and grasses; that is, a fairly complex shrub layer existed below the trees. All necessary ingredients for making ropes, canoes, and wooden slides for statues could be obtained from forest products. One of the Easter Island palms, the Chilean wine palm, can grow to over 80 feet high and 6 feet in diameter—more than enough tree for a seafaring canoe. The mainstay of the Polynesians' diet was, of course, fish. With large canoes, offshore fishing was not a problem. The inhabitants could easily have traveled hundreds of miles to rich fishing areas in the region. According to Steadman (personal communication), most of the ancient garbage dumps consisted of fish bones. From 900 A.D. to 1300 A.D., porpoise made up part of the protein in the typical diet; however, porpoise in the Polynesian diet is rare elsewhere. Porpoises were found far offshore and therefore must have been hunted in sturdy canoes.

Also in the diet were seabirds. Easter Island must have been an ideal site for nesting seabirds. Without terrestrial mammals, the ground-nesting birds could raise their young with higher success rates. Diamond (1995) suggested Easter Island was one of the richest seabird breeding islands in the Pacific, with at least 25 nesting species. The surrounding waters must have provided much fish for the birds. The islanders ate both birds and fish, based on findings in the garbage dumps.

The historical picture now is beginning to become clearer. Sometime around 400 A.D., the seafaring Polynesians landed on Easter Island, a veritable subtropical paradise in a very remote part of the Pacific. The island was forested, the soil fertile, and the sea rich with fish. Some of the visitors decided to stay. For hundreds of years, they survived and prospered. The human population slowly expanded. Trees were used for canoes and fuel. By 800 A.D., pollen cores show that the destruction of the island's forest was becoming serious and that woody shrubs and grasses were being used for fuel. After 1400 A.D., the palm trees were gone and what once was a forest was now a grassland. Rats had prevented any trees from germinating by consuming seeds. All the native land birds such as rails were caught and eaten. Shellfish were overexploited and wiped out.

Around 1500 A.D., porpoise bones disappear from the dumps. We can conclude that the seafaring canoes were no longer available. Chicken became a mainstay of the diet, and chickens eat seeds. About this time, human bones show up in the garbage dumps. Archeologists confirmed that these bones are not from people dying of old age. Instead, humans began eating humans for food. Cannibalism became a common practice. By the 1600s, warriors with stone spears raided villages and took captives for food. The decline and fall of life in paradise had bottomed out. With the natural resources gone, people turned to hunting people. Since there were no canoes, there was no escape.

Mesopotamia

Most of us have heard of the Cradle of Civilization, the valley of the Tigris and Euphrates rivers, and the Sumerian culture. These rivers flowed highest in the spring due to snowmelt at their sources. This was not ideal for crops that needed more water in the late summer; however, rains in the higher valley acted as a buffer. Lower in the valley, there were fewer rains and higher summer temperatures, thus making irrigation essential to sustain agriculture. Because the land was flat, especially lower in the valley, and the water table was higher, waterlogging and surface salt accumulation were problems. As long as agriculture was restricted to the higher valleys, it was sustainable and the impact on the entire ecosystem was reduced.

The limited amount of easily developed land and an increasing population meant that lower parts of the valley had to be brought under agricultural production. No thought was given to long-term sustainability. Human life spans argued against planning for a distant future. It was in this valley that the first literate society developed, a great achievement for our species.

With literacy came historical records, detailed accounts of a society's rise to prominence and its eventual, and inevitable, fall. So successful were farmers that a military class was created to defend against other rising city-states. Excess food produced by farmers was used to defend, protect, and establish the domination of Sumerian society. While production increases came easily at first, the lower in the valley agriculture extended, the more problems arose. Eventually, environmental degradation through centuries of irrigation and excess production caught up with society.

By 2500 B.C., wheat cultivation was being replaced with barley as a means of keeping production consistent. Barley, it seems, did better in the saline soils that were now common in the lower valley. Slowly over the centuries, wheat production declined up the valley and was replaced by barley. By 1700 B.C., no wheat was grown at all. Throughout the entire region, crop yields decreased.

The human population continued to grow, armies fought other armies for control of the land, and new land was desperately needed. Continued advances in agricultural technology were not enough to sustain the society. Burdened by an army, bureaucracy, and clergy who only consumed, the farmers could not support the society. As early as 2300 B.C., conquests coincided with crop failures. With the Babylonian conquest in 1800 B.C., Mesopotamian society left the Cradle of Civilization for good. Crop yields throughout the valley were less than one-third of those obtained nearly 500 years earlier. The ground was white with mineral salts. The great society was no more.

Mesoamerica

The Mayan civilization is also well known. It rose and fell in the lowland tropical forests of Mexico, Guatemala, Honduras, and Belize. The time period coincides with the rise of Sumerian society. The earliest records date back to 2500 B.C. The population grew, and a complex society developed, with temples, stone pyramids, plazas, and other elaborate architectural and scientific wonders. Astronomical observations led to the development of a calendar and the capability to predict positions and phases of the moon. To support a class of builders, astronomers, and other nonproducers, surplus agriculture was necessary.

By 600 A.D., the Mayan empire flourished. In many population centers, pyramids and inscribed stone pillars used for commemorative purposes were erected. The current view of Mayan civilization differs radically from past beliefs. Today, the stone pillars are recognized to be monuments to the rulers of the various cities scattered throughout the empire. They also contain the written history of one city's conquest over another and thus confirm the existence of armies. The deceptive belief of a peaceful religious society, wholly wrapped up in its astronomy and calendar, was flawed. Archeological work has confirmed that an elite class supported by armies and surplus food conducted almost continuous warfare against neighboring cities.

Around the central city and its great temples were masses of laborers who supported the elite. Elaborate agricultural works have been unearthed which show intensive cultivation of the surrounding lowlands. Extensive hillside terraces were used to contain soil erosion. Raised fields in swampy areas also existed. Complex drainage systems were used to control surface water, with material from the ditches used to heighten fields. This is similar to many agricultural areas of south Florida.

Crops of maize and beans sustained the population, while cotton and cacao were also cultivated. "The agricultural system was the foundation for all the achievements of the Maya" (Ponting 1992). Increased warfare, a swelling human population, and ever-more elaborate construction of monuments led to demands on the agricultural system that could not be sustained. Tropical soils are notoriously poor and easily eroded. Cities were clustered around sites of fertile soil, but once the forests were removed, the soils were easily eroded in the tropical rains.

Forests were cleared for the usual reasons: fuel, building material, and to make room for agricultural fields. The Mayans also had no domestic stock to turn weeds into manure for the fields. As the population expanded, more marginal areas were forced into production. Deforestation followed by soil erosion, siltation of the river systems, and general environmental degradation was the predictable outcome.

Human skeletons unearthed from grave sites show increasing nutritional deficiencies. Within decades, the civilization collapsed. The environment of the tropical forest was unable to support a massive human population. Today, the cities are engulfed by tropical rain forest.

Other ancient societies in China and the Mediterranean also collapsed due to long-term persistent environmental degradation. Are human societies now suffering from a similar environmental decline? In times long ago, transportation systems were poorly developed, and most material goods and supplies were made, created, or grown locally. In today's global economy, goods and services can be grown or extracted in one place, assembled or processed in another, and sold in yet another. This has allowed the human population to expand; however, global transportation systems have made possible global exploitation. Our reach has extended around the globe, and exploitation is usually efficient. The local problems faced by ancient societies are similar to the problems modern society faces today. Now, however, we are able to import necessary commodities and thus allow others to sort out the local difficulties. If the producers become too expensive or too inefficient, consumers simply switch to another producer. Does the consumer know or even care that a producer has gone out of business because the local environment is degraded?

Whereas Mesoamericans and Mesopotamians depended on agriculture, modern societies have shifted their dependence to fossil fuels to sustain their economies. Is dependence on a nonrenewable resource such as oil a strategy for a sustainable future?

WILDLIFE ISSUES IN A CHANGING WORLD

Estimates indicate that humans consume nearly 50% of all primary production of the planet. Naturally, while humans are getting their half, all the other creatures of the world are splitting the rest. As the human population increases, the percentage of primary production consumed by humans will increase, leaving less for all other creatures. Advanced technology will enable people to extract more resources and increase primary production; however, it is likely that the human population will increase and consume any excess. An increasing global human population seems a certainty. Wildlife issues are becoming ever-more common across a rapidly shrinking resource base. Rich nations will be able to cope with their problems by purchasing natural resources such as oil from other nations. Third World nations, whose populations are growing most rapidly, will eventually dismiss their wildlife issues as irrelevant. One answer to these problems is global education. What we learn must be translated and retaught to others as efficiently as possible. The Electronic Age offers us an unprecedented opportunity to reach billions of people without using fossil fuels.

EXERCISES

3.1 Identify wildlife issues from all 50 states by reading various local newspapers. Is there a common thread that links the issues regionally?

3.2 Use your favorite search engine to locate wildlife issues on the World Wide Web. Identify issues of concern from other countries.

3.3   Across much of Africa, wildlife is equated with food. Bush meat is routinely eaten. Monkeys are a favorite bush meat. Is there a wildlife issue here? If so, identify the vested interests.

3.4   Are there any wildlife dealers advertising on the World Wide Web? What are the products for sale?

3.5   Visit your local ethnic grocery store and look for interesting foods, medicines, and treatments that are composed of wildlife.

3.6   Do a more thorough investigation of the bear trade. What is the status of the Asian bear today?

3.7   What is the status of the panda in China today?

3.8   Perhaps the most powerful weapon used to solve wildlife crimes is the U.S. Fish and Wildlife Service Forensics Laboratory in Ashland, Oregon. From poisoned bald eagles to dried seal penises, from women's shoulder bags made from cane toads to piles of illegally obtained skins, the lab receives packages from all 50 states and 125 countries that are signatories to CITES. Find out more about the laboratory using the World Wide Web, the library, and by writing for more information.

3.9   Obtain more information about TRAFFIC–USA from the World Wildlife Fund in Washington, D.C. Summarize its activities. Who provides funding for the work carried out by TRAFFIC–USA?

3.10 Trace the history of early Mediterranean societies. What environmental problems caused the collapse of their social systems?

3.11 The northern spotted owl is a famous wildlife issue of the Northwest old-growth forests. Summarize what happened. Discuss the so-called "God Squad's" decisions regarding old-growth timber, logging, and the future of the spotted owl.

3.12 Track down some statistics in the international trade of snakes. What is the demand for snakes, and where does this demand originate? How is it satisfied?

3.13 Vicunas are South American mammals prized for their fine, warm hair which makes wonderful sweaters. What is the outlook for this species today?

3.14 The international cat trade, including tigers, is threatening many species of cat. What is happening to snow leopard populations in Asia and ocelot and margay populations in Central and South America? Compare the trade in spotted cats versus cats without spots.

LITERATURE CITED

Brower, L.P. 1995. Understanding and misunderstanding the migration of the monarch butterfly (Nymphalidae) in North America: 1857–1995. *Journal of the Lepidopterists' Society* **49**(4):304–385.

Caughley, G., N. Shepard, and J. Short. 1987. *Kangaroos: Their Ecology and Management in Sheep Rangelands of Australia.* Cambridge University Press, Sydney, Australia.

Colclough, C. 1996. Guatemala's macaws on verge of extinction. The Associated Press, April 30.

Culbert, T.P. 1973. *The Classic Maya Collapse.* University of New Mexico Press, Albuquerque.

Diamond, J.M. 1995. Easter's end. *Discover* August:63–69.

Dold, C. 1996. Shark therapy. *Discover* **17**(4):51–57.

Fitzgerald, S. 1989. *International Wildlife Trade: Whose Business Is It?* World Wildlife Fund, Washington, D.C.

Hughes, J.D. 1973. *Ecology of Ancient Civilizations.* University of New Mexico Press, Albuquerque.

Littell, R. 1992. *Endangered and Other Protected Species: Federal Law and Regulation.* The Bureau of National Affairs, Washington, D.C.

Mills, J. 1992. Milking the bear trade. *International Wildlife* May–June:38–45.

Owens, M. and D. Owens. 1984. *Cry of the Kalahari.* Houghton Mifflin, Boston.

Ponting, C. 1992. *A Green History of the World. The Environment and the Collapse of Great Civilizations.* St. Martin's Press, New York.

Rzoska, J. 1980. *Euphrates and Tigris: Mesopotamian Ecology and Destiny.* W. Junk, The Hague, Netherlands.

Sachs, A. 1991. A grisly and illicit trade. *Time* April 18:67–68.

Safina, C. 1995. The world's imperiled fish. *Scientific American* November:46–53.

TRAFFIC-USA 1995. **4**(2):1–2.

TRAFFIC-USA 1996. **15**(2):1–20.

Williams, T. 1996. The great butterfly bust. *Audubon* **98**(2):30–37.

WHAT IS WILDLIFE?

4

For the purposes of this book, we define *wildlife* to mean any living non-human, undomesticated organism in the kingdom Animalia. Although wildlife issues are not limited to any particular taxonomic group, we will emphasize nonhuman, undomesticated members of the animal kingdom. Our reasons for this bias stem partly from history (the term "wildlife" has traditionally been limited to certain types of animals) and partly from our own expertise and experience. One could further argue that the vast majority of humans are more interested in the fates of animals than in, say, the fate of an endangered bryophyte. This is not to suggest that bryophytes are unworthy of our attention. There was a time historically when the term "wildlife" was applied only to so-called *game* species. Game species were those species (almost exclusively birds, mammals, and fish) that people hunted (we consider fishing to be a modified form of hunting). Of course, if one is willing to classify a handful of species as being "game" species, then the remaining species must all be *nongame* species. In any event, if the term "nongame" refers to any species that is not a game species, then we must define wildlife as any species of living organism. That is to say, we must include all the lower life forms such as unicellular species, as well as the evolutionarily primitive species such as the bacteria.

The brand "wildlife" is no longer limited to birds, mammals, and fish; it now also includes so-called protists, bacteria, fungi, and plants. There may be a bias toward only classifying vertebrate (see below) species as wildlife, but this is a prejudice to which we do not subscribe. Perhaps the main reason for our willingness to relax the boundaries of our definition is the number of issues that arise involving nonvertebrate species. Many of these issues are detailed in this chapter.

CLASSIFYING WILDLIFE

We have already seen that the term wildlife encompasses a vast number of species. Clearly, a classification scheme of some sort is required. A main reason for this lies in developing management plans. If one can group species in a logical fashion, it simplifies, or at least has the potential to simplify, the implementation of management plans or policies. Thus, we might set size limits for harvesting finfish (vertebrates) but not shellfish (invertebrates).

No matter how one chooses to categorize species, the system is based on shared characteristics. Secondarily, the characters should be mutually exclusive. Recall that in our discussion of the scientific method, we stated that hypotheses should be formulated so as to be mutually exclusive. The use of mutually exclusive characteristics here is no accident, as it mirrors the scientific method. Thus, edible species share obviously at least one characteristic, as do inedible species. Inedible species are simply "not edible"; thus, our characteristic indeed has two mutually exclusive states. Note also that we could go further by imposing the standard "tastes good" versus "does not taste good." But this would be futile, because some people like some foods that others simply refuse to eat. Therefore, it is immaterial how the food tastes because human preferences vary. The characters involved here may be shared, but they are rather superficial.

Biologists have long classified species on the basis of observable traits. Such a scheme is referred to as a phenetic scheme. A phenetic scheme can be a powerful system if properly applied but can lead to some very funny conclusions if misapplied. Consider the set of a mosquito, hummingbird, snake, earthworm, and bat. Clearly, we could group species by virtue of their having wings (mosquito, hummingbird, and bat) or not (snake and earthworm). The grouping here ignores evolutionary relationships that, for the sake of argument, we know would place the hummingbird and bat together, and then these two with the snake. This group then would be near the earthworm, and at the other extreme would be the mosquito. Why does the phenetic classification scheme fail? Clearly, it fails because the characters were not weighted. This means that the characters are all assumed to be of equal importance. From an evolutionary perspective, this is weak because some characters are likely more important than others in classifying species. Thus, biologists would consider it more important that the snake, bat, and hummingbird are all vertebrates than that the mosquito, hummingbird, and bat can all fly. In other words, the characteristics associated with being a vertebrate are "weighted" more than the character of mode of locomotion.

In the course of developing our understanding of what species comprise wildlife, we will follow the five-kingdom classification. We will see that there are wildlife issues at nearly all the levels or taxa (singular is taxon). This system is based on shared characteristics, and the characters are mutually exclusive.

The categories used in the scheme are presented below in the form of an example of how to classify an American crow:

Kingdom	Broadest grouping that identifies species that share fundamental similarities in cell structure (e.g., Animalia)
Phylum	Chordata (animals with backbones)
Class	Aves (birds)
Order	Passeriformes (songbirds)
Family	Corvidae (crows, jays, and magpies)
Genus	*Corvus*
Species	*brachyrhynchos*

SPECIES

The genus and species make up the so-called Latin or scientific name. We follow the system of binomial (two names: the genus and the specific name) nomenclature established by Linnaeus. Many species also have a common name; for *Corvus brachyrhynchos,* it is the American crow. A point of caution involving common names: We must be careful today when dealing with some phylogenetic groups of species because many of them have more than one common name; this makes it difficult in some cases to identify a species (i.e., figure out its scientific name) just by the common name.

Here, when we use the term species, it will imply a number of individuals that are reproductively isolated from all other individuals in other populations. Reproductive isolation means the individuals are not able to interbreed with individuals from other populations and produce viable offspring (i.e., their offspring cannot produce offspring of their own).

It is important to note that because it is often unclear whether or not a population is a true species or a subspecies, certain provisions of the Endangered Species Act may apply to just one of several subspecies of a species. The term species is not easy to define either biologically or politically. Because of the difficulty in defining the term, the Endangered Species Act is extended to include subspecies. We simply cannot in every case use the condition of reproductive isolation to identify species versus subspecies (subspecies are geographically distinct subpopulations of a species).

Several wildlife issues are based on the possible confusion over deciding if populations are species or subspecies. The following are examples of some fuzzy cases where this is important:

1. *Red wolf* versus gray wolf (the red wolf is indigenous to the southeastern United States, whereas the gray wolf lives elsewhere)
2. *Florida panther* versus Texas cougar
3. *Concho* water snake versus Harter's water snake

The italicized names may refer to subspecies rather than species, since there is no reproductive isolation except by geography and no general agreement among scientists let alone politicians!

In each of these cases, there are people who claim that these populations are actually different species, not just subspecies. In other words, the red wolf and gray wolf may be different species and not just different subspecies.

THE FIVE KINGDOMS

- Kingdom Monera
- Kingdom Protista
- Kingdom Fungi
- Kingdom Plantae
- Kingdom Animalia

The following are shared characteristics among all these groups:

1. All are made up of cells
2. All have deoxyribonucleic acid (DNA) with the same genetic code

Differences among kingdoms are as follows:

1. **Kingdom Monera**—Comprised of prokaryotic cells, do not have either an organized nucleus or membrane-bound organelles (all other kingdoms have eukaryotic cells, which means they do have an organized nucleus and membrane-bound organelles)
2. **Kingdom Fungi**—Cell wall (chitinous), no chloroplasts, multicellular, no true alternation of generations (see below), heterotrophic
3. **Kingdom Plantae**—Cellulose, cell wall, have chloroplasts, multicellular, true alternation of generations, autotrophic

4. **Kingdom Animalia**—No cell wall, multicellular, no true alternation of generations
5. **Kingdom Protista**—Unicellular, eukaryotes

As stated at the beginning of this chapter, we take the position that any living organism may be considered wildlife. With this in mind, we now consider characteristics of each of the five kingdoms, with special attention devoted to the animals.

KINGDOM MONERA:
BACTERIA AND BLUE-GREEN ALGAE

The importance of bacteria to wildlife populations is typically indirect. Humans may go to the field to observe birds, mammals, reptiles, or fish, but typically not bacteria. Nevertheless, bacteria play an important role in wildlife biology in at least three important ways.

First, different bacteria are responsible for numerous wildlife diseases (Forrester 1992), such as Lyme disease, which is transmitted by ticks. The bacterium that causes Lyme disease is found in various species of vertebrates such as rodents and larger mammals, as well as birds. Another example of a bacterium that threatens humans is *Vibrio vulnificus,* which is found in shellfish (e.g., oysters) and may be lethal to humans.

The second effect of bacteria on wildlife populations is through their role as decomposers. After a recent red tide event in Florida, a large fish kill was caused by bacteria. Decomposing fish can also be a source of disease. Indeed, the relationship of bacteria to red tides may be even more complex, as some authors have suggested that the toxins produced by red tide algae (see Kingdom Protista) may themselves be produced by symbiotic (see below) bacteria that live in close association with the algae (Anderson 1994).

The ability of various types of bacteria to decompose organic polymers and other complex man-made molecules also aids in indirect wildlife concerns, such as cleaning up toxic chemical and petroleum product spills (bacteria are involved in eating petroleum products).

The third effect involves the transfer of nutrients to plants. In particular, certain types of blue-green algae (or, as they are often called today, blue-green bacteria) play an important role in fixing gaseous nitrogen. Plants need nitrogen, just as animals do, for building proteins. Unfortunately, elemental nitrogen exists (albeit abundantly) in the atmosphere as a gas. Certain blue-green algae and fungi are able to "fix" this gaseous nitrogen into a more usable form for plants to take up. Of course, once plants take up the nitrogen, animals can harvest it from the plants (see Chapter 9).

KINGDOM PROTISTA:
UNICELLULAR ALGAE AND PROTOZOANS

Red Tides

Our discussion of red tides is based on the work of Anderson (1994). Red tides have long captured the imagination of humans. Recently, red tides and the resultant toxins have been associated with deaths of marine mammals (humpback whales) as well as finfish. Red tides are the result of algal blooms (an algal bloom is basically a population explosion of a type of algae which is a single-celled eukaryote). Red tides can result in massive fish kills due to toxins and are often (but not always!) characterized by a change in water color. Red tides are caused by toxins produced by dinoflagellates and a few other groups of algae. The term red tide is a misnomer since the water may not turn red; it may turn brown or not change color at all. Moreover, some algal blooms cause a change in water color but are not associated with production of any toxin.

There are three types of red tides:

A. Indirect toxic
   1. Dinoflagellates (or associated bacteria) produce a toxin.
   2. The toxin is consumed by shellfish.
   3. Humans eat the shellfish and suffer poisoning (paralysis, diarrhea, neurotoxicity, amnesia).
B. Direct toxic
   1. Dinoflagellate bloom (explosive increase in the number of algal cells in a local area) occurs.
   2. Fish swim by and rupture the dinoflagellate cells.
   3. The toxin released from the cells attacks gills and kills the fish.
C. Ciguatera
   1. Dinoflagellates live in seaweed.
   2. Fish eat the dinoflagellates.
   3. Humans eat fish and suffer ciguatera fish poisoning. Here, the toxins are accumulated in fatty tissues of the fish. The bigger the fish, the more intense the poisoning.

Notice the interesting situation where a unicellular organism is able to cause ecosystem-level changes.

The following factors may favor red tides:

1. **Low resources**—Certain dinoflagellate species revert to sexual reproduction when resources are scarce. As a result of sexual reproduction, these organisms produce hard-coated cysts which can be translocated around the world by ship ballast or hurricanes.
2. **Pollution**—Some evidence suggests that red tides are positively correlated with nutrient pollution. More nutrient pollution leads to more red tides.

Protozoan Parasites

A number of protozoan parasites play an important role in wildlife ecology. For example, avian malaria is caused by a protozoan in the genus *Plasmodium*. In Hawaii, bird-watchers can see native forest birds (that are not yet extinct) only on islands that have high elevations (Kauai, Molokai, Lanai, Hawaii, Oahu, Maui). The reason for this may be related to the distribution of mosquitoes that transmit the *Plasmodium* organism to the birds. At lower elevations, temperatures are high enough for mosquitoes to thrive; however, at higher elevations, mosquitoes apparently cannot survive the chilling rain-soaked nights. Warner (1968) argued that the distribution of native forest birds was the opposite of mosquitoes. He suggested that avian malaria was killing the birds that invaded lower elevation habitats. To test his idea, he conducted the following experiment; note the use of the scientific method.

- **Observation**—Native forest birds largely do not occur at lowland elevations.
- **Null hypothesis**—Native forest birds cannot live at lowland elevations (presumably because of exposure to malaria-transmitting mosquitoes).
- **Test**—Place native birds in cages with food and water at lowland elevations and monitor them. If mosquito exposure (malaria) is responsible, then the birds should die from malaria.
- **Result**—The birds died from malaria.

What was Warner's control? What other factors could have been responsible?

KINGDOM FUNGI: MUSHROOMS, BLIGHTS, RUSTS, AND MOLDS

This kingdom includes some 70,000 species. Major characteristics are as follows:

1. Heterotrophs ("hetero" = other and "troph" = eats; therefore, heterotrophs eat other food, as opposed to auto [= self] trophs, which are able to produce their own food via photosynthesis. Ecologists call plants "producers" and animals "consumers.")
2. Cell wall made of chitin
3. Many are symbiotic

Symbiosis: The Art of Living Together

Symbiosis refers to the situation where two or more species live together in close association. There are so many examples of symbiotic relationships in nature that it is difficult to know where to start in describing them. Perhaps the best approach lies in describing the three main forms of symbiosis: competition, where both species suffer; predation/parasitism, where one species suffers and the other benefits; and mutualism, where both species benefit from the interaction. Note that there are costs and benefits here, which means that the decision as to which species suffers and which one benefits is determined by the net effect of the interaction on population size or growth rate. When we say a population suffers, we mean that its population size or growth rate is reduced as a result of the interaction with its symbiont.

The following are examples of symbiotic relationships:

1. **Lichens**—A mutualistic relationship between a fungus and an alga
2. **Mycorrhizae**—Fungus associated with trees roots; the fungus helps with ion transfer, while the tree supplies the fungus with carbohydrates

Fungi play an important role in ecosystems, mostly as decomposers. Decomposers are species that break down tissues into simpler organic molecules that can be used by other organisms. However, certain symbiotic forms such as mycorrhizae play an important role in fixing gaseous nitrogen for plants. Fungi often attack plants and human feet (athlete's foot) and are essential for making important human foods, such as bread and beer. In some countries (e.g., France), various fungi are national delicacies. Finally, as described below, fungi may play an important role in detecting environmental dangers, much the same way that miners once relied on canaries to warn them of reduced oxygen levels.

The Fall of European Fungi and Atmospheric Pollution

In the fall of 1990, mycologist Eef Arnolds of the Netherlands gave a presentation at the Fourth International Mycological Congress in Germany in which he showed evidence of a catastrophic decline in abundance and diversity of ectomycorrhizal fungi—mushrooms (Jaenike 1991). The basic observation reported by Arnolds was that ectomycorrhizal fungi (mushrooms) had suffered major declines in northern Europe between 1970 and 1985. These losses included (1) a loss of species (60% decline in number of species collected per year) and (2) a decline in the abundance of remaining species.

The fungi involved in these declines are associated with trees in a mutualistic (symbiotic, where both species benefit) relationship. The fungi facilitate ion transport (chiefly phosphates and immobile ions) for the trees, and the trees provide the fungi with carbohydrates. Both the fungi and the plants benefit. Scientists now think that with increased nitrogen in the form of deposits from pollution, trees divert carbohydrates to growth of aboveground biomass rather than to roots (for nutrient uptake). Thus, the fungi that are associated with the roots get less and therefore decline.

Other explanations include overharvesting by humans and habitat loss. If overharvesting by humans were the cause, we would predict that only the more desirable species would decline. However, we can reject this hypothesis because of comparable declines in inedible species. Arnolds also rejected the habitat loss hypothesis as being, at best, only a partial explanation.

Other Fungi Important to Wildlife Issues

Chestnut blight reduced hard mast crop bears, and other species that depend on chestnuts were reduced. Ganoderma affects transport tissue. This fungus has attacked palms in Gainesville, Florida. Convulsive ergotism is a condition caused by consuming rye flower that is infested with the fungus *Claviceps purpurea*.

The Salem witchcraft affair is the story of a very localized outbreak in a small area of Massachusetts and one county in Connecticut (Matassian 1982). The symptoms of "bewitchment" were described in Essex County, Massachusetts, in 1692. A high percentage of cases (24 of 30) showed convulsions. Many individuals had the sensation of being pinched or bitten. Other symptoms included temporary blindness, deafness, loss of speech, burning sensations, visions of a "ball of fire," the sensation of flying through the air, and spells of laughing and crying.

Most victims were children or teenagers. Original records suggested that the affair was really just an outbreak of convulsive ergotism caused by a fungus that grows on rye in particular. We now know that ergot (*C. purpurea*) is a source of LSD. This fungus is especially common in plants grown in wet areas that have recently been converted to cultivation. Other records showed that due to environmental conditions, specifically wet weather, the people had to rely on rye when other crops produced inadequate yields.

Many mushrooms produce narcotics or toxins. Psilocybin is produced by the fungus *Psilocybe mexicana*. *Amanita phalloides* produces the deadly toxin amanitin, which selectively inhibits mammalian RNA polymerase. Fungi often attack plants. In Florida, there has been much controversy surrounding the use of Benlate, a fungicide manufactured by DuPont that protects certain crops. Unfortunately, it may break down to form compounds that kill plants. The main roles fungi play in ecosystems are as mutualists, as in mycorrhizae, and as decomposers.

In the three kingdoms we have covered so far (Monera, Protista, and Fungi), the categories used to establish relationships among species are based on highly technical, nonintuitive characters. However, in plants and animals, these categories (below the kingdom level) are more familiar to the average person and are also useful in talking about different issues.

KINGDOM PLANTAE

Evolutionary History of Plants

The evolutionary history of plants is fascinating and no doubt played a critical role in the evolution of animals. We must recall that it was only after plants invaded the land and began to produce molecular oxygen that air-breathing animals could survive. The discussion below generally follows Raven et al. (1986).

Kingdom Plantae, which contains 275,000 species, is distinguished by two main characteristics:

1. Cellulose cell wall
2. True alternation of generations:
   - Sporophyte generation (diploid—2N)
   - Gametophyte generation (haploid—N)

The relevance and importance of plants are as follows:

1. As autotrophs, plants fix gaseous carbon from CO_2 into simple sugars via photosynthesis and thus ultimately produce food for themselves, animals, and fungi. For this reason, plants are called primary producers.
2. Plants produce all the oxygen we need, as a by-product of photosynthesis
3. Habitat, at least for terrestrial species, invariably depends on plants.

The nucleus of typical eukaryotic cells contains, among other things, the chromosomes. We say typical because some eukaryotic cells, such as the red blood cells of mammals, have no nucleus. Chromosomes are comprised of DNA and various proteins. Chromosomes are important because they are the bodies that harbor the genetic information that (1) is needed for cells to operate on a day-to-day basis and (2) enables an individual to be different from all other individuals. Chromosomes are the means by which genetic information is passed between generations.

Every cell needs enzymes and other proteins to function properly. A different and highly specific portion of DNA positioned along the chromosomes includes the instructions for an elaborate cell machinery to produce each of the different enzymes and proteins. These portions of DNA are called genes. Every organism has many genes to code for the different proteins. As a simple example, consider the genes that code for the human ABO blood group. The gene has three forms or *alleles*: one for Type A (IA), one for Type B (IB), and one for Type O (i). The ABO blood group gene occupies a position on a specific chromosome. Since each individual inherits chromosomes in pairs (one from each parent), each of us has two alleles for the ABO blood group gene.

Mitosis and Meiosis

There are two ways in which the nuclei of cells may divide: mitosis and meiosis. In mitosis, the chromosomes duplicate and form sister chromatids. The chromatids condense and line up on the equator of the cell. An elaborate apparatus of microtubules, called the spindle, forms and attaches to the chromosomes and draws them apart toward the ends or poles of the cell. As the chromatids converge at the two poles of the cell, the cytoplasm divides (called cytokinesis), and two daughter cells are formed, each identical to the parent cell (before the chromosomes duplicated to form the chromatids).

Let's say that the parent cell had 20 chromosomes (= 10 pairs). When these duplicate, they form 40 chromatids (= 20 pairs of chromatids). Remember, each chromatid is an exact copy of the original parent cell chromosome. In each daughter cell, there are 20 chromatids (one from each of the original 20 pairs of chromatids). Therefore, the parent cell had 20 chromosomes and the daughter cells have 20 chromosomes each. (Chromatids is a term that refers to the duplicated chromosomes.)

Now we would say that the parent cell had a diploid number of 20 and a haploid number of 10. (The diploid number refers to the *total* number of chromosomes in a cell, and the haploid number refers to the number of *pairs* of chromosomes in the cell.) Notice that we began with a diploid cell and ended up with two diploid cells. Mitosis refers to the basic form of nuclear division that nearly all cells use to make new cells.

Meiosis is a very special form of nuclear division that is only involved in one process: sexual reproduction. Consider a cell with the same diploid number as the parent cell above (2N = 20). Not every cell can undergo meiosis, because meiosis is a type of nuclear division reserved for special circumstances.

In meiosis, the chromosomes duplicate as before to form sister chromatids. However, these chromatids will undergo two divisions after this duplication, instead of one as in mitosis.

Before we continue, let's assume one thing about chromosomes: they are all different in size. Now we can identify the chromosomes by numbering them from largest to smallest; thus, the largest chromosome is number 1 and the smallest is number 10. Each individual has chromosomes in pairs and thus has two number 1 chromosomes, two number 2 chromosomes, and so on. The reason for the pairs is that each individual inherits one chromosome of each chromosome pair from each parent. In mitosis, all the pairs of *chromatids* for all 20 chromosomes line up on the equator. In meiosis, the pairs of *chromosomes* (each with its own pair of chromatids) line up on the equator. Thus, we see the ten pairs of chromosomes, each with two chromatids, lined up. These are called homologous pairs, because the pair of number 1 chromosomes is together, as is the pair of number 2 chromosomes, and so on. Each homologous pair is comprised of four chromatids. Sometimes these homologous pairs are called bivalents.

In the first division in meiosis, the homologues separate. Thus, two initial daughter cells form, each with both chromatids of one chromosome of each chromosomal pair. Therefore, each cell would get, for example, a chromosome number 10 with both its chromatids. A daughter cell might get the chromosome number 10 that the parent cell inherited from its father and the chromosome number 2 from its mother. There are many ways of dividing up the homologues: There are 2K ways of dividing up K pairs of homologous chromosomes. Thus, if you had ten homologous pairs of chromosomes, there would be 210 = 1024 ways of dividing up the homologues.

In the second division, the homologues in each daughter cell line up on the equator of the cell and the chromatids separate as in mitosis.

In the end following meiosis, a cell with 20 chromosomes (10 pairs) produces 4 cells each with 10 chromosomes. These chromosomes are not paired, but rather there is a representative of each chromosome. Thus, meiosis begins with a diploid cell and ends with four haploid cells.

Alternation of generations basically consists of an alternation of meiosis and fertilization (which involves fusion of two cells to produce one cell with twice as many chromosomes).

Sporophytes and Gametophytes

In plants, the two stages of the life cycle are the sporophyte (meaning "spore plant") and the gametophyte (meaning "gamete plant"). The sporophyte makes spores (via meiosis), whereas the gametophyte makes gametes (via gametogenesis). Each of these is a distinct free-living entity.

All cells in a typical sporophyte are said to be diploid. This means that all cells have two sets of chromosomes, or at least their chromosomes occur in pairs or multiples of pairs. The sporophtye produces haploid spores (through meiosis). Haploid spores disperse and grow into gametophytes, and gametophytes produce gametes via mitotic cell division. The gametes are the so-called egg and sperm cells. The gametes fuse (fertilization) to produce a zygote, which develops again through mitotic cell division to form the sporophyte.

Sporophyte generation dominates in more evolutionarily advanced species. Gametophyte generation dominates in more primitive species. In flowering plants (the most evolutionarily advanced plants), the male gametophyte is just the pollen grain.

Classification of Plants

Plants can be distinguished as nonvascular or vascular:

1. **Nonvascular**—No xylem or phloem
2. **Vascular**—Xylem and phloem (specialized tissues, cell types, for transporting water and nutrients) are needed for true roots, stems, and leaves

Vascular plants are more advanced in an evolutionary sense than nonvascular plants.

Nonvascular Plants

Bryophytes—This is the single group of nonvascular plants excluding the algae, which in the present classification scheme are considered to be protistans. Mosses (14,500 species), liverworts (9000 species), and hornworts (100 species) are the three subgroups of bryophytes. Bryophytes are characterized by having a free-living gametophyte and no vascular tissue.

Vascular Plants

Seedless Vascular Plants—This group includes the lycophytes, horsetails, and ferns. Although these species have vascular tissue, they are primitive plants. Vascular tissue is needed for a plant to develop true stems, leaves, and roots, which allow a plant to become tall. Nevertheless, the lycophytes (about 1000 species) are all small, as are the horsetails (15 species). Although present-day horsetails are mostly small herbaceous plants, in the Paleozoic era they reached their maximum abundance and diversity and were represented by enormous (15 m in height) ancestors.

The ferns, represented by nearly 12,000 species, are by far the most successful of the seedless vascular plants. Ferns are mostly tropical, and some species are truly tree-like in size. Spectacular tree fern forests occur on islands such as the Hawaiian Islands.

Seed Plants—The seed plants are clearly the most evolutionarily advanced group of plants. The least advanced seed plants are the so-called gymnosperms (= naked seed). Among the gymnosperms are the conifers, cycads, gingko, and gnetophytes. Conifers are familiar to most people in regions of the temperate Northern Hemisphere, whereas cycads or sago palms are chiefly tropical or subtropical. The gingko is endemic to China, whereas the gnetophytes occur in deserts and the tropics. There are 550 species of conifers, 100 species of cycads, 1 gingko, and about 70 gnetophytes. As a characteristic, gymnosperms grow quite slowly.

Angiosperms—The angiosperms (= vessel seed), or flowering plants, are the most advanced of all plants. The flowering plants are also highly diverse in structure and have invaded a number of habitats, both aquatic and terrestrial. The angiosperms are subdivided into two groups: monocots and dicots. Monocots are characterized by having flower parts in threes, a single cotyledon, usually parallel leaf venation, scattered vascular bundles, and no vascular cambium to allow for secondary growth. Dicots are characterized by having flower parts in fours or fives, two cotyledons, net-like leaf venation, vascular bundles in a ring, and a vascular cambium.

Major Plant Evolutionary Events

Although we will discuss the history of life on earth in Chapter 6, a brief history of plants is provided here:

1. Invasion of land during the Silurian period of the Paleozoic era.
2. Evolution of vascular tissue (specialized cells for transport) begins in Upper Silurian ("upper" refers to not as deep in the sediments and therefore means more recent).
3. Evolution of "naked" seeds, or gymnosperms ("gymno" means naked and "sperm" means seed), during the Permian period of the Paleozoic era. Seeds are an adaptation for a cooling and drying environment.

4. Gymnosperms dominate throughout the Mesozoic era.
5. Angiosperms (covered seeds) evolve in the Cretaceous and dominate the Cenozoic era. These are the most highly evolved plants and are called flowering plants.

Plant Conservation

There are many endangered species of plants in Florida (e.g., Florida scrub species). Many plant species are pests, such as the melaleuca, which was introduced from Australia. Plants can produce some interesting compounds, including drugs and toxins. Castor beans produce ricin, a deadly toxin. Johnsongrass produces glucocyanoside when there is an early freeze. Bracken ferns produce an anticoagulant.

KINGDOM ANIMALIA

The two main characteristics of animals are as follows:

1. Cells lack a cell wall
2. No true alternation of generations

Animals, like fungi, are heterotrophs (i.e., consumers). The diets of primary consumers are as follows:

- Herbivores—eat herbaceous food
- Granivores—eat seeds
- Frugivores—eat fruit
- Foliovores—eat leaves

The diets of secondary consumers are as follows:

- Carnivores—eat primary consumers
- Piscivores—eat fish

The animals can be divided into two main groups: vertebrates and invertebrates. Our goal in the following discussion is to provide a brief description of some of the major groups in terms of wildlife ecology and to use these groups as benchmarks for other taxa mentioned elsewhere in the text.

Invertebrates

The invertebrates comprise numerous phyla, of which we will consider only a handful. The invertebrates are quite diverse. They range from taxa that are little more than aggregations of cells, such as the metazoans and sponges, to such advanced forms as the arthropods and chordates.

Phylum Proifera

These are the sponges. Sponges are very primitive filter feeders that principally occur in marine environments. Sponges are little more than collections of cells. The body is not symmetrical and there are no organs. There are some specialized cells in the bodies of sponges, along with intercellular structures called spicules. Spicules are hard (in some groups they are silicon based, whereas in other taxa they are calcareous) structures that reinforce the body. Water is drawn through openings in the body wall by specialized cells called choanocytes. Food particles are swept into the central canal, and amoebocyte cells engulf and then digest them.

Despite the fact that most sponges are filter feeders, there is an apparent exception: carnivorous sponges (Vacelet and Boury-Esnault 1995). These sponges occur in deep water and live in an environment that is poor in resources. The central adaptations seen in these sponges include loss of choanocytes and the presence of filaments that are efficient at capturing small crustaceans (see below). The captured crustaceans struggle for hours until the filaments that ensnared them shorten and thicken while new slender filaments grow over them. Digestion is completed in a few days.

Phylum Cnidaria

The phylum Cnidaria comprises species with a defined two-cell-layer body and a gut that consists of a blind sac. Among the species in this group are the jellyfish, corals, sea anemones, and their relatives. The specialized cells seen in this group are the nematocysts. These cells are also known as stinging cells. Anyone who has run into a flotilla of sea lice or other such "beach blobs" (also known as moon jellyfish) is probably familiar with the extreme discomfort that nematocysts can inflict.

In the fall of 1994, an invasion of large moon jellyfish (*Aurelia aurita*) hit the west coast of Florida in the vicinity of St. Petersburg. Moon jellyfish are typically pelagic forms, which means they occur in the open ocean far from the coastal or littoral zone. The invasion was apparently a product of heavy storms and accompanying winds that simply blew the jellyfish toward the coast.

Other cnidarians of major importance to wildlife are the corals. Coral reefs are essentially mountains built from calcium carbonate deposited by the reef-building corals. Coral reefs only form when strict environmental conditions of water clarity (for penetration of sunlight), salinity, and temperature are met. Thus, coral reefs are restricted to the zone between 30 degrees north and south latitude (Hickman et al. 1990). Coral reefs provide habitat for numerous species of marine invertebrates, such as sponges, echinoderms, and vertebrates, including numerous bony and cartilaginous fishes.

Phylum Mollusca

The mollusks include several important wildlife species. Among these are the bivalves, which include the scallops, oysters, mussels, and clams; the gastropods, which are comprised of snails and slugs; and the cephalopods, which include squids and octopus.

Many species, including humans, are known to feed on bivalves. Bivalves are filter feeders that typically live in the littoral zone and play an important role in ecosystems as detritus feeders.

The majority of gastropods are herbivores that rasp plant material from surfaces. However, some are predatory and some are scavengers. Snails have long been a food source for humans. In France, native snails have recently been depleted, and imports from Turkey have replaced them in escargot. The important point here is that this may represent an example of human overexploitation of a wildlife resource.

The cephalopods include the octopus and squid, which are active predators. Their prey species include small fish, crustaceans, other mollusks, and worms.

Phylum Echinodermata

The echinoderms include several groups among which are the sea urchins and the sea stars (also known as starfish). Sea stars is a preferred name, as these animals are certainly not fish (which are vertebrates). Sea stars are important predators of other invertebrates on coral reefs. Keep in mind that many invertebrates are either sessile or quite slow (such as bivalves) and thus cannot move away when a sea star approaches. Paine (1966) showed that removal of sea stars from reef plots could result in the loss of total species. His argument was that the sea stars fed heavily on species that were superior competitors to other species. Once the sea stars were removed, the control on the populations of the superior competitors was released, and these species quickly outcompeted (for space) several of the other species, forcing them off the reef.

As a group, the echinoderms are along the evolutionary pathway that led to the vertebrates. Other groups along this line are minor and are not covered here.

Phylum Arthropoda

The arthropods are arguably the most advanced invertebrates. There are numerous taxonomic groups in the phylum, but they can be placed into one of three subphyla: subphylum Chelicerata, subphylum Crustacea, and subphylum Uniramia.

The chelicerates include the spiders, scorpions, horseshoe crabs, ticks, and mites. Spiders are important predators that feed on insects and other small animals, as do the scorpions. Visitors to eastern shores of the United States are undoubtedly familiar with horseshoe crabs, the bodies of which litter many beaches by the thousands.

Many crustaceans are important wildlife species. This group includes the shrimps (several species), crabs, crayfishes, and lobsters. All these species are harvested commercially and recreationally.

The uniramians (unbranched appendages) include the insects, centipedes, and millipedes. The centipedes are predators and may be quite large (several centimeters). Often, they may be found under rocks or rotting logs. They feed on earthworms, insect larvae, and other small animals. The millipedes are herbivorous and, like centipedes, prefer dark places under rocks or logs.

The insects are the most diverse (in number of species) group of animals. Insect species have diverged in a number of interesting ways biologically. Many insects are serious agricultural pests and require the use of massive quantities of insecticides for control. Indeed, one of the worst environmental disasters of modern times involved the use of DDT (dichlorodiphenyltrichloroethane), an insecticide. DDT has been linked to egg shell thinning and the collapse of numerous wildlife populations. Another chemical used as an insecticide is methyl bromide, used by the ton in agricultural enterprises around the world. If nothing else, insects have led humans to impact wildlife populations in indirect but nevertheless severe ways.

Phylum Chordata

Phylum Chordata includes both invertebrates and vertebrates. The invertebrates are important in evolutionary studies, but are nowhere near as important as the vertebrates.

Vertebrates

Vertebrates are not nearly as numerous as invertebrates (of the 1.5 million species of animals, only 50,000 are vertebrates), but they include several important taxa. The best way to view the vertebrates is to look at them in an evolutionary context. In this view, there are two main groups: the jawless forms (agnathans, which includes lampreys and hagfishes) and the jawed forms, which includes all other vertebrates.

Class Agnatha includes the hagfishes and lampreys. Hagfishes are bottom feeders in marine environments. They are scavengers and are only rarely encountered by humans other than marine biologists. Their impact on wildlife populations is not thought to be great at this time.

Lampreys are somewhat more important to wildlife ecology. Several species of lampreys are parasitic. Each is equipped with an oral disk armed with sharp and hard denticles. The lamprey attaches itself to a fish and rasps away at the body wall, ingesting the precious bodily fluids of the host fish. Once sated, the lamprey drops off the hapless fish, which may later die. The sea lamprey (*Petromyzon marinus*) invaded the Great Lakes of the United States and Canada when the Welland Canal around Niagara Falls was deepened in 1918. Invading lampreys, along with overfishing, led to the collapse of the lake trout fishery in Lakes Huron, Erie, Michigan, and Superior in the 1950s.

The classes of vertebrates with jaws are briefly introduced here. More indepth discussion is provided elsewhere in the text where relevant.

1. **Chondrichthyes**—These are the cartilaginous fishes, sharks, skates, and rays. As far as wildlife issues are concerned, recreational fishers have long pursued sharks, and now commercial fishers are after them, too.
2. **Osteichthyes**—These bony fishes comprise the largest group of vertebrates in terms of number of species (>24,000 species). Bony fishes of several species are heavily exploited for food by humans.
3. **Amphibia**—This class includes salamanders, frogs, and toads (4000 species). They are the most primitive of the so-called tetrapods (= four feet). They form the basal group for all the remaining groups (reptiles, mammals, and birds). Amphibians are found on all continents except Antarctica.
4. **Reptilia**—These are the snakes, lizards, turtles, and crocodilians (7000 species). They have epidermal scales and an amniote egg (self-contained egg, no longer dependent on water for reproduction). Reptiles form the basal group for birds and mammals. Both birds and mammals evolved from reptiles in the early part of the Mesozoic era. Both groups began to diversify in the Cretaceous era and continued to diversify in the Cenozoic era, as did flowering plants (angiosperms).
5. **Aves**—This class includes only birds. Members of the class Aves have feathers. Birds are homeothermic and have diverse diets. Some species feed chiefly on fruit, some on seeds, some on other animals, and one (one of Darwin's finches) even takes blood from the bodies of other

birds as they sit on their nests. There are approximately 9000 species of birds, and they occur worldwide in all surface habitats (they do not occur, for example, on the sea floor, but several pelagic species spend much of their lives miles from land).

6. **Mammalia**—The mammals include about 4200 species and are extremely diverse in both morphology and ecology. Like birds, mammals evolved from reptiles and are homeotherms. In place of feathers, most mammals have hair (an exception is the naked mole rat). Mammals are comprised of several important wildlife species, including ourselves. As humans, we have hair, are homeothermic, nurse our young, and possess a number of other characters found only among the mammals.

Mammals are more diverse in size than birds. Welty (1975) noted that the shrew *Microsorex* (the smallest mammal, which weighs just a few grams) and the blue whale (which weighs several tons) as opposed to the hummingbird (the smallest bird) and the ostrich (the largest bird). Even though mammals are more diverse in this sense, there still are more than twice as many species of birds as there are mammals. The birds have a much longer evolutionary history.

EXERCISES

4.1 Birds have a much longer evolutionary history than mammals. When did the first flighted birds appear? When did the first flightless birds appear?

4.2 Bats are flighted mammals. When did bats arrive on the evolutionary scene? Are birds or bats more committed to flight?

4.3 There are many examples of convergent evolution, where similar forms evolved completely independently of each other. For instance, birds and bats are flighted yet have a separate evolutionary history. What other mammals are convergent with kangaroo rats?

4.4 Make a list of the extant flightless birds. Where are they located?

4.5 Why are there so many insects? Should we include insects in a textbook on wildlife issues?

4.6 How many different trees are there? Where do trees reach their highest diversity?

4.7 Explain the relationship between wasps and fig trees. How did this relationship evolve?

4.8 What is a euphorbia and what has it converged with?

4.9 Hornbills are found in the Old World (Asia, Africa, Europe). What have they converged with in the New World (North and South America)?

4.10 Are feral hogs, cats, dogs, and other escaped pets wildlife? Are they of any special concern?

4.11 Are zoo animals wildlife?

4.12 Should we establish reserves to protect rare fungi?

4.13 How many different beetle species are there? How many have been identified with Latin names?

4.14 Why should we care if the panda becomes extinct? If a species is the unit of conservation biology, isn't a mite the equivalent of a panda? Should we be just as concerned for the mite as the panda?

4.15 Only 2% of the people in Florida purchase a hunting license. Couldn't voters outlaw hunting if they wanted to? Would outlawing hunting be a good idea?

LITERATURE CITED

Anderson, D.M. 1994. Red tides. *Scientific American* August:62–68.

Forrester, D.J. 1992. *Parasites and Diseases of Wild Mammals in Florida.* University Press of Florida, Gainesville.

Hickman, C.P., Jr., L.S. Roberts, and F.M. Hickman. 1990. *Biology of Animals,* 5th edition. Mosby, St. Louis.

Jaenike, J. 1991. Mass extinction of European fungi. *Trends in Ecology and Evolution* **6**(6):174–175.

Matossian, M.K. 1982. Ergot and the Salem witchcraft affair. *American Scientist* **70**: 355–357.

Paine, R.T. 1966. Food web complexity and species diversity. *The American Naturalist* **100**:65–75.

Raven, P.H., R.F. Evert, and H. Curtis. 1986. *Biology of Plants,* 4th edition. Worth Publishers, New York.

Vacelet, J. and N. Boury-Esnault. 1995. Carnivorous sponges. *Nature* **373**:333–335.

Warner, R.E. 1968. The role of introduced diseases in the extinction of the endemic Hawaiian avifauna. *Condor* **70**:101–120.

Welty, J.C. 1975. *The Life of Birds,* 2nd edition. W.B. Saunders, Philadelphia.

GENETIC DIVERSITY 5

s a panther from British Columbia, Canada, the same as a panther from Venezuela or a panther from Tierra del Fuego, at the tip of South America? If you were a zoo manager in charge of carnivore conservation and wanted to have a panther exhibit, what would you do? Would you choose a panther from Canada and one from the tip of South America to breed, or would you choose two from the same region? If you had to maintain these animals in a zoo for possible reintroduction later, what would your strategy be? Indeed, many Siberian tigers in zoos today are actually the result of crosses with Bengal tigers. Few are pure-bred Siberian tigers, and this raises serious questions regarding the very survival of this beautiful species. For many reasons, knowing the lineage of the animal is critical to maintaining the offspring's pure genetic makeup. It's little wonder genetics is playing a more important role in conservation today.

INHERITANCE AT THE INDIVIDUAL LEVEL

To better appreciate conservation and wildlife issues, we must understand the difference between genetic diversity and species diversity. Gregor J. Mendel (1822–1884) is credited as the father of modern genetics. Sadly, Charles Darwin did not know of Mendel's work, although the two were contemporaries. During Darwin's time, *blending inheritance* was the accepted principle by which inheritance occurred. Darwin doubted blending inheritance but had no other mechanism to explain how traits were passed from parents to offspring. Blending inheritance can be likened to the mixing of traits, as if some averaging was occurring. For instance, if a single species of rose has white flowers on some plants and red flowers on other plants, then breeding or crossing the two plants would lead to a rose with pink flowers—a sort of intermediate form. Mendel knew that this was not the case in nature, however.

Mendel performed hybrid experiments with garden peas and discovered the basis of what is now called *Mendelian inheritance*. Although Mendel had no idea of the mechanism of genetic inheritance, his experiments showed that inheritance was particulate. That is, Mendel demonstrated that traits from the red-flowered plants could be combined with white-flowered plants to produce red-flowered and white-flowered plants, but not pink-flowered plants. Furthermore, red-flowered plants could be crossed to produce both white- and red-flowered plants.

Mendel's concept of *unit characters* was deduced from solid experimental evidence gained from pea plants. The physical traits of his plants were, he reasoned, controlled by what we now call *genes,* which are present in all living organisms. Each gene has two or more *alleles,* which are different forms of a gene that determine alternate characteristics such as color, height, or shape. Thus, if a gene has two alleles, one for red color and one for white color, either color, but not both, can be passed on to offspring in the form of a *gamete*. The separation of the pair of alleles is referred to as *segregation,* which occurs during the maturation of the reproductive cells, called gametes. Since both parents each contribute one gamete, only one allele from each parent can be passed to their offspring. Another way of looking at this is that each new plant has one gamete and hence one allele from each parent. Note that this allele might control the color of the flowers, which could be white or red. Also, gametes carry many alleles for different traits. We say that the *genotype* determines the *phenotype*. This means that the gene, the genotype, determines what you see, the phenotype. Let's look at a simple example.

First, let's assume that the plant is *true breeding*. This means that red-flowered plants that are interbred always produce offspring with red flowers. Similarly, white-flowered plants that are bred with white-flowered plants produce only white-flowered plants. Furthermore, we will assume that the red color is *dominant* over the white color. That is, if a red allele from one parent is combined with a white allele from another parent, offspring will have only red flowers. The allele for the red color of the flower, labeled **R**, is dominant over the allele coding for the white color, labeled **r**. The allele coding for the white-colored flower is *recessive*. The parents will each be *homozygous*. That is, each will contain two identical alleles, as opposed to being *heterozygous,* which means having two different alleles. The first-generation plants in our experiment are usually labeled F1 plants, and the offspring of those are labeled F2 plants, for second-generation offspring.

Now let's cross or mate a red-flowered plant with a white-flowered plant, just as Mendel did, and then cross the resulting offspring, again as Mendel did. What happens?

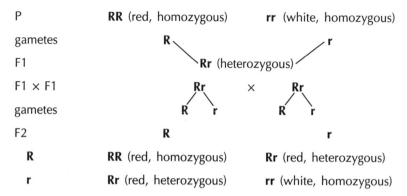

P   **RR** (red, homozygous)   **rr** (white, homozygous)

gametes   **R**   **r**

F1   **Rr** (heterozygous)

F1 × F1   **Rr**   ×   **Rr**

gametes   **R**   **r**   **R**   **r**

F2   **R**   **r**

R   **RR** (red, homozygous)   **Rr** (red, heterozygous)

r   **Rr** (red, heterozygous)   **rr** (white, homozygous)

Thus, there will be three red-flowered plants for each white-flowered plant produced in the second generation. The red allele **R** was dominant, as you recall. Furthermore, there will be two homozygous plants and two heterozygous plants. We say that the probability of being heterozygous is thus one-half. The probability of being red is three-fourths, and the probability of being white is one minus three-fourths, or one-fourth. Each parent contributed a single gamete to the *zygote* that forms in the offspring. We know from genetic theory that the contribution from each parent is a random and independent event. The probability of two independent events is equal to the product of the independent probabilities.

Taking the F1 × F1 step above, we know that the probability of an offspring receiving a dominant allele **R** was one-half from each parent. Thus, the probability of being **RR** is $(1/2) \times (1/2) = 1/4$. The probability of receiving an **r** and being recessive or white-flowered is also one-fourth. Thus, the probability of being homozygous is the sum of the two probabilities, or one-half, as shown above. Hence, the probability of being heterozygous is one minus one-half, or one-half.

What is the probability of being red-flowered? Apparently, this is also the probability of receiving at least one **R**, since only one **R** allele is required to express a red flower. But this is simply one minus the probability of receiving no **R**. If the plant receives no **R**s, then it must receive two **r**s. Thus, the probability of a plant having a red flower is just one minus the probability of being white-flowered, or one-fourth, and hence is three-fourths.

This is a simplified example of a very complex problem. We could have used animals instead of plants with no loss of generality. Generally, gametes carry many, many traits. Some traits even skip a generation and are known to be sex linked. Long ago, people noticed traits in a father but not in any of his children, male or female. When these children had families, the trait would surface again in the male offspring. This distinctive pattern suggests that the particular trait is sex linked. Sex-linked traits, such as the inability to recognize the color green, are well studied. It is known that females must receive two recessive alleles to have deficient color vision, while males need receive only one. This is because the male sex chromosome carries no color vision *locus,* whereas the female chromosome does. Thus, defective green color vision is a heritable trait. Exactly what traits are heritable?

Many morphological, behavioral, physiological, and life history traits are heritable (Ridley 1993). Indeed, in studies of common houseflies (*Drosophila*), Roff and Mousseau (1987) showed that longevity, fecundity, development time, locomotion, mating activity, and the number of abdominal bristles were all heritable traits. Morphological traits were more heritable than behavioral traits, and life history traits were less heritable. However, chances are if your grandparents and your parents live long and healthy lives, then you might as well plan on living a long and healthy life yourself. That is, you have the right genes.

Genetic diversity within a species is immense; among living creatures, it is incalculable. Nearly every creature is unique in its genetic makeup. Obviously, this raises serious questions with respect to zoos and the species they contain. With so much genetic uniqueness, can we really hope to save any of it in zoos? If a species is experiencing an extinction spasm, such as the whooping cranes of North America, genetic diversity decreases. Alleles are more likely to be bred out of a population simply by chance alone. Keeping a few in zoos is little more than a last-gasp effort, since genetic diversity will almost certainly have been lost forever through the loss of some alleles. Unless we can save the species *in situ*, that is, in its place, and protect its habitat from destruction, then perhaps it is better to limit human involvement altogether. This is most certainly a wildlife issue!

INHERITANCE AT THE POPULATION LEVEL

Animals respond to their environment but do not adapt to it. The English sparrow (*Passer domesticus*) provides an excellent example. Johnston and Selander (1971) and Gould and Johnston (1972) studied male house sparrow body weights across North America. They discovered that the smallest sparrows were found near San Francisco and Central America, while the heaviest sparrows were found in Minnesota and central Canada. Animals tend to be larger in colder regions because a larger body mass helps to maintain body heat. To be sure, the sparrows were adapting to their environment. As we will see later, there must have been differential survival among the sparrows. In cold regions, more heavier sparrows survived than lighter ones. Since life history traits are heritable, the heavier sparrows were more likely to survive and thus produced more offspring than the lighter sparrows in these regions. This led to the average sparrow being heavier in the colder regions. We can presume the opposite effect occurred in the southern and western portions of the sparrow's range. Of course, this assumes that no other complicating forces were at work, such as predation. However, there are many similar examples of population diversity.

Geographic variation across a gradient, such as winter temperature, in a species like the house sparrow produces a *cline*, or gradient of continuous variation in a genetic or phenotypic character. Indeed, the two North American meadowlark species are ubiquitous, being found from the east coast to the west coast. During the ice age 22,000 years ago, a great sheet of ice protruded into the northern United States, splitting the population of meadowlarks in half. At that time, meadowlarks were probably one species. Now we recognize two species of meadowlarks, the western meadowlark (*Sturnella neglecta*) and the eastern meadowlark (*S. magna*), although they appear nearly identical. The ice sheet apparently lasted long enough for the single species of meadowlark to separate into two populations that followed divergent evolutionary paths and *speciated*, that is, became separate and distinct species unable to interbreed in nature (Lanyon 1957). Although similar in appearance, the two species are reproductively isolated due to differing behavior and different songs. As the keeper of an aviary for a zoo, you would naturally want both species of meadowlark represented; 22,000 years ago, you would only have required a single pair. Hence, genetic diversity as it changes in time is certainly a consideration in deciding some wildlife issues. Once again, we are using a scientific basis to decide how many and what kind of birds we need for the aviary.

Whereas individuals have genes, populations have gene frequencies. Similarly, individual genotypes are expressed phenotypically, but populations have genotype and phenotype frequencies. One way of analyzing populations of a single species is to assume that the genotype frequency is in equilibrium and then look more closely at deviations from that equilibrium. We say that a population is in Hardy–Weinberg equilibrium (H-WE) when mating in the population is purely random and no evolutionary forces (see below) are acting on any phenotype in the population (Starr and Taggart 1992). What does H-WE mean in terms of genotype frequencies?

We will again consider only one gene with two alleles, R and r, that have a frequency of p and q, respectively, where p + q = 1. Note that p and q summed must be unity because R and r are the only two alleles we are considering. Under the assumptions of random mating, and a very large population existing without immigration or emigration, p^2 is the frequency of RR, 2pq is the frequency of **Rr**, and q^2 is the frequency of **rr**, just as we saw above in the example using flowers. Note also that $p^2 + 2pq + q^2 = (p + q)^2 = 1$. Now, the claim is that if the population is in H-WE, the next generation genotypic frequencies will be stable over succeeding generations, although the genotypic frequencies may have changed initially.

To see this fact, pick out two random individuals from the population and consider the offspring produced. The chance that the offspring will have an **R** is just p, and the chance of having an **r** is then q. Producing an offspring that is an **RR** genotype has probability p^2. Similarly, the chance that the offspring will have an **r** genotype has probability q^2, and finally, the probability of an offspring being **Rr** or **rR** is $pq + qp = 2pq$. Thus, the allele frequencies did not change in the offspring when compared to the parents. The population is in H-WE. When are populations not in H-WE?

EVOLUTIONARY FORCES

Populations undergoing evolutionary pressures have changing genotype frequencies, not because the members of the population are adapting to their environments but because there is differential survival among the offspring due to natural selection acting on the population. Other evolutionary forces such as mutation, inbreeding, and genetic drift may also be acting. The fundamental property of natural selection is described eloquently by the scientist who first used the term. Darwin (1859) described evolution as follows:

> As many more individuals of each species are born that can possibly survive; and as, consequently, there is a frequently recurring struggle for existence, it follows that any being, if it vary however slightly in any manner profitable to itself, under the complex and varying conditions of life, will have a better chance of surviving, and thus be *naturally selected*. From the principle of inheritance, any selected variety will tend to propagate its new and modified form.

Darwin goes on to tell us:

> Natural selection acts solely through the preservation of variations in some ways advantageous, which consequently endure.

Later, he adds:

> Natural selection, it should never be forgotten, can act solely through and for the advantage of each being.

First, Darwin tells us that organisms produce more offspring than can survive. Second, any offspring that is ever so slightly better prepared for its environment will have a better chance of living and reproducing. The organism is naturally selected to reproduce more successfully. The other organisms that do not reproduce as successfully are selected against. Note that, as Darwin tells us, natural selection acts on the individual, that is, on the phenotype and not the genotype. Third, by the rules of inheritance that were discussed previously, the organism will tend to pass on its traits to its offspring, and presumably some of them, but not necessarily all of them, will be differentially naturally selected. The results of natural selection surround us in our everyday lives. Every living organism, from the house sparrow to the elephant, is the product of natural or human-induced selection. Entire orders of organisms have survived for 100 million years and then disappeared from the fossil record forever, never to be seen on earth again. As Darwin said:

Natural selection will not produce absolute perfection...

All living organisms are subjected to natural selection. But what is natural selection acting on? A population can change due to natural selection acting on mutation, inbreeding, migration, and genetic drift—natural evolutionary forces changing allele frequencies within a population. Gene mutation can be beneficial, neutral, or harmful depending upon whether or not the organism's *fitness* increases. Fitness is a measure of reproductive success. An organism that reproduces more viable offspring is considered relatively more fit than another organism of the same species. Thus, if one bull elephant sires 20 young in his lifetime and another bull only 10 young, then the first elephant has greater fitness.

Any mutation that increases reproductive success is considered beneficial. In the worst case, if a mutation causes the organism to die at birth, then the mutation can be safely presumed to have been harmful. Most mutations are either neutral or harmful, with the result that the organism is either unaffected or dies before it reproduces. Mutations that increase fitness can be spread through a population because of the differential success of organisms with the mutation. Eventually, the mutated gene will be found more and more often in successive generations.

Thus, gene mutation does not necessarily connote a negative gene transformation. Indeed, natural selection acts on the phenotype produced by the mutant gene and determines the result. If the gene produced a malformed heart in a zebra calf, the chance of the calf surviving to mate would be small. On the other hand, if the calf's chance of survival increased because the mutant gene increased the size of its heart ever so slightly, natural selection would favor this individual. However, the rigors of everyday life are stacked against the zebra calf. Even with a possibly important mutation, the young zebra would have to survive day after day on the African plain, as perfect prey for hungry lions and hyenas. Natural selection has acted on the predators as well.

Genetic migration acts to decrease differences between populations by *gene transfer*. When an individual or group of individuals migrate from one population to another that is geographically separated and reproduction occurs, the transfer of genes acts to make the two populations more similar. Geographically separate populations of organisms must respond to different environmental variables and thus develop their own gene frequencies that are representative of their ancestral populations. When these populations interbreed, the genetic differences between them eventually disappear, assuming that preferential mating does not take place. Returning to the opening paragraph of this chapter, populations of panthers in British Columbia, Venezuela, and Tierra del Fuego are genetically quite different. If populations from these three very different regions were to interbreed, the resulting panthers would be genetically unique as well. Also, if we were to sample body sizes of panthers along the geographic cline from north to south, we would find variations, with the large body size in the highest latitudes and the smallest in the lower latitudes. These differences in body size could be correlated with certain allele frequencies, although body size is probably accounted for by multiple alleles interacting.

Random genetic drift also occurs in populations, and the smaller the population, the more likely it is that random genetic drift will have a significant impact. Allele frequencies in a population can change simply by chance alone. Furthermore, genetic drift always acts to reduce variability in a population. This means that there is an increase in homozygous individuals in the population at the expense of heterozygous individuals. Recall that homozygous individuals carry two copies of the same allele. In smaller populations, genetic drift reduces heterozygosity and hence variability. Indeed, theoretical studies show that if there are only four individuals in a population, after some 70 generations almost all heterozygosity will have been lost, whereas a population of 1024 individuals will have lost only 8% of its heterozygosity. Obviously, genetic drift can have a powerful impact on small populations. All too often these days, we hear the expression "the vortex of extinction." What this means is that the population of an organism has decreased so that genetic drift is causing the increasing loss of genetic diversity (i.e., genetic variability). Alleles can disappear from a population. When genetic diversity is lost, a population's ability to adapt is reduced simply because the number of alleles expressing phenotypic variation for natural selection to act upon has decreased. With increasing loss in genetic variability, extinction probability nears unity.

Inbreeding is similar to genetic drift and results when a departure from random mating occurs. That is, if individuals with a certain gene tend to mate with other individuals with the same gene, then nonrandom mating occurs. Inbreeding becomes increasingly more likely with a smaller population. Also, inbreeding is associated with *deleterious* or harmful genes being perpetuated in a population. For instance, according to Maehr (1997), there are about 50 Florida panthers living in south Florida in the Big Cypress State Preserve. Many of the panthers are purported to show the effects of inbreeding. That is, many of the panthers have a crooked tail, and breeding males sometimes have but one descended testicle (having none makes them infertile). One breeding female only weighed 40 pounds, about half the size of a normal female at reproductive age. These phenotypic characters are blamed on inbreeding and are believed to be deleterious. A multimillion-dollar introduction of Texas panthers into the Florida population has been undertaken to increase genetic variability. Certainly, over the short term, genetic variability will increase in the Florida panthers. The long-term effects of the introduction program cannot be known with any certainty. A project manager was quoted as saying, "Any panther born in Florida is a Florida panther." Is this a wildlife issue?

QUANTITATIVE GENETICS

Complex traits such as height and weight that vary continuously in a population are known as *quantitative* traits (Strickberger 1990, Futuyma 1986). Most quantitative traits involve the contribution of different alleles from many genes. Furthermore, phenotypic variation can be caused by environmental factors above and beyond that caused by genotypic differences. For instance, suppose there are four different alleles controlling height, A, a, B, and b. Individuals in the population must have two of these alleles. There would be individuals that are AA, Aa, aa, AB, aB, Ab, ab, BB, Bb, and bb. The effects of multiple alleles are usually additive. Hence, the phenotype of the individual is determined as the entire contribution from all the alleles. If each one codes for a different increment in height, for example, then total height would be some base height plus the additive effect of each allele. Imagine the possibilities if a trait were controlled by ten different alleles.

Frequency graphs are most often used to understand quantitative traits. The *x* axis, or *abscissa,* usually represents the trait, such as height, weight, or other characters that can be measured. The *y* axis, or *ordinate,* records the frequency of the trait in the population. The effects of natural selection acting on a trait such as height can be understood by inspecting a frequency graph.

When different rates of survival and reproductive success exist because of some inherited characteristic, natural selection can operate on that character. Natural selection will differentially favor the individuals that exhibit the trait that permits greater reproductive success, and hence greater fitness. On average and over the long term, those individuals with the favorable trait will survive and reproduce more often. Note also that not necessarily all such individuals with the favorable trait will survive. Indeed, the challenges of life demand more than favorable genetic makeup. Generally, natural selection can be stabilizing, directional, or disruptive, and these effects show up in the frequency graphs of the trait that is acted upon.

Stabilizing selection acts to reduce variability of a trait or character. Here, the average individuals in a population have higher fitness than individuals at the extremes. Natural selection is acting against any change from the average. Human birth weight is a classic example of stabilizing selection. Over the course of human evolution, birth weights have stabilized at about eight pounds. Babies born under- or overweight by two pounds have a less than 95% chance of survival, while babies born near the mid-range of the birth weight frequency curve have a better than 98% chance of living to age four weeks.

Directional selection acts to push a phenotypic character in one direction or another over time. The most famous example of directional selection comes from the experiments of H.B.D. Kettlewell with peppered moths (*Biston betularia*) in England. The most common form of the moth in preindustrial England was the black-and-white peppered form or *morph*. The moths rest by day on trees covered with lichens of the same peppered color, camouflaged against visual predators such as birds. With industrialization, the lichens on trees were killed, and hence the peppered morph of the moth was subjected to intense selective pressure by birds. About this time, a black or melanistic morph of the moth appeared more frequently. Camouflaged against the dark tree bark, the melanistic morph was safe from predatory attack. Meanwhile, the peppered morph became rare. Birds were the agents of natural selection. When industrialization was reduced, the lichens responded and once again covered trees. Now the melanistic morph of the moth was subjected to predation, and the peppered morph once again increased in numbers. Here, natural selection was directional, first against the peppered morph and then against the melanistic morph. In both instances, both morphs existed at the same time but in different proportions within the population. Neither was completely removed. Kettlewell's (1973) use of experimentation to test hypotheses remains an excellent example of the scientific method to test evolutionary theories.

Disruptive selection acts to increase variation in a character or trait and is the opposite of stabilizing selection. The phenotypic character at either end of the range is naturally selected for, while the intermediate form is selected against. For instance, disruptive selection can occur when mating is nonrandom. Body size is an inherited characteristic and usually varies smoothly in populations. Suppose that small individuals tended to mate with small individuals and large individuals tended to mate with similar sized individuals. Individuals at the extreme ends of the character frequency curve would have a selective advantage over the intermediate forms in the middle of the curve. Over time, the population might segregate into two groups with different body sizes. Indeed, over a very long time, the groups might become reproductively isolated and speciate.

Why is maintaining variability in a population important? Why haven't all populations stabilized by now? Simply put, maintaining biodiversity means maintaining genetic variability within and between populations of organisms. From plate tectonics to climate change, to variability in the weather, to the competitive advantage of a similar species, organic adaptation is essential to the long-term viability of a species. Recall that individuals do not evolve; populations evolve. Although natural selection acts on the individual, an individual cannot evolve or adapt. The offspring produced will, however, be ever so slightly better adapted to local conditions or, as Darwin put it:

> ...if some of these many species become modified and improved, others will have to be improved in a corresponding degree or they will be exterminated.

If conditions change, there will most likely be differential survival. Indeed, the environment might change, requiring some special feature that most organisms in the population do not possess. Selection can act on these special features, leading to more offspring being produced by the favored organism. Genetic variability will have increased, leading to a population of organisms being better suited to their environment. Genetic variability is the key to the long-term survival of a species, but even species with high genetic variability have no guarantee of surviving forever. We can say that the loss of genetic variation is one step on the road to extinction.

THE VALUE OF GENETIC THEORY

In 1982, the International Whaling Commission (IWC) voted to impose a moratorium on commercial whaling. However, some members of the IWC have continued to take whales for scientific or subsistence purposes, both permitted activities under the ban. A spot check of Japanese retail markets (Baker and

Palumbi 1994) showed that whale products available were not all from species hunted and traded in accordance with the IWC international treaty. The hard evidence came from the genetic analysis of mitochondrial DNA (mtDNA) and demonstrates the ability of genetic techniques to identify not only particular species but also the geographic locations where the species were taken.

Meat products ranging from dried and salted strips of meat, marinated in sesame oil and soy sauce, to unfrozen meat sold as sashimi were purchased from open markets in Japan and analyzed using a portable laboratory. Whale tissue samples from many known species and populations had also been analyzed to construct a library. Of 17 Japanese samples that matched the library specimens closely, 14 samples were from minke whales, found close to Australia. One sample of marinated meat showed both minke whale and humpback whale sequences. Two samples matched the family Delphinidae, which includes killer whales, dolphins, and pilot whales. One humpback whale sequence was an identical match with known humpback wintering grounds in Mexican, Hawaiian, and Japanese waters. One fin whale sequence was identical to fin whales sampled near Iceland.

Arguments about sustainable whaling are based on the assumption that only abundant species will be hunted and that depleted populations and species will be left to recover. These findings show that such assumptions are ill advised and suggest that active enforcement of IWC rules will be essential if continued whaling is practiced. "Without an adequate system for monitoring and verifying catches, however, history has shown that no species of whale can be considered safe" (Baker and Palumbi 1994).

EXERCISES

5.1 At what level does conservation start? Do we begin conservation at the individual level for birds and mammals and at the species level for insects, for instance?

5.2 Give some examples of the uses of modern genetic techniques in conservation.

5.3 Investigate how zoos ensure against inbreeding of captive populations.

5.4 Is it true that small, highly inbred populations have a necessarily dim long-term survival outlook? Consider the American bison and musk ox populations in the far north.

5.5 Use a search engine on the World Wide Web to identify the range of problems genetics can address.

5.6 Browse the Web for information on zoos. What functions do zoos serve in international conservation?

5.7 Do zoos maintain genetic variability within species?

5.8 Look for Mountain View Farms Breeding and Conservation Centre on the World Wide Web. Comment on its program. How does its strategy for conserving biodiversity differ from that of your local zoo?

LITERATURE CITED

Baker, C.S. and S.R. Palumbi. 1994. Which whales are hunted? A molecular genetic approach to monitoring whaling. *Science* **265**:1538–1539.

Darwin, C.R. 1859. *On the Origin of Species.* John Murray, London.

Futuyma, D.J. 1986. *Evolutionary Biology.* Sinauer Associates, Sunderland, Massachusetts.

Gould, S.J. and R.F. Johnston. 1972. Geographic variation. *Annual Review of Ecology and Systematics* **3**:457–498.

Johnston, R.F. and R.K. Selander. 1971. Evolution in the house sparrow. II. Adaptive differentiation in North American populations. *Evolution* **25**:1–18.

Kettlewell, H.B.D. 1973. *The Evolution of Melanism.* Oxford University Press, Oxford, U.K.

Lanyon, W.E. 1957. *The Comparative Biology of the Meadowlarks (Sturnella) in Wisconsin.* Publication of the Nuttal Ornithology Club No. 1. Cambridge, Massachusetts.

Maehr, D.S. 1997. *The Florida Panther: Life and Death of a Vanishing Carnivore.* Island Press, Washington, D.C.

Ridley, M. 1993. *Evolution.* Blackwell Scientific, Boston.

Roff, D.A. and T.A. Mousseau. 1987. Quantitative genetics and fitness: lessons from *Drosophila. Heredity* **58**:103–118.

Starr, C. and R. Taggart. 1992. *Biology.* Wadsworth, Belmont, California.

Strickberger, M.W. 1990. *Evolution.* Jones and Bartlett, Boston.

ORIGIN AND DISTRIBUTION OF BIODIVERSITY

6

E .O. Wilson (1988) wrote about the amount of biological diversity. There are about 750,000 insects, 43,000 vertebrates, and 250,000 plants that have been classified. Of the vertebrates, 4200 are mammals, 9000 are birds, 6300 are reptiles, 4200 are amphibians, 18,000 are bony fishes, and 900 are jawless fishes and cartilaginous fishes. Add to this number sponges, nematodes, algae, fungi, bacteria, mollusks, and arthropods and the total comes to about 1.5 million known species. We have already seen that within-species diversity is enormous as well. What is even more astounding is that Wilson estimates there may be from 5 million to 30 million species worldwide. In short, humans have only begun to appreciate the number of species that co-exist on planet earth.

These numbers seem bewildering, especially to people living in temperate North America. Whereas some forests in North America have perhaps a dozen or so species of tree, on the island of Borneo there may be 500 species of tree *per hectare*. And since each different tree has its own peculiar community of insects, the number of animals inhabiting these highly diverse forests is as yet unknown. How do we measure such biodiversity? How do we begin to organize our understanding? The patterns of distribution and abundance we see in nature now are the result of the past history of life on earth.

Raven (1994) suggests that biodiversity is the sum total of genetic variation and the interactions of all living organisms on earth, or in a particular area. The term biodiversity is a way of encompassing, in a single word, all of life's history and complexity on earth. Raven estimates that between two-thirds and three-fourths of all species on earth occur somewhere in the tropics, and more than half live in tropical rain forests. Yet rain forests occupy only 7% of the earth's terrestrial surface, and this amount is decreasing.

Many oddities confound the modern observer of wildlife distributions. For instance, fossilized leaves of the so-called *Glossopteris* flora (Schopf 1970, Brown and Gibson 1983) have been found in Permian deposits in Antarctica, South America, Africa, India, Australia, and Papua New Guinea. Hence the evidence suggests that either this ancient conifer-like form somehow rafted to many southern continents and India or that these land masses were once much closer together. Indeed, we now know that these land masses were at one time a single continent called Gondwana. To understand more clearly the present distribution of life on earth, we must understand the distribution and abundance of life throughout the history of the earth (McKinney 1993). From life's earliest beginnings, the dynamic nature of the earth has carried life along, and life has modified the land, sea, and atmosphere so as to sustain and promote life. In short, we cannot fully understand and appreciate the present distribution and abundance of living organisms without knowing how things got to where they are today.

THE ORIGIN OF BIODIVERSITY

Paleozoic: Old Life

The famous paleontologist G.G. Simpson (1969) wrote: "The existence of a natural community, or ecosystem, is the result of evolution." Evolution is responsible for biodiversity. Certainly evolution is driven, in part, by the dynamic geological phenomenon of plate tectonics. After all, *Glossopteris* flora rode into the Northern Hemisphere on India. Here, we are interested in understanding the evolution of life and how past changes in the distribution of organisms have determined present distributions. Indeed, today even the most casual observer notices that the distribution of most living organisms is limited. No species, save humans and their parasites, are cosmopolitan.

Life is hypothesized to have started about 3.5 billion years ago (BYA), when the atmosphere of the earth consisted of methane, carbon dioxide, and gaseous nitrogen compounds—an atmosphere that would poison life today. Bacterial communities consisting of cyanobacteria that could photosynthesize are known today as fossilized stromatolites. Limestone stromatolites are found on nearly every continent. Those in Australia are found along the seacoast and look like large stone toadstools. These cyanobacteria expelled poisonous oxygen as a waste product. About 2 BYA, iron began to rust as the oxygen content of the primitive atmosphere began to rise. Bacteria were oxidizing the earth, beginning with the ocean. Once the iron supply of earth was reduced, the atmospheric oxygen content rose further. By the time earth was 2.5 billion years old, life had transformed the atmosphere and the stage was set for an explosion of biological activity. From simple bacteria, more complex life forms developed.

About 800 million years ago (MYA), the supercontinent of Pangaea, which had merged from separate land masses, began to break up. The earth's surface has never stopped moving. The earliest animals are 580- to 560-million-year-old Cambrian rocks that hold fossils of arthropods, brachiopods, mollusks, and others known as metazoans. From these beginnings arose the first life forms that can be easily recognized with the unaided eye. An explosion in shells and skeletons of calcium carbonate and silica occurred as life continued to evolve. Some 500 MYA, the Cambrian ended with marine organisms having diversified into gastropods that today include snails, cephalopods with their single shell (whose only example today is the nautilus), and echinoderms, such as sea lilies, on stalks attached to the sea bottom in the inner tidal zones.

The Ordovician land mass locations were nothing like the land mass locations today. Laurentia was an isolated continent. Gondwana included Australia, Antarctica, India, Africa, and South America. Other massive islands existed where none are found today. Marine life expanded its ecological range and continued to diversify. Hinged shells evolved during this time. Starfish evolved as well, preying on shelled mollusks attached to rocks. Sponges, corals, snails, bryozoans, and many other creatures populated nearshore coasts. However, the end of the Ordovician was marked in the fossil record when some 70% of marine species disappeared. For sea creatures, this was a mass extinction, perhaps the second most all-encompassing setback. Entire reef-constructing communities disappeared from the fossil record. The emerged land masses were on the move into less favorable positions on the globe. The megacontinent of Gondwana had now drifted over the South Pole. With so much of the earth's land mass crowded around the South Pole, a great ice sheet formed, the seas cooled, and polar faunas grew within ten degrees of the equator. The expansive ice cap sequestered water and global sea levels fell, exposing Laurentia's surrounding coral reefs. Toward the end of the Ordovician, the polar ice caps melted as land masses drifted away from the South Pole. Sea levels again rose and drowned coastal flora and fauna.

The Silurian began about 438 MYA and lasted only 30 million years. The first plants moved onto land during this time. They were simple forms, such as *Cooksonia,* that had vascular conducting strands to carry water and nutrients through the stem. Fish began to diversify, and toward the end of the Silurian, the fossil record contains jawless armored fishes less than 20 cm in length. Sea life was beginning to explore greater depths.

The Devonian began about 408 MYA with the reappearance of reef communities of corals, sponges, brachiopods, bryozoans, and echinoderms. According to Cox and Moore (1994), separate floras and fish faunas can be distinguished in different parts of the world as it was then. Amphibians, the first land vertebrates, emerged during the late Devonian, and the first reptiles evolved from amphibians. The key to success on land was the evolution of the hard-shelled egg that freed the land-dwelling reptiles from the necessity of having to return to water to lay their eggs. Fossils suggest that these land creatures first evolved on the continent of Laurentia, which had been located on the equator. The Devonian also saw an explosive radiation of land plants, no doubt affecting the global climate by reducing *albedo*, or reflectance, of the earth's surface and also reducing the CO_2 content of the atmosphere further still. The Devonian closed with another mass extinction.

The next period in the history of life began about 360 MYA with the Carboniferous period and lasted until 286 MYA. During this time, terrestrial life forms diversified and multiplied. Trees, ferns, insects, dragonflies, and amphibians were common. However, life had not been freed from its aquatic past. All amphibians must return to the water to lay their eggs. The land masses of the Carboniferous were coming together again. Africa and South America were connected, and North America abutted Africa. India, Antarctica, Australia, and Europe were all close to Africa and each other. Africa was traveling northward to its present position on the globe. The Carboniferous forests were lush and green in the lowland swamps and estuarine areas that were routinely inundated by the sea.

Cockroaches, spiders, scorpions, and centipedes were thriving in the green regions of earth. The southern part of the Gondwana was glaciated extensively. Central and Southern Africa, most of South America, and all of Australia were covered with ice at one time during the Carboniferous. Our vast coal deposits of today date back to the Carboniferous period. Major coal deposits today are found in what is called Laurasia—the eastern United States, Europe, Siberia, China, and part of Australia—which is presumed to have been more tropical. These tropical forests provided suitable habitat for the amphibians and reptiles.

Laurasia lay on the equator, and the seas were rich with life. Sharks hunted the waters; starfish, gastropods, crinoids, corals, and sea urchins would be familiar to us. The bony fishes were evolving as well and competing for space. Indeed, these bony fishes make up 90% of all fish today. Most of the fish at this time were bony fishes, but a second group, the lobe-finned fishes, were also present. Although unimportant during the Carboniferous, the lobe-finned fishes had great evolutionary significance. They were the vanguard for the invasion of land. Those lobed fins were supported structurally with bone and were the forerunner to legs. Several reptile groups are known from the Carboniferous as well.

The Paleozoic era closed with the Permian period, which began 286 MYA and ended 245 MYA. During this time, the continents were moving closer together than in the Carboniferous period. Laurasia and Gondwana merged to become Pangaea. The vast ice cap that had formed during the Carboniferous began to shrink. The northern parts of Pangaea became hotter and drier, and the great wet forests of Laurentia began to dry out. The mosses, horsetails, and other wet-adapted flora were replaced by conifers, cycads, ginkgoes, and seed ferns—gymnosperms. The Gondwana ice cap completely disappeared during the Permian and *Glossopteris* spread throughout the southern land mass. Rich beds of fossil amphibians and reptiles are found in the Permian basin of Texas.

The end of the Permian period and the Paleozoic era was also the end of ancient life on earth. The Paleozoic ended with the most devastating mass extinction ever. Nearly 75% of all amphibians and reptiles were erased from the annals of life. Trilobites, perhaps the most familiar fossils, were never to occur again. Some 50% of all marine families were wiped out. Many of the forms that survived have never fully recovered their former glory. The assembly of the land masses of earth into the supercontinent Pangaea is believed to have precipitated the mass extinctions. Volcanoes, falling sea levels, global climate change, and the creation of a single land mass giving rise to a single global ocean all have been suggested as possible causes of the mass extinctions. Using the number of extant families as a measure, life was set back to mid-Ordovician times—a setback of some 300 million years.

Mesozoic: Middle Life

The Triassic period lasted from 245 MYA to 208 MYA. The supercontinent Pangaea was at its maximum contiguous size at about the middle of the Triassic. With the exception of ferns, most of the forests from which coal would eventually form were gone. The extant reptiles of the Triassic are the thecodonts, the precursors of the dinosaurs. The earliest recognizable frogs (order Anura) from the early Triassic are found in Madagascar and became more common in the fossil record throughout time. The Karroo, an arid region in South Africa, is known for the number of Triassic fossils discovered there. Among the fossils are the therapsids, mammal-like reptiles of both carnivorous and herbivorous forms.

Since the continents abutted, there existed a single global ocean. A large finger of ocean, the Tethys Seaway, penetrated the land. On the vast shoreline and in the warm seas, sea life slowly recovered. Large reptiles, such as ichthyosaurs, efficient swimming reptiles, caught fish, and placondonts, with their specialized jaws and teeth, grazed on mollusks by first stripping the creatures from the rocks and then crushing them with powerful jaws and plate-like teeth. The fish so common in today's waters were but an unimportant group in the Triassic. Significantly, it was during the Triassic that the archosaurs were undergoing a transformation from what Gould (1993) refers to as "a sprawling gait," such as that of a modern crocodile, to "an erect gait," like that of the much later Tyrannosaurs. Changes in the skeletal structure allowed some groups to walk upright.

The reptiles we refer to as dinosaurs appeared in the latter half of the Triassic, along with several other important groups. The first turtles appeared, as well as the first crocodilians, pterosaurs, and mammals. The Paleozoic reptiles were gone, and many of the forms we see today have their origin in the late Triassic. Birds, however, had to wait another 40 million years to appear.

From 208 MYA to 114 MYA, during the Jurassic and all through the Cretaceous, ending 65 MYA, the dinosaurs flourished. Cycads, ferns, and ginkgoes diversified while seed ferns disappeared. Conifers also diversified, and the angiosperms (the flowering plants) radiated gloriously about 150 MYA during the Jurassic, replacing the gymnosperms as the earth's dominant vegetation. Crocodiles of the Jurassic would be familiar to us, but other inhabitants would not. The theropods (*Tyrannosaurus rex*), sauropods (*Brontosaurus*), ornithopods (duck-billed dinosaurs such as Tsintaosaurus), ceratopsians (*Triceratops*), stegosaurs (*Stegasaurus*), and ankylosaurs (heavy club tails) were true dinosaurs that spread across the earth.

During the Jurassic, Pangaea started breaking up, and by the mid-Cretaceous (100 MYA), the land masses were separate and their trajectories were carrying them into their present positions. South America had just separated from Africa and was on a westward trajectory. Africa was moving slowly northward and rotating slightly into its place on the present globe. India was accelerating northward but was still just north of Antarctica. Europe had barely separated from North America. Pangaea had broken up—again. There is little doubt the continents will once again come together and give rise to another Pangaea in the distant future.

The flying reptiles, pterosaurs, are known from early and middle Jurassic fossils. These were the first vertebrates to be freed from terrestrial life. The oldest known reptilian flyer was *Eudimorphodon*, a fish-eating form with a 1-m wingspan. The pterosaurs consisted of two highly differentiated groups. The rhamphoryncoids had a tail and a toothed beak, while the pterodactyloids lacked these features. Some of these flying reptiles were adapted to catching insects on the wing.

Perhaps the best-known bird fossil ever found is of Archaeopteryx, discovered in 150-million-year-old limestone in Bavaria, Germany. Archaeopteryx had more dinosaur affinities than modern bird attributes. For instance, it had teeth, a heavy skull, a long slender tail, and long fingers, which modern birds lack. Modern birds also have large wishbones; Archaeoptryx had a reptilian wishbone. Indeed, it was classified as a meat-eating reptile for many years before it was reexamined and found to have feathers.

The explosive radiation of the angiosperms about 100 MYA during the Cretaceous gave rise to an increase in insect diversification. With the large variety of plants and flowers, and differing climatic conditions, the insects speciated. Bees are known to have originated during the Cretaceous, for instance, with the radiation of flowering plants.

Mammals were an unimportant group during the Jurassic and Cretaceous. They arose from therapsid reptiles, mammal-like reptiles in the Triassic. The earliest true mammals were the morganucodontids. These diminutive creatures were as small as present-day shrews, probably nocturnal, and insectivorous. They are presumed to have laid eggs. Before the end of the Cretaceous, mammals evolved into three groups that still exist today: monotremes, marsupials, and placentals. Some groups such as the multituberculates, for instance, radiated and existed for 160 million years before disappearing. The most primitive group, the monotremes, still lays eggs. However, even by the end of the Cretaceous, many mammals were delivering live young.

The end of the Mesozoic era and the Cretaceous 65 MYA came with yet another mass extinction. No dinosaur fossils are known beyond this time. Other forms such as flying reptiles and large marine animals died out as well. Compared to the Permian extinctions, only 70% of all species died out. All the surviving animals were small, including the mammals. Many floral species passed into extinction.

What caused the so-called K-T (Cretaceous–Tertiary, the next geological period) extinction? Alvarez et al. (1980) suggested that a meteorite 10 km in diameter struck the earth and led directly to the mass extinction. The fossil record shows that within a few years, 70% of all species vanished forever. The evidence was based solely on the discovery of an iridium layer in Cretaceous rock in Italy. Iridium is exceedingly rare on earth but is a common element found in meteorites (that is, meteors that strike the earth). Alvarez and his team suggested that this iridium layer would be found around the world and was caused by an extraterrestrial object impacting the earth.

Since 1980, evidence has continued to accumulate which indeed shows that iridium is found in a layer around the earth. If a meteorite of the size Alvarez and his team suggested did strike the earth, an enormous amount of dust and debris would be thrown into the atmosphere. In a few days, most of the heavier particles would have fallen back to the ground; however, the lighter aerosols would have remained suspended in the upper atmosphere. High-aerosol, low-latitude volcanoes such as Mount Pinetubo reduced temperatures worldwide. The K-T meteorite would have had a similar but extensive effect. The dust cloud would have blackened the sky, especially in the tropics. Photosynthesis would have been interrupted, causing the death of plants in a few weeks. Global temperatures would have dropped, more so in the tropics than in the higher latitudes, but most life is in the tropics. There would have been a crash in animal populations as large animals died out first across the globe. Smaller flesh eaters would have survived somewhat longer, feasting on the dead.

Other meteor impacts have been recorded without causing global extinctions. However, these impacts were in Canada and Siberia, both high-latitude locations. Similarly, Mount Saint Helens exploded, throwing smoke and ash into the atmosphere. Obviously, the earth's global inhabitants did not suffer. However, volcanoes like El Chichon did lower global temperatures. El Chichon was a high-aerosol, low-latitude volcano. In contrast, Mount Saint Helens was a high-latitude, low-aerosol volcano. In any case, most scientists agree with the meteorite hypothesis, and the impact crater is believed to be located on the coast of the Yucatan Peninsula in the Gulf of Mexico. The mass extinctions of the K-T boundary once again wiped away a majority of the earth's inhabitants and species. The continents were separated and moving apart. Whatever survived the K-T boundary would be the precursors of life on the separate continents for 65 million years.

Cenozoic: New Life

The Cretaceous period lasted nearly 80 million years and ended abruptly with an extinction spasm. The Tertiary period of the Cenozoic era began 65 MYA with the Paleocene epoch and lasted a mere eight million years. Evidence of possible mammals as far back as the Triassic period (about 220 MYA) shows that these proto-mammals were small creatures—much smaller than the dinosaurs with which they co-existed. Certainly, true mammals existed by 210 MYA in the Triassic. True birds evolved about 60 million years after mammals, around 150 MYA. However, with the K-T extinctions, the playing field was wide open. The multituberculate mammals had made it thus far.

Mammals that made it through the K-T boundary now radiated to fill new niches left by their extinct competitors. Marsupials, placentals, multituberculates, and monotremes all were extant forms. The continents were moving into their present positions; however, India and Australia were noticeably out of place.

The global climate was warmer and damper than today. Most mammals, although strange, would be recognizable herbivores, carnivores, or insectivores. Regional differences began to show in the fossil record. Since the continents were moving apart, it is not surprising that different forms were evolving according to differing climates and other ecological opportunities. The Eocene epoch began 57 MYA and lasted until 34 MYA. India was midway on its journey to Asia, and Australia and New Guinea were approaching their present positions. Many living orders and even families of mammals trace their ancestors to the Eocene. Indeed, Gould (1993) points out that three living orders of today arose in the Eocene. The Chiroptera (bats) arose in the early Eocene, while Cetacea (whales) and Sirenia (sea cows and manatees) arose in the late Eocene. Early primates also appeared in the Eocene. The important point to realize is the continents had split apart and were moving into place. Mammals were radiating from archaic stocks *in situ*. Whereas Australia had many marsupials, primates occurred in North America.

Climate change in the late Eocene forced tropical vegetation to the lower latitudes, and conifer forests spread in the high latitudes in response to colder temperatures. The late Eocene saw the loss of many ancient forms of mammals. The multituberculates died out, ending a 160-million-year existence—by far the most successful mammalian group ever to have existed. Not all groups retreated, however. Rodents, for instance, radiated and in sheer numbers of species comprise one-fifth of all present-day mammals. In the Oligocene epoch, 34 MYA to 23 MYA, the earth was cooler than previously, and the temperate forests ruled the high latitudes in both hemispheres. Strange camel-like mammals roamed North America but later became extinct. The earliest seal comes from the Oligocene. Around 23 MYA, the Miocene began during a period of mountain building. India was crashing into Asia, creating the Himalayas, which are still growing today.

A warming trend expanded the tropical forests, and mammals thrived in the rich coastal waters. Whales radiated during this time. Africa made contact with Europe, and mammals crossed between the two continents. The Proboscideans existed in three groups: deinotheres, mastodonts, and gomphotheres. Ancient horses radiated spectacularly and spread across the Northern Hemisphere during the Miocene. Cats such as the saber-toothed machairodontids were present, and fossils have been found in North America. The Miocene epoch lasted until only 5 MYA. About 11 MYA, North America was much like the Serengeti of Africa is today, with an array of different herbivores and carnivores.

The Pliocene epoch lasted from 5 MYA to 1.8 MYA. During this time, the earth grew cooler and drier. Grasslands grew more extensive and more treeless plains came into existence. Mammal diversity decreased due to a less diverse environment. North America lost most of its Miocene fauna and during the Pleistocene epoch lost some 54 large mammals over 45 kg. The megafauna of North America were gone before modern humans could enjoy them. The Florida peninsula was about three times as wide during the glacial period as it is today. It was also an area of high vertebrate biodiversity. An alternating sequence of glaciers now grips the earth, and we are currently enjoying an interglacial period. Both the Indian subcontinent and Africa support a large complement of megafauna similar to what North America contained only 12,000 years ago. Now that the polar ice cap has retreated, two-thirds of what was Florida is underwater.

The disappearance of the North American megafauna from across the continent is perplexing. Some have suggested that climate change is responsible. Others note that around this time, humans spread across the continent and into South America. For whatever reason, the megafauna—the mastodon and mammoth, the tapir and camel, the cheetah and lion, the giant ground sloth, the giant beaver, the four-horned pronghorn antelope, the short-faced bear, two species of bison, the saber-toothed cat, and the teratorn (a giant bird)—all are gone from North America. Some lament their passing.

MODERN DISTRIBUTION AND ABUNDANCE

Looking outside your window, you may see trees and squirrels or perhaps cactus and birds. They did not just happen by one day. Their distribution is the result of the entire past history of life, at least 600 million years of life, before they arrived as ancestors of the riders on the continents before them. The distribution and abundance of living things are controlled by many factors. For instance, Birch (1957) suggests that "weather is a component of the environment of animals which effectively determines the limits to distribution and the abundance of some species." There is no doubt that weather plays a vital role in affecting the geographical limits of species from plants to animals. Competition between species also plays a role, as do predation, food resources, a place to live and reproduce, and a host of other biological and physical factors. Huston (1994) summarized landscape patterns of species distributions and biological diversity. The fastest rates of change result from mortality-causing disturbances, such as fires or volcanic eruptions. Moderate rates of change result from vegetation changes. Slow change occurs as a result of gradual processes, such as natural climate change, or soil development processes. Finally, the slowest changes are caused by geologic processes of plate tectonics and the movement of land masses. Currently, the earth's biota is undergoing a rapid change.

Today, wildlife across the globe is under immense pressure, principally from direct and indirect human causes. For many species, extinction is a real possibility as their populations become ever more fragmented and isolated. Ecological processes such as the monarch butterfly migration are also threatened. As we have just seen, there have been many extinctions in the fossil record. Indeed, many periods and eras in the earth's history are divided by these so-called mass extinctions, yet life has always recovered and continued to diversify, always reaching higher levels of previous diversity. In one instance, 300 million years was required to attain the former diversity, but life, if nothing else, is tenacious. After all, it led to the evolution of humans in spite of mass extinctions and catastrophic losses of biodiversity. Indeed, it is because of these mass extinctions that we even have humans on earth! Extinction is, without doubt, the ultimate fate of all species, including humans.

Today, we read about this or that species going extinct. We do not even have a verb form of the adjective "extinct." It must be that humans have little interest in the final event of the last member of a species. We will never see the passenger pigeon again, yet flocks turned the sky black as recently as 80 years ago. We will never hear the high notes of the Carolina parakeet, yet just 100 years ago it was a common winter migrant to Florida. The ivory-billed woodpecker must have been a spectacular bird to see, its bright ivory-white bill gleaming in the sun. The dusky seaside sparrow is another bird gone in our lifetime. These are just a few of the creatures we will never enjoy. But extinction is normal. Indeed, between 98 and 99% of all species ever to have lived are extinct! What, then, is the big deal? It is the rate of extinction around the world that has attracted attention.

Pimm et al. (1995) suggest that recent extinction rates are 100 to 1000 times their pre-human levels in well-understood diverse groups of organisms from many different environments. The ultimate fate of every species is to evolve into a new form and hence become extinct, but Wilson (1988) warns that:

> The current reduction of diversity seems destined to approach that of the great natural catastrophes at the end of the Paleozoic and Mesozoic eras—in other words, the most extreme in the past 65 million years. In at least one important respect the modern episode exceeds anything in the geological past. In the earlier mass extinctions, which some scientists believe were caused by large meteorite strikes, most of the plants survived even though animal diversity was severely reduced. Now, for the first time, plant diversity is declining sharply.

Some argue that since extinctions have been part of life's history, we should not be concerned about modern extinctions. Some insects, such as mosquitoes and malaria-causing flies, and some viruses, like the HIV and Ebola viruses, they say, should be driven to extinction. Certainly both camps can agree that extinction is a wildlife issue.

Humans depend on biodiversity to sustain them. Indeed, humans arose in a diverse world, not in a depaurate one. The fruit and grains we consume are but a tiny fraction of all that we might one day utilize. Babbitt (1995) suggests that we do not need more laws to protect biodiversity. "Eco-showdowns" must be avoided because they polarize people, making a compromise difficult to achieve. In the same article, Babbitt, as secretary of the interior, announced the formation of the National Biological Service. In late 1995, the U.S. Congress demoted the service to a department under the U.S. Geological Survey. The new mission for the service was to "examine the status and trends for all U.S. wildlife habitats." Its budget was cut by 25% to $146 million, a fraction of what it takes to run a single government weapons laboratory.

There are about 270 million people living in the United States. Before Europeans arrived, there were about 12 million Native Americans. Since European arrival, more than 90% of the tallgrass prairies, 55% of wetlands, 26% of all forests, 50% of tropical forests, and 75% of all old-growth forests have been destroyed. Since 1620, over 500 species and subspecies of native plants and animals have been erased forever. According to Raven (1988), the earth may be losing 100 species per day. One way to preserve biodiversity and mitigate wildlife issues is through private sector initiatives (Edwards 1995). We will discuss this further in Chapter 20.

EXERCISES

6.1 Do a search on the World Wide Web using the keywords "biodiversity" and "biological diversity" to look for interesting wildlife issues. Summarize your results.

6.2 Look for the Econet site on the World Wide Web. Identify countries that seem to have an interest in wildlife issues. Which countries are represented on the World Wide Web?

6.3 Run a search on "Pangaea" and "Gondwana" using your favorite search engine. What comes up?

6.4 Go to the science library on campus and browse through the journal section. Are there any journals devoted to studying biogeography or plate tectonics? If so, select an interesting article to report on.

6.5 Run a search on "tectonics" or "plate tectonics" on the World Wide Web. Report on your findings.

LITERATURE CITED

Alvarez, L.W., W. Alvarez, F. Asaro, and H.V. Michel. 1980. Extraterrestrial cause for the Cretaceous–Tertiary extinctions: experiment and theory. *Science* **208**:1095–1108.

Babbitt, B. 1995. Protecting biodiversity. *Nature Conservancy* January/February:17–21.

Birch, L.C. 1957. The role of weather in determining the distribution and abundance of animals. *in Population Studies: Animal Ecology and Demography, Cold Spring Harbor Symposia on Quantitative Biology,* Vol. XXII. The Biological Laboratory, Cold Spring Harbor, New York.

Brown, J.H. and A.C. Gibson. 1983. *Biogeography.* C.V. Mosby, St. Louis.

Cox, C.B. and P.D. Moore. 1994. *Biogeography.* Blackwell Scientific, Boston.

Edwards, V. 1995. *Dealing in Diversity.* Cambridge University Press, New York.

Gould, S.J. 1993. *The Book of Life.* W.W. Norton, New York.

Huston, M.A. 1994. *Biological Diversity.* Cambridge University Press, Cambridge, U.K.

McKinney, M.L. 1993. *Evolution of Life.* Prentice-Hall, Englewood Cliffs, New Jersey.

Pimm, S.L., G.J. Russell, J.L. Gittleman, and T.M. Brooks. 1995. The future of biodiversity. *Science* **269**:347–350.

Raven, P.H. 1988. Our diminishing tropical forests. *in* Wilson, E.O. (Ed.). *Biodiversity.* National Academy Press, Washington, D.C.

Raven, P.H. 1994. Defining biodiversity. *Nature Conservancy* January/February:11–15.

Schopf, T.J.M. 1970. Relation of flora of the Southern Hemisphere to continental drift. *Taxon* **19**:657–674.

Simpson. G.G. 1969. The first three billion years of community evolution. *in Diversity and Stability in Ecological Systems.* Brookhaven National Laboratory, Upton, New York.

Wilson, E.O. 1988. The current state of biological diversity. *in* Wilson, E.O. (Ed.). *Biodiversity.* National Academy Press, Washington, D.C.

THE EVIL QUARTET 7

The history of life on earth is punctuated with extinctions. Yet each time, life rebounded, diversified, and took new forms. The most successful mammalian radiation occurred with the multituberculates and lasted 160 million years. Today there are none. Based simply on the number of species extant today, biodiversity has never been higher in spite of the stunning fact that 98 to 99% of all species that ever lived are now extinct. Biological organisms are a force to be dealt with. The fact of the matter is that all species evolve or become extinct. Another way of putting it is that all species disappear one way or another. Why are some people concerned about extinction when it is as natural as life itself?

THE WAY THINGS WERE

A few vignettes will set the stage for our examination of the present dilemma. George Cavendish Taylor visited Florida to collect birds in the spring of 1861 and wrote about his trip (Taylor 1862). Most of the wildlife he observed remains common today. He suggested that Florida be called the "Mockingbird State" because mockingbirds (*Mimus polyglottus*) were common. They remain so today. Other birds Taylor saw are not as common. His list included the ivory-billed woodpecker (*Campephilus principalis*) and the Carolina parakeet (*Conuropsis carolinensis*). Today, the Carolina parakeet is extinct and the ivory-billed woodpecker is believed to be extinct (Lammertink and Estrada 1995). Perhaps Taylor's sightings were just luck.

Taylor's trip aboard the steamship *Darlington* took him up the St. Johns River. South of Lake George, the captain of the ship took several rifles to his post on the top deck and proceeded to shoot alligators sunning on the banks of the river. The captain suggested that alligators had greatly decreased in number in that part of the river in recent years. However, above Lakes Jessup and Harney, higher up the St. Johns, alligators were more common and "cannot do harm in a country where the population is so scanty as in Florida." The captain shot at some turkeys. Both turkeys and deer were noticeably diminishing in numbers by this time.

The steamship reached its destination, Lake Monroe, in a day. The captain shot another alligator. Taylor observed ivory-billed woodpeckers from his room. The next day, Taylor traveled by horse-drawn cart to New Smyrna, near the Atlantic coast. Along the way, he happened by country settlers referred to as "crackers." "One of them told me that there were a 'smart' of Bears, Wolves, and Turkeys about. The Wolves had been 'bad' on his hogs, and he had killed a good many of them with strychnine."

Near some lagoons, Taylor again sighted a flock of parakeets (Parroquets in his text). He raised his gun to bag them, but the sun prevented him from getting a clear shot and the parakeets made their escape. The next morning, he bagged a few Florida scrub jays (*Aphelocoma coerulescens*). That afternoon, he again saw parakeets and shot two. He observed another ivory-billed wood-pecker. His companion bagged an opossum but left it behind. Taylor stated that the barred owls were so tame he could approach within a few feet of them before shooting them.

Quail were abundant around Smyrna, but Taylor was without a dog to hunt them. Taylor met two Englishmen visiting on a sporting outing. The two had just returned from killing a female bear, but the cub they shot at had escaped. That morning, the two had killed four deer, a "good many alligators," and enjoyed harpooning sawfish with three-foot saw-like snouts. They also saw parakeets and "large" woodpeckers but did not shoot any.

Taylor wrote about a Sir Francis Sykes who spent three winters hunting in Florida. "One winter they killed as many as thirty-five bears, principally on Merritt's Island near Cape Canaveral, where they were camping out. A few years ago Roseate Spoonbills were plentiful down by the Indian River, but of late their numbers have greatly diminished, owing to their being shot for the sake of the wings, which are greatly in demand for the purpose of making fans." Later, Taylor saw the ground full of holes, "like large rabbit-burrows, made by the Land-tortoises, here called Gophers (*Testudo carolina* = *Gopherus polyphemus*). These tortoises are extracted from their burrows by hooks with long handles, and are, I believe, used for food."

Later, Taylor shot "a fine Pileated woodpecker" and shot at "a pair of Swallow-tailed kites" but could not see the result. The parakeets were nesting at this time of the year and "generally breed in the cypress-swamps. They roost in company, making use of a hollow tree as their nesting-place. I am told that some live-oak-cutters, up the Halifax River, saw a flock go one evening into a hole in a tree to roost; next day, while the birds were absent, they cut the trunk of the tree nearly through, only leaving just enough uncut to keep it standing. After the Parroquets had gone in to roost, they felled it with a few blows of the axe, and secured them all."

In a letter to Taylor, one Dr. Bryant, who wrote "Notes on the Birds of Florida," which was published as Volume VII of the Proceedings of the Boston Society of Natural History, expressed regret that although the eggs of the ivory-billed woodpecker would be an important contribution to science, he was never able to collect any. "There are some birds, common enough, the eggs of which I have never succeeded in getting, such as the Ivory-billed Woodpecker, which would be a most important acquisition to science, and a most ornamental and curious egg, *never likely to be common*, and growing rarer every year as the bird does."

Later, on the same steamship with the same trigger-happy captain, Taylor described with disgust the shooters aboard. "Soon after the steamer has left Lake Monroe and has arrived in the narrow river, we see a flock of Turkeys on the right bank, consisting of a 'gobbler' and a few hens. Captain Brock and several others have their rifles ready, and blaze away at them, unfortunately wounding the 'gobbler.' This I consider shameful destruction, and unsports-manlike in the extreme." More alligators were killed. "One of the gunners shoots an unfortunate Fish-Hawk [osprey] while bearing a fish to its nest. It drops the fish, falls in the woods, and perishes uselessly."

One day on Deep Creek, a tributary of the St. Johns, Taylor spent time in dense swamp with many large cypress trees and apparently many ivory-billed and pileated woodpeckers. "I saw the Swallow-tailed Kites coming nearer, and gliding just clear of the tops of the trees, where, no doubt, they find a good supply of lizards, tree-frogs, and insects. At last one came within shot, and I killed it....I obtained a specimen of Bonaparte's Gull, which I shot while it was running along the shore of the river."

"Red-bellied and Red-cockaded Woodpeckers were common around Orange-Mills, and easily obtained....I saw a single flock of ten Parroquets, and next day went again to the same locality in search of them, but without success." Apparently, killing in the name of science was an acceptable alternative to the wanton destruction of the gunners. Nevertheless, the result was the same.

Just 130 years ago, millions of wading birds inhabited the Everglades and Big Cypress. One roost area contained an estimated one million birds. First, the plume hunters decimated roosting sites. Later, water management policies destroyed what the hunters could not. Today, estimates are that fewer than 50,000 wading birds live in the Everglades. What have we lost in Florida? Perhaps we should ask what there is left to lose in Florida.

One problem is that most people do not know what used to be in Florida or in any other place. We have all read about, heard about, or seen on television the great wildlife of Africa. By comparison, North America appears almost devoid of wildlife. However, the situation has not always been so. Imagine a Florida inhabited by more wolves, panthers, bears, and deer than humans. Try to imagine wading bird flocks in the hundreds of thousands coming in to roost. Imagine 60 million American bison roaming the interior of the United States. A roadside information sign in Kansas tells us that a person could stand at the site and watch for *five days* as a great bison herd moved by. Few people know that American elk were plains animals and retreated into the high-elevation forests as humans increased their hunting pressure on them. A mere 800,000 American pronghorn antelope survive today in the western plains. Imagine how many roamed the West just 200 years ago. What have we lost? Many people suggest we have lost far too much.

The majority of Americans do not appreciate what has been lost and do not understand that life has not always been the way it was just yesterday. Humans are an infinitely adaptable species with almost no memory of the past. Indeed, most people do not know that Lake Powell in Utah is the result of a hydroelectric power plant concrete and steel dam placed in a narrowing of the canyon walls between the Grand Canyon and what used to be Glen Canyon. Were it not for the full-page advertisements placed in major metropolitan newspapers across the United States, a series of dams would exist in what is now Grand Canyon National Park. How many people born since 1970 would suggest that the Glen Canyon dam be removed? Not many. Why? The answer is that *the dam has always been there.*

DOCUMENTED EXTINCTIONS

Islands have been the scene of many extinctions, particularly bird extinctions, since most remote islands did not have native mammals. Human-caused extinctions began thousands of years ago on islands in the tropical Pacific. Steadman (1995) estimates that perhaps as many as 2000 species of land and sea birds were exterminated by prehistoric human activities. Such a figure represents nearly 20% of all bird species. Native birds across a vast area vanished as human invaders cleared forests and introduced mammals. Most of these birds evolved in the absence of mammalian predators and probably had not learned to fear humans before they were driven to extinction. Many extinctions have been well documented. One such story is told by Bailey (1956).

The Hawaiian Island chain extends northwestward from the big island of Hawaii to Kauai, and on to Laysan, Pearl, Midway, and Kure islands. Midway, as the name implies, is centrally located in the mid-Pacific between China and San Francisco and is about the same distance from Australia. Laysan is between Midway and Kauai.

Bailey reported that "Laysan was considered one of the most remarkable bird islands of the world because of the five species of indigenous birds which had evolved there, and because of the thousands of sea birds assembling during the nesting season." Captain John Paty annexed Laysan Island to the Hawaiian Kingdom on May 1, 1857. Despite being only 25 to 30 feet high, about 3 miles long, and 1.5 miles wide, with a lagoon about 1 mile long and a half mile wide, Paty reported that Laysan was quite literally covered with birds whose numbers he estimated at 800,000. He also recorded that seals and turtles were tame and too numerous to avoid stepping on.

In 1859, Captain N.C. Brooks visited Laysan and harvested 1500 Hawaiian monk seals (a federally endangered species today). The piles of bird guano, however, doomed the island fauna and flora. On March 29, 1890, Laysan was leased to the North Pacific Phosphate and Fertilizer Company for a 20-year period. Guano digging commenced in 1894 and ended in 1904. Despite these mining activities, the five endemic species survived, but it was only a matter of time.

Around 1903, the manager of the mining operation, Max Schlemmer, introduced rabbits and guinea pigs to Laysan to act as his food supply and as a profitable business venture. The animals thrived in the absence of predators and "literally every green leaf on the island was devoured except for a tobacco patch." With no vegetation left, the island quickly became an almost "uninhabitable desert," and three species of endemics became extinct. Drifting sands buried tens of thousands of birds alive. In 1909, more than 200,000 albatrosses and other birds were killed by Japanese feather hunters. However, it was the rabbit and guinea pig invasion that finally did in Laysan's avifauna.

As for Midway Island, Captain N.C. Brooks first discovered the island on July 8, 1859. In 1902, Midway was visited by the naturalist W.A. Bryan. He reported on the activities of the Japanese feather hunters who killed hundreds of thousands of birds for the fashion industry. Bryan found the island covered with bones. On other nearby islands, the same destruction occurred from the Japanese feather trade. On the island of Lisianski, an estimated 300,000 albatrosses and other seabirds were killed before the feather hunters were interrupted in their work. As a result of these grizzly activities, President Theodore Roosevelt set aside all the islands as bird sanctuaries in 1909.

A tiny grayish-brown warbler, the so-called miller bird (*Acrocephalus familiaris*), endemic to Laysan and Nihoa, became extirpated when rabbits destroyed the vegetation on Laysan. In 1902, Fisher (1903) reported that "the miller bird is one of the most abundant of the strictly land birds peculiar to Laysan." Bailey (1956) reported that the birds were abundant on Laysan in 1913 and the warblers were common visitors to the buildings on the island. They were so tame that "they would land upon our heads" in search of millers (a variety of moth) and caterpillars. By 1923, they were extinct.

The Laysan honey-eaters (*Himatione sabguinea freethii*) were beautiful scarlet-vermillion birds "with slender down-turned beaks and eyes of reddish-yellow." Fisher (1903) wrote that "this brilliant little bird is found all over the island, but is most abundant in the interior among the tall grass and low bushes, bordering the open stretches near the lagoon, where all the land birds seem fond of congregating." Bailey (1956) cites Palmer's diary wherein he described securing one. "A most touching thing occurred: I caught a little red Honey-eater in the net, and when I took it out the little thing began to sing in my hand. I answered it with a whistle, which it returned and continued to do so for some minutes, not being the least frightened."

Dill (1913) visited Laysan in 1911, shortly after the rabbits began the destruction of the vegetation there. He estimated that 300 honey-eaters remained. "The honey-eater is not common on Laysan." Bailey and Willett arrived in 1912, and the brightly colored birds greeted them. Because of the rapidly disappearing vegetation, "the birds we noted about the island were few, and were confined to patches of wild tobacco, and the few remaining clumps of *Scaevola* and bunch grass." The sad end of the Laysan honey-eater came in 1923, when the rabbits had completely denuded the entire island.

Bailey writes, "The last naturalists to see these birds alive were the members of the Tananger Expedition, and the late Donal Dickey was probably the last to take photographs when he secured his dramatic motion picture of a male in full song on a coral rock twig—one of three survivors of the race which was doomed to extinction in swirling dust a few days later." Wetmore (1924) pulled the curtain on three of the native birds on Laysan when he wrote:

> In spite of the rabbits, a few dozen Laysan finches still end their sprightly songs above the buildings or hopped among the rocks near the lagoon. Three individuals alone of the little Honey-eater remained on our arrival: these perished during a three-day gale that enveloped everything in the cloud of swirling sand. The Miller-bird had disappeared entirely, and of the Laysan Rail but two remained.

The Laysan rail (*Porzana palmeri*) was unique to Laysan. Munro (1946) reported that the rails were "plentiful all over the island" in 1918. Fisher (1903) wrote, "The rails were everywhere on Laysan in great numbers" and photographed them. Dill (1913) reported them to be abundant during his visit to Laysan. Wetmore (1925) was probably the last to see the living bird. None were seen during a visit to Laysan in 1936. Although the rails survived on Midway until 1942, their destruction was rapid. Rats succeeded in getting ashore on Midway during World War II.

The Laysan finch (*Telespyza cantans cantans*) was the hardiest of the three endemic perching birds of Laysan, and it was able to find food even after the destruction of the vegetation. It even outlasted the rabbits and was there when new growth returned. On Midway, however, it was no match for the rats and eventually succumbed.

Who cares about a few birds on remote islands where few people live? And if we don't care about a few birds, why, for instance, should we care about whether or not a wasp becomes extinct? The answer is, of course, that we do not know all the possible ramifications species have in the natural world. Indeed, we cannot know. After all, mammals were given their chance with the extinction of the dinosaurs brought about by a chance meteorite impact. An example from Borneo illustrates what can happen when this balance is upset.

Harrison (1968) tried to illustrate the interrelationships with an example given to him by a biologist who had spent five years in Borneo. The island of Borneo was suffering an outbreak of mosquitoes. To control the pests, the World Health Organization decided to spray heavily with the pesticide DDT. This was before the long-lasting detrimental effects of DDT were understood. Much DDT was used to deal the mosquitoes a severe blow. For the local people, however, their problems were not solved—they were only beginning. Unfortunately, the DDT killed a large fraction of the wasp population as well. As a result, the caterpillars, whose main threat was the wasps, were devouring the thatched roofs of the local houses, causing them to collapse.

The houses were now open to flies, and further DDT spraying was done to reduce their population. The gecko lizards, which were a common feature in every home, also controlled the flies indoors. Now the bodies of the flies were laced with DDT. The geckos consumed the flies and accumulated the toxins in their bodies. When the fatal toxic level was reached, the geckos died, and their corpses were, in turn, consumed by the other common pet, a house cat. Soon the house cats also died as they, too, consumed the poisons.

With the death of the cats, another outbreak was visited upon Borneo. Freed from the terror of house cats, rat populations exploded; they invaded the houses and consumed food. Now the circle was complete, and the attacker became the attacked. The rats were "potential plague carriers." In the end, house cats were airlifted by the World Health Organization to remote villages to control the rats. The ecological web is a complex one, and no one can say for certain what will happen when this or that species is lost. All we can be sure of is that nature is complex, and humans must find a way to live with—not without—nature. As C.S. Holling points out (*Natural History* 1968), "We do know that, despite the arguments for the delicate balance of nature, natural systems are profoundly resistant to change." Humans evolved in a natural system. Can we now live without them?

Mammalian extinctions have also occurred. The vast herds of elephants across Africa are no more. Rhino are extremely rare. The Bengal tiger is once again under intense human pressure. Of the Siberian tigers, only 200 remain in the wild. The last Tasmanian wolf died in captivity in 1938. In the last 100,000 years or less, some 70% of North American, South American, and Australian large mammals and 40% of the genera of large mammals in Africa have become extinct (Martin and Klein 1984). In the next 50 years, we could well witness the extinction of the African Cape hunting dog (*Lycaon pictus*), the panda (*Ailuropoda melanoleuca*), and a host of others. The consequences of these extinctions are unimaginable.

Most modern extinctions have been caused by humans. The mechanisms of (1) overexploitation, (2) habitat fragmentation and destruction, (3) impacts of introductions, and (4) chains of extinctions were referred to by Diamond (1984) as the "Evil Quartet" and are the mechanisms by which modern extinctions occurred. In Florida, habitat destruction is a main cause of the extinction of wolves and ivory-billed woodpeckers, as well as the near elimination of panthers. Diamond (1989) suggested that overhunting by humans may have caused the Pleistocene extinctions of as many as 45 species of large mammals in the New World, although this is still debated today.

The islands of Hawaii are world famous for the extraordinary number of successfully introduced exotic species and for the resulting chains of extinctions that eliminated native plants and birds. As Hafernik (1992) points out, 870 nonnative plants have been successfully transplanted to Hawaii, and that approximates the number of native plants that once existed on Hawaii. Some 2000 invertebrates and 81 vertebrates have also been introduced in Hawaii. Cattle, pigs, rats, mongooses, goats, and birds of all sorts have been established on Hawaii, and their effects on the native flora and fauna have been nothing short of devastating. Indeed, a number of brown tree snakes have been recovered at the airport. These are the same snakes that devastated the island of Guam.

EXTINCTION RATES

Ehrlich (1995) claims that extinction rates are so high that "a major extinction episode is clearly underway." We already know that five major extinctions

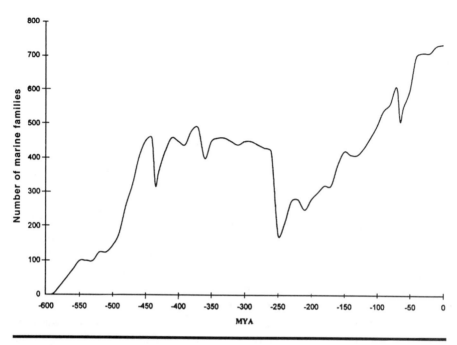

FIGURE 7.1 Approximate number of families of marine body fossils. (Adapted from Droser and Fortey 1996.)

occurred in the past: the Ordovician 440 million years ago, the Devonian 365 million years ago, the Permian 250 million years ago, the Triassic 210 million years ago, and the Cretaceous 65 million years ago—all of which were catastrophes for life (see Figure 7.1). How does the modern "extinction" compare with these past setbacks?

Four of the previous mass extinctions eliminated 65 to 85% of all animal life in the ocean. The Permian extinction was far more devastating, erasing 95% or more of all marine species. These facts led Jablonski (1995) to suggest that "extensive as today's species losses probably are, they have yet to equal any of the 'Big Five' mass extinctions." Moreover, he points out that over 90% of all past extinctions occurred at times other than during the "Big Five" mass extinctions. Rebounds from mass extinctions, Jablonski points out, are "geologically rapid and ecologically slow," and biodiversity recovery might happen in only five to ten million years. What is there to worry about? Since 1600, we have only lost 1% of the known bird species, and most of these losses occurred on islands. Whales are making a comeback. Whooping cranes introduced in Florida are surviving. Are we coming back from the brink?

The second mechanism in the Evil Quartet, habitat fragmentation and destruction, is the single most important mechanism causing extinctions today. In New Mexico, for instance, habitat fragmentation and destruction is the most common threat to threatened and endangered mammals (60% of species), birds (91%), amphibians (67%), fish (100%), and invertebrates (81%). Only among threatened and endangered reptiles is the first mechanism in the Evil Quartet I, overexploitation, more important (Bailey 1996).

Is the human population the problem? Not directly. There are about 5.5 billion humans. The state of Alaska is 586,400 square miles (375,296,000 acres). If the average household worldwide is four people, then every household on earth could have more than one-quarter acre and live in Alaska. However, humans have to eat. Humans require about 40% of all primary production to maintain themselves. Crops and livestock require huge amounts of land. When land is converted to food crops, habitat loss for most other species occurs. There is little doubt that habitat fragmentation continues to be a threat to biodiversity. Many species can survive in fragmented habitat; however, their long-term viability is questionable. Indeed, human impacts affect species well beyond human population centers and where habitat fragmentation is a major factor (Nott et al. 1995). Furthermore, Nott et al. point out that freshwater mussels and freshwater fish in North America, mammals in Australia, plants in South Africa, and amphibians worldwide are suffering rates of extinction 1000 times higher than extinction rates of the past.

The rate of the modern extinction event is cause for concern. According to Steadman (1995), one-fifth of all bird species worldwide have been eliminated in the past 2000 years, a rate far higher following human expansion. Another 1000 are globally threatened. Additionally, migratory songbird populations in the eastern United States have fallen 50%. Fully 20% of the world's freshwater fish species are extinct or close to it. Even in the United States, plants and freshwater and land mollusks are threatened. Many are already listed as extinct. Others are declining. In many instances, we know what the problem is, but we lack the will to carry out the solution. For some, enough has been lost and another loss is intolerable. For others, species losses are part of progress. This begs the question, "What do we mean by progress?"

Sadly, it seems inevitable that we will look back in 50 years and ask why nothing was done to save this or that creature from extinction. Why wasn't the manatee fully protected? Why does the Alaska Fish and Game Department shoot wolves? Why wasn't something done to save habitat for ivory-billed woodpeckers? Some resent not being able to see this giant woodpecker. How can we sit by and watch rhinos become extinct in the wild without doing something to help? Extinction is more than a wildlife issue. It is the single most important issue of our time. The legacy of extinction will be found in museum drawers across the world for all future generations to see.

The road to extinction for a species is usually, but not always, a long one. Clearly, rarity is a prelude to extinction. However, species do not have to be rare for very long before they become extinct. The passenger pigeon was so ubiquitous that harvesters were convinced the species could be harvested forever. Populations of American bison are estimated to have been as many as 60 million during the early 1800s. Within half a century, the population was reduced to less than 10,000. Modern-day extinctions are no less dramatic. On the continent of Africa, the large ungulate population has been reduced by an estimated 95%. Populations of both species of rhinos and the African elephant have been reduced to a tiny fraction of their former glory, although rhino populations in Africa are now increasing.

Many instances exist, however, that indicate our true ignorance of present situations. No one knows how many African okapi exist. We have no clue how many snow leopards haunt the high Himalayas. What is true is that these animals are habitat specialists and they are rare. Perhaps they have been rare throughout the history of their species. Certainly we would expect top predators to be rare. After all, if they were common, their prey would be even less common and hence could not support them. But how rare is rare, and when does a rare animal demand special attention?

EXERCISES

7.1 Do a search on "extinction" and "extinct" on the World Wide Web. Summarize the results.

7.2 Look for "manatee" on the World Wide Web. Is there an organization dedicated to ensuring the survival of manatees?

7.3 Look for "rhino," "elephant," and "penguin" on the World Wide Web. How many species and subspecies of these animals are recognized? What is the status of each?

7.4 The extinction of the dusky seaside sparrow was a sad day in modern wildlife history. Many subspecies of sparrows existed in Florida as a result of the Pleistocene glacial retreat. Look for an article on Florida's sparrows in the journal *Ecology*. Use the library and the Web to construct a history of these subspecies. Predict the future of these subspecies, given global climate change.

7.5   How many pandas are left in the wild and how many exist in zoos around the world? What is it about the breeding biology of the panda that lessens the species' chances of surviving the next century? How many pandas have been bred in zoos? If the panda is lost to extinction in the wild, will zoos be able to reintroduce pandas?

7.6   Categorize the threatened and endangered species in your state using the Evil Quartet. What cause most affects these species and what can be done to mitigate the cause?

7.7   Run a keyword search on the World Wide Web for "threatened" and "endangered." Summarize the results.

7.8   Do a search for "CITES," "IUCN," and "WCMC." Are there databases that contain information on ranges of endangered species?

7.9   What biological parameters are most often shared by threatened and endangered species? How can this information be used to plan conservation strategies?

LITERATURE CITED

Bailey, A.M. 1956. *Birds of Midway and Laysan Islands.* Denver Museum of Natural History, Denver.

Bailey, J.A. 1996. Questions & answers. *New Mexico Partners Conserving Endangered Species* 1(4)3–4, 9.

Diamond, J. 1984. "Normal" extinctions of isolated populations. pp. 191–246 *in* Nitecki, M.H. (Ed.). *Extinctions.* University of Chicago Press, Chicago.

Diamond, J.E. 1989. Overview of recent extinctions. pp. 37–41 *in* Western, D. and M. Pearl (Eds.). *Conservation for the Twenty-First Century.* Oxford University Press, Oxford, U.K.

Dill, H.R. 1913. The albatross of Laysan. *American Museum Journal* **13**:185–192.

Droser, M.L. and X.Li. Fortey. 1996. The Ordovician radiation. *American Scientist* March–April:122–131.

Ehrlich, P.R. 1995. The scale of the human enterprise and biodiversity loss. pp. 214–226 *in* Lawton, J.H. and R.M. May (Eds.). *Extinction Rates.* Oxford University Press, Oxford, U.K.

Fisher, W.K. 1903. Notes on the birds peculiar to Laysan Island, Hawaiian group. *Auk* **20**:384–397.

Hafernik, J.E. 1992. Threats to invertebrate biodiversity: implications for conservation strategies. pp. 171–195 *in* Fiedler, P.L. and S.K. Jain (Eds.). *Conservation Biology*. Chapman & Hall, New York.

Jablonski, D. 1995. Extinctions in the fossil record. pp. 25–44 *in* Lawton, J.H. and R.M. May (Eds.). *Extinction Rates*. Oxford University Press, Oxford, U.K.

Lammertink, M. and A.R. Estrada. 1995. Status of the ivory-billed woodpecker *Campephilus principalis* in Cuba: almost certainly extinct. *Bird Conservation International* **5**:53–59.

Martin, P.S. and R.G. Klein. 1984. *Quaternary Extinctions: A Prehistoric Evolution*. University of Arizona Press, Tucson.

Munro, G.C. 1946. Laysan Island in 1891. *Elepaio* **8**:24–25.

Natural History. 1968. Ecology: the new great chain of being. **77**(10):8–16, 60–69.

Nott, M.P., E. Rodgers, and S. Pimm. 1995. Modern extinctions in the kilo-death range. *Current Biology* **5**(1):14–17.

Steadman, D.W. 1995. Prehistoric extinctions of Pacific Island birds: biodiversity meets zooarchaeology. *Science* **267**:1123–1131.

Taylor, G.C. 1862. Five weeks in the peninsula of Florida during the spring of 1861, with notes on the birds observed there. Part I. *The Ibis* **IV**:127–142, 197–207.

Wetmore, A. 1925. Bird life among lava rocks and coral sand. *National Geographic* **48**:77–108.

Wilson, E.O. 1992. *The Diversity of Life*. W.W. Norton, New York.

EVIL QUARTET 1: OVEREXPLOITATION

8

Before the advent of agricultural systems 12,000 years ago, humans were hunters and gatherers. Shortly after humans reached Australia about 50,000 years ago, the continent lost most of its large animals with the exception of kangaroos. Paleoecologists also believe humans hunted many of Australia's large snakes, reptiles, and most of the flightless birds to extinction. North and South America lost 73 and 80%, respectively, of their large mammals 11,000 years ago. Human hunters are suggested as the cause of these extinctions, although this was a time of climate perturbations as well. Shortly after the Polynesians reached New Zealand 1200 years ago, more than ten species of large, flightless birds called moas became extinct. Across the islands of the Pacific humans exterminated literally thousands of bird species that occurred nowhere else. Human hunters drove to extinction more than ten large endemic lemurs on the island of Madagascar.

Since 1600, there have been 115 extinctions of mammal species and subspecies (Diamond 1989). Of 29 recent mammalian extinctions in Africa, Asia, and Europe, 26 were due to human hunting. Flightless birds on islands evolved in the absence of terrestrial predators and were easy victims. Plants, too, suffered from overexploitation. Logging of sandalwood trees drove several species to extinction.

Overexploitation is still a factor in the extinction process today; the ocean's fisheries are an example. Although there remain many species that could be further exploited and driven to extinction, such as rhinos, elephants, whales, walruses, and some seals, we now realize the importance of preserving species. Although the outright killing of the last individuals of a species is recognized as objectionable, other forms of the extinction process are equally insidious and no less effective.

Overexploitation of species to extinction is believed to have taken place in North America. Mammoths, mastodonts, and other large mammals were hunted by early humans. Although we will never see live mammoths again, we can see bison that were nearly hunted to extinction in the 1800s. Efforts are now under way to reestablish bison populations in formerly occupied areas. A recent example of human exploitation of rare tortoises for the pet trade is discussed later in this chapter. Collecting wild-caught animals to supply the pet trade is the modern equivalent of overhunting.

BISON IN NORTH AMERICA

Many animals once present in North America were comparatively recent immigrants from Asian stocks. Among these species were mammoths and several species of bison, one species with two subspecies of which still survive today. During the Pleistocene Ice Ages, land bridges across what is now the Bering Strait appeared and disappeared as the ice sheets grew and later melted. This is because water was sequestered in the great North American and European ice sheets and sea levels were reduced some 100 m below present levels. Because this land bridge was under the influence of a cold climate, only cold-adapted animals could pass from Asia to North America. Bighorn sheep (*Ovis*), mountain goats (*Oreamnos*), musk ox (*Ovibos*), and several species of bison (*Bison*) were able to withstand the cold and migrated to North America (Vaughan 1986). Morphological divergence occurred in the bison, and several species may have occupied the same habitats for brief periods. Some Pleistocene bison, such as *B. latifrons* from California, stood 2 m at the shoulder and carried horns that spanned 2 m. The steppe bison of Asia (*B. sivalensis*), from which North America's bison (*B. bison*) arose, was also much larger than bison of today and had long upcurved horns.

To understand bison, we must understand their evolutionary history, which began in Eurasia. Bison (*Bison* spp.) are the northern analog of cattle (*Bos* spp.), both phylogenetically and ecologically. Indeed, modern cattle and bison are so closely related in morphology that their skulls can be confused (the bison eye orbital protrudes more) (Guthrie 1970). Both formerly occupied the large bovid grazing niche across Eurasia, with *Bison* in the north and *Bos* in the south. In the far south, buffalo such as the African Cape buffalo and the Asian buffalo occupied analogous habitats. Moving from south to north geographically, buffalo, cattle, and bison appear. Intermediate zones between these three genera existed and contained two species. Interestingly, from south to north, there is an increasing tendency to remain in open areas throughout the day and to rely on flight or the protection of the herd rather than seclusion for safety.

When *Bison* colonized North America, they did not encounter *Bos* and thus were not excluded from extending their range farther south. The Bison evolution was more dramatic in the New World, as they did not compete with similar species. Indeed, with the exception of the yak, buffalo and cattle were not able to extend beyond their northern ranges, as this was the habitat of the bison, a cold-adapted species. The yak, a *Bos,* was adapted to cold and did manage to colonize but not persist in Alaska in the early Pleistocene.

The Orthogenetic Theory, Horn Core Attrition Theory, Wave Theory, Synthetic Theory, Dispersal Theory, and Clinal Theory are competing theories of bison evolutionary history. McDonald (1981) presents a modern treatment of bison evolution. Bison are presumed to have arisen during the early Pliocene, about four million years ago (MYA), in eastern Asia. The earliest bison is *Bison sivalensis,* from what is now Beijing, China, and Pinjor, India, and was widespread throughout much of eastern Asia in the early Pleistocene 1.8 MYA. *B. sivalensis* dispersed into North America in the early or middle Pleistocene, during glacial periods when the Beringia land bridge between Alaska and Asia existed.

B. sivalensis gave rise to *B. priscus,* and from *B. priscus* a shorter horned species, *B. alaskensis,* occurred in the middle and late Pleistocene in Eurasia. *B. priscus* evolved first as a savanna-adapted species and speciated to *B. alaskensis*, a forest-opening- or woodland-adapted species. Later, *B. sivalensis* also gave rise to *B. latifrons* and *B. antiquus*, both of North American origin. *B. antiquus* led to the two species we have today in North America.

B. sivalensis, with medium-sized horns, occurred in northern Eurasia and Alaska and seems to have dispersed into North America during the Yarmouth or Illinoian glacial period that preceded the Sangamon Interglacial and Wisconsin glacial maximums. The Illinoian glacial period lasted from 500,000 to 125,000 years ago and was the longest, harshest glacialization for Europe, Asia, and Alaska during the Pleistocene. The Sangamon Interglacial lasted from 125,000 to 75,000 years ago. Small populations of *B. sivalensis* dispersed across Beringia and into North America, south along western lowland Canada, and into mid-latitude North America. Entering central North America, *B. sivalensis* encountered a large, complex ungulate community but not one competitive enough to prevent its colonization (Guthrie 1970). The native grazing community in North America was dominated by various horses (*Equus*). Once established in North America, bison increased in body and horn size as they successfully exploited the grasslands.

These small, dispersed populations of *B. sivalensis* evolved into *B. latifrons* and *B. antiquus* during the Illinoian glacial maximum (Figure 8.1), around 200,000 years ago. *B. antiquus* was a smaller bodied arid- and open-savanna-adapted species of the southwest. *B. latifrons* was the largest horned species of

FIGURE 8.1 During the Illinoian glacial period, 450,000 years ago, much of North America was covered by a large ice sheet. *Bison priscus* and *B. alaskensis* (circles) occupied Beringia (Siberia, Alaska, and the land bridge between them), while *B. latifrons* (squares) and *B. antiquus* (triangles) would evolve from *B. sivalensis* stock on lands south of the ice sheet. No corridor existed between the populations. All are extinct today.

bison to exist. *B. latifrons* persisted into at least part of the interglacial period between the Illinoian and Sangamon glaciers. Fossils of *B. latifrons* from 160,000 and 25,000 years ago are known. Although fossils have been found in Florida, none are known from the eastern United States. By the late Wisconsin, some 18,000 years ago, *B. latifrons* was extinct.

During the Sangamon Interglacial period, there was no land bridge between Asia and North America, Beringia was underwater, and when the ice sheet retreated, *B. priscus* and *B. alaskensis* extended their ranges southward to meet with *B. antiquus* (Figure 8.2). During the last glacial maximum of the Wisconsin, 20,000 to 18,000 years ago, a glacier separated the two extant bison species (Figure 8.3). *B. antiquus* of the Great Plains was the dominant grazer of the Great Plains ungulate community and was also the mainstay of humans during that time.

During the period of maximum ice extent, 18,000 years ago, many large animals lived south of the ice sheet, including mastodons, two species of mammoths, elk, deer, several species of bison, bighorn sheep, two species of musk-ox, caribou, a smaller horse, a camel species, wolves, panthers, bobcats, grizzly bears, black bears, and short-nosed bears. In Europe, horses, wooly rhinoceros, bears, lions, and other large animals that no longer occur were found in abundance. During this time, the Cordillean Corridor was closed (Figure 8.4). As the ice sheets melted and the Cordillean Corridor opened, *B. antiquus* invaded northward and replaced *B. alaskensis*. The land bridge between Asia and North America was severed again. Between 4000 and 5000 years ago, *B. antiquus* speciated into *B. bison bison*, the plains bison of central North America, and *B. bison athabascae*, the wood bison of northern Canada, which we have today.

Previous studies have shown that *Bison* and *Bos* are predominantly grass and grass-like plant foragers, and few differences in their preferences exist (Larson 1940). However, *Bos* did not naturally occur in North America. Indeed, cattle (*Bos*) were brought from European stock in recent times. The various species of bison inhabited grasslands, and their grazing habits and body structure developed in association with the grasses they consumed. One essential difference, however, was that bison were naturally nomadic whereas domestic cattle were bred to be sedentary. Bison could go without water and find it on their own. Domestic cattle required water nearly daily. Whereas bison were adapted to the harsh winters of the Great Plains, domestic cattle must be stabled or artificially fed during severe weather. Wild bison and domestic cattle are thus fundamentally different creatures behaviorally.

With the complete retreat of the great ice sheets, there remained only one species of bison, *Bison bison*, in North America (Figure 8.5). Estimates of its population size ranged from 40 to 60 million strong just 300 years ago. Bison

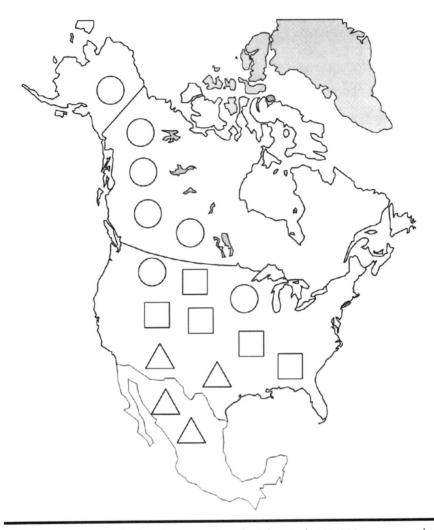

FIGURE 8.2   During the late Sangamonian glacial period, 100,000 years ago, the great ice sheet had retreated. Beringia was now underwater. *Bison alaskensis* (circles) migrated south to commingle with *B. latifrons* (squares). *B. antiquus* (triangles) were farther south. All are extinct today.

roamed the great grasslands of North America from coast to coast and from Canada to northern Mexico. The grasses they grazed upon had co-evolved with them. Early European explorers and military troops reported that "the plains were black and appeared as if in motion," so dense were the herds (Callenbach 1996).

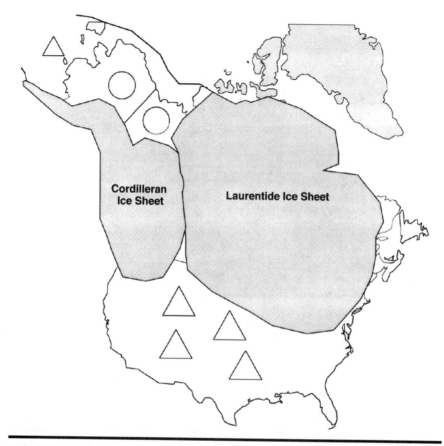

FIGURE 8.3 Two great ice sheet covered North America 18,000 years ago during the Wisconsinian glacial period: the large Laurentide Ice Sheet and the smaller Cordilleran Ice Sheet. The grassland bridge of Beringia extended across the Bering Sea to Asia. The rest of North America was largely grassland as well. *Bison alaskensis* (circles) occupied Beringia, while *B. antiquus* (triangles) occupied lands south of the ice sheet. *B. latifrons* was extinct. No corridor existed between the ice sheets. All are extinct today.

Since the time people crossed the land bridge into North America from Asia, people have hunted animals, including the bison. Well-documented hunting sites across North America contain bison bones and spearheads that were used to kill them. Before the 1750s, native hunters hunted in groups on foot using stealth to get close enough to their prey. Imagine, for a moment, being covered by a bison hide, with the skull atop your head and the empty skin

FIGURE 8.4   The Cordilleran Ice Sheet had retreated by 10,000 years ago and the great Laurentide Ice Sheet had begun to melt, creating a string of lakes. The grassland bridge of Beringia that once extended across the Bering Sea to Asia had disappeared. *Bison antiquus* (triangles) had evolved earlier and was speciating into the two extant subspecies of *B. bison* we have today.

forelegs dangling in front of your arms. In a crouched position, you are attempting to approach a small herd of bison in the open plains. A few 2000-pound bulls and several cows and calves are feeding placidly. Several of your companions are sneaking forward from other directions. Suddenly, one of your hunting party stands, draws his bow, and sends an arrow into one of the bulls. The small herd of bison panics and stampedes—toward you.

Evidence that early humans hunted bison is abundant (McDonald 1981). Interestingly, some populations of bison showed morphological abnormalities

FIGURE 8.5 The range of the bison (*Bison bison*) before the arrival of Europeans.

suggesting that genetic drift was occurring, possibly because these bison existed in small, highly inbred populations. Numerous dental abnormalities, facial bone differences, deviant horns, and certain limb abnormalities have been attributed to inbreeding (McDonald 1981). Small populations could have been isolated on the periphery of the vast bison range and, as a result of such isolation, become highly inbred over many hundreds of years. Isolated populations of bison are all we have left today!

Plains Indians made a living off bison, using nearly every part of the animal in some way. The meat provided food and was dried for lean periods, weapons were made from bones, glue was made from ground hooves, and hides were made into ropes, blankets, clothing, and shelters. Bison dung was dried and burned to cook meals and heat shelters. Just as the bison were nomadic, so, too, were Native Americans. Native Americans lacked the technology to wipe out the bison and lived sustainably from them for thousands of years. One advance, however, took place near the close of the 18th century, when Native Americans began using horses to hunt bison. Because the time required to field dress a bison and the logistics involved in transporting the meat, bones, and hide back to the campsite was considerable, Native Americans took only what they could use. Rarely was a resource destroyed.

By the 19th century, vast changes conspired to redefine the relationship between humans and bison. As a result, humans nearly caused the complete demise of the bison. Following the famous expedition of Lewis and Clark, fur traders acted as the vanguard of invasion into the plains early in the 1800s because of the availability of furs and hides along North Dakota's Missouri River. Soon the commercial slaughter of the bison began in earnest (Haines 1970).

Weak demand for beaver skins drove trappers to the plains in search of marketable alternate furs such as bison. Bison meat was also sold to passing wagon trains heading west. Waterways such as the Missouri River provided the most efficient route back to markets in the East. After 1850, however, railroad companies invaded the Midwest. First, railroad construction crews had to be provided with food, which, naturally, came from local bison herds. Railroads and the army contracted with hunters like Buffalo Bill Cody to supply buffalo meat for employees and soldiers. Bison hides were shipped east. By the late 1800s, bison were gone from the western plains.

Next to follow were Native Americans of the Great Plains. Stolen from them was the very means that supported their societies. Tribe after tribe surrendered to the U.S. government. Most were put on reservations and became completely dependent on the government. Productive people were turned into beggers. Soon railroads became the delivery service of choice to expanding markets. Bison meat and hides from more easily accessible herds kept prices down even though demand was high. Needless to say, the lives of Native Americans changed forever, and society as a whole is still paying the terrible price.

Much has been written about the almost total elimination of the bison from North America. Native Americans also suffered a similar fate. The absurd destruction of wildlife and the relentless pursuit of Native Americans across America for no other reason than because they were there is a dark side of American history. The carnage was at times both laughable and sad and bordered on the surreal. For example, during the winter of 1872–1873, the peak season for hide shipments, one firm reported handling 200,000 hides, more than 1.6 million pounds of meat, and $2.5 million worth of bones. Records show that approximately 1.5 million hides were shipped in 1872 and 1873 and more than 160,000 in 1874—a total of well over 3 million hides sent to market in only three years (Haines 1970).

As usual, certain individuals stand out and their names live on. An extreme example was provided by Thomas C. Nixon of Kansas, who in 1870 killed 120 bison in 40 minutes, a record that will likely stand forever (Dary 1974). Senseless killing and bizarre sport were the order of the day as trains routinely stopped in the middle of dwindling bison herds so passengers could kill indiscriminately without leaving their seats.

Many Anglo-Americans advocated extermination of the bison as a way to starve Native Americans into submission (Callenbach 1996). Once the bison were gone, Native American tribes could be more easily subdued. In 1876, U.S. Representative James Throckmorton of Texas said, "I believe it would be a great step forward in the civilization of the Indians and the preservation of peace on the border if there was not a buffalo in existence" (Hodgson 1994). The extermination of the bison was an example of a successful attempt to eliminate a species for the purpose of achieving a political victory over an enemy that could not otherwise be defeated.

By 1880, bison were doomed, and the Native American tribes that relied on bison were forced to submit to the political will of Anglo-Americans or cease to exist at all. The race by big game hunters to secure a trophy bison head and perhaps win a place in history by killing the very last bison was on.

Two bison-hunting expeditions undertaken in consecutive years brought the reality of the dwindling populations into sharp focus for naturalists and so-called sportsmen alike. Upon discovering that the U.S. National Museum had no bison skins in its inventory that were acceptable for display, the museum's chief taxidermist and a staunch wildlife advocate, William Hornaday, set out for Montana on a bison-hunting excursion in 1886. Eight weeks in the field yielded Hornaday a handsome crop of 25 animals (McHugh 1972). Only one year later, a contingency from New York's Museum of Natural History launched an expedition to the same area for the same reason, but aborted the mission after three months without a single bison sighting.

The last wild bison in the East was killed in western Pennsylvania just after 1800 and the last in the eastern United States before 1825. By 1900, only two wild herds were left, one in Yellowstone and one in Wood Buffalo National Park in Alberta, Canada.

The relentless final pursuit of the last of the bison had begun. Killers tracked down bison even in areas that should have provided protection for bison. Poachers infiltrated Yellowstone National Park in droves despite regulations prohibiting hunting. Lax punishment for illegal hunting combined with the enticing profitability of a successful hunt conspired against the bison. When an 1894 inventory in Yellowstone reported only approximately 20 live bison, public opinion forced Congress to pass the National Park Protective Act, which imposed a penalty of a stiff fine or imprisonment for the offense of bison poaching (McHugh 1972). It was too little too late. As the Yellowstone herd struggled for survival, unprotected bison were hunted down and exterminated. In 1897 in Lost Park, Colorado, poachers killed four bison—in all likelihood the last free-ranging, unprotected herd in the United States (McHugh 1972).

Subsequent to his post at the U.S. National Museum, William Hornaday headed up the New York Zoological Park and from this position put into effect several measures designed to revive bison populations. In 1905, he sent 12 of the park's bison to a new 8000-acre reserve in Kansas as a gift to the United States for the establishment of a national bison herd. Later that year, with the help of President Theodore Roosevelt, Hornaday and naturalist Ernest Harold Baynes founded the American Bison Society. By coordinating gifts of land, money, and animals, the society slowly nursed bison populations back from the verge of extinction.

Despite last-ditch efforts to save the species, only an estimated 1000 bison survived to see the dawn on the 20th century on a continent where just 100 years before some 60 million had inhabited the land. A hearty, adaptable species, winner of the evolutionary race, stood at the brink of extinction after a century of heavy exploitation. Poignantly dramatic is the bison's rapid demise and its crushing impact on native cultures; both remain on our collective conscience as Americans. America's past cannot be changed, but its future rests in your hands. Are you one who wants to bring back the bison?

Brucellosis (*Brucellosis abortus*) and Bison

Brucellosis is a bacterial disease carried by bison; when carried by domestic cattle, it causes them to abort their fetuses. Naturally, cattlemen fear that bison carrying brucellosis might infect their cattle. There is, however, no known instance of such a case. Indeed, there has never been a single verified case of brucellosis transmission from free-ranging wildlife to cattle. Transmission has been demonstrated only in experimentally confined animals.

In 1989, 810 cattle from 18 different herds where Yellowstone bison ranged were tested twice for brucellosis infection. No cattle tested positive for exposure to brucellosis. This is because brucellosis is transmitted during reproductive events. Although there is a very small chance that bison bulls can transmit the bacteria to female cattle during a cross-mating between the two, the Greater Yellowstone Interagency Brucellosis Committee stated that transmission from bison to cattle is almost certainly confined to contamination during a birth event by adult female bison.

Bison usually do not commingle with cattle during the calving season. Cattle can be controlled during the time they are receptive. Therefore, it is reasonable to conclude that nonpregnant cows, calves, and male bison pose virtually no risk of disease transmission. In fact, the most likely method of transmission to cattle may be from exposure to an aborted fetus, placenta, or fetal fluids of a bison cow. That is, if a cow licks fetal fluids or eats an aborted fetus or placenta of a bison cow infected with the bacteria, then the cow can become infected. Uninfected young can also be infected from infected milk, because the bacteria can be passed through the mother's milk. Specifically, if a cow calf takes milk from a bison female that is infected with the disease, the calf can become infected. The bacteria has limited viability outside its host and is quickly killed by direct sunlight, although it can remain viable for longer periods of time in cold weather. Moreover, a vaccine exists for cattle that is 75% effective and a new, more effective vaccine has been developed. No safe and effective vaccine exists for bison, however.

Yellowstone's bison have probably evolved a tolerance for brucellosis and have not been observed to abort often. Only a few abortions due to brucellosis by Yellowstone's bison have been documented. The Yellowstone bison population would continue to grow if thousands had not been shot, *prima facie* evidence that brucellosis does not limit bison population growth.

Brucellosis is no longer a threat to humans. In the past, humans contracted brucellosis by ingesting unpasteurized milk or by handling infected carcasses. If adequately cooked, even the meat from infected animals can be safely consumed. Because of the widespread pasteurization of milk and near elimination of the disease from cattle, the number of reported human cases dropped from 6500 in 1940 to 70 in 1994. The Centers for Disease Control and Prevention in Atlanta no longer considers brucellosis a reportable disease. In summary, brucellosis has been nearly eradicated in the United States. There are only 57 cattle herds left in the United States that test positive for brucellosis. However, in Montana, bison are guilty of brucellosis by association alone.

Some 2000 bison have been killed in the area surrounding Yellowstone National Park since 1984 under the guise of brucellosis control. As of 1996, there were some 3500 bison in Yellowstone National Park, 600 of which were in what was referred to as the northern herd. As snow accumulated in Yellowstone, bison began to migrate north into Montana and also west into Wyoming. Those that left the park for Wyoming were tested for brucellosis and destroyed if they were found to carry the disease. Bison that entered Montana were shot on sight whether or not they carried the disease. A federal judge refused to stay the execution of the migratory northern herd. Cattlemen in Montana were concerned their cattle would be boycotted elsewhere in the United States if they were suspected of carrying a disease.

By late 1996, 200 bison had been shot near West Yellowstone, Montana, and another 350 that left the park near Gardner, Montana were sent to slaughter in a so-called "disease management" program sponsored by the state of Montana. The Yellowstone bison scientist estimated that the population would crash as the combined result of natural mortality because of severe winter conditions and the number sent to slaughter. Montana cattle ranchers successfully thwarted plans to stop the slaughter, and the executive officer of the Montana Department of Livestock, Laurence Peterson, said his office was opposed to the plan to relocate bison. "We do not want an animal from an infected herd in the state of Montana," for quarantine or other purposes (Robbins 1997). In January 1997, with more than 700 bison dead, Governor Marc Raciot of Montana asked President Clinton to stop the slaughter. Only 1600 bison remained alive in Yellowstone.

Curiously, brucellosis is also found in elk, but no one has yet suggested killing all elk outside national parks. The prevalence of brucellosis is higher for the southern Yellowstone elk herds because elk are fed during the winter on the National Elk Refuge north of Jackson, Wyoming, and on 22 feeding grounds operated by the state of Wyoming. The Jackson bison herd also spends the winter on the National Elk Refuge.

The ultimate consequences of human conflicts are often difficult to predict. Bison represent what North America used to be and are especially important to many Native American peoples. Like fire, bison represent deep-rooted spiritual values. Bison were a source of food (sometimes the only source), shelter, and clothing for many tribes. When Native Americans learned of the killing of bison outside Yellowstone, their anger motivated them to take positive action, the consequences of which we can only hope to live to see come true.

In 1990, a coalition of tribes formed the InterTribal Bison Cooperative (ITBC). One purpose of the ITBC was to foster bison reintroduction onto tribal lands as an alternative to raising cattle. In 1996, the ITBC, acting as a coalition of more than 40 Native American tribes, and the National Wildlife Federation (NWF), the largest conservation group in the United States, tried to halt the slaughter of bison outside Yellowstone. They asked the National Park Service to stop killing wild bison and instead use them as seed stock to reestablish free-roaming herds throughout the America West. In late 1996, members of the Cheyenne River Sioux joined the effort after viewing the butchering of

Yellowstone bison shot by Montana officials. The Sioux agreed with other tribes that killing the bison showed complete disrespect for one of North America's best-known wildlife symbols and for Native American values. Although park officials back the ITBC proposal, Montana officials do not.

Bringing Bison Back

On January 22, 1997, a joint press conference was held in Denver, Colorado, to announce a historic agreement aimed at stopping the government-sanctioned slaughter of bison in and around Yellowstone National Park and restoring the wild species to the range in its historic habitat on Native American and eventually public lands throughout the West. In the first wildlife management agreement signed by a national conservation group and a group of Native American tribal leaders, the NWF and the ITBC agreed to a comprehensive plan to restore wild bison to tribal lands and ultimately to public lands in the American West. Fred DuBray, president of the ITBC, spoke for all Native Americans: "While the world around us has changed, our spiritual and cultural link to buffalo is eternal. By working together with NWF's millions of members, we will reestablish healthy buffalo populations on Indian lands, and reestablish hope for the Indian peoples." Note that Mr. DuBray used the words bison and buffalo interchangeably.

The Memorandum of Understanding between the NWF and the ITBC will:

1. Reestablish management of North American bison as one of the premier wildlife species of the West from their current status as domestic livestock
2. Restore buffalo to those tribal and public land habitats capable of supporting their long-term propagation as one of the preeminent wildlife species in North America
3. Enhance availability of wild bison for Native Americans for cultural and subsistence uses

In the future, this means that the NWF and ITBC will cooperatively advocate the capture and quarantine of Yellowstone bison that migrate to private lands outside of the park. Bison that pass quarantine will then be made available to the tribes for reintroduction to tribal lands. Coupled with limited sport and subsistence hunting conducted on lands outside the park, this strategy will halt the "shoot-on-sight" policy of the Montana Livestock Board.

The NWF and the tribes have also pledged to work toward the eventual reestablishment of bison herds on public, nontribal lands in the West. To support this effort, the tribes have agreed that a percentage of the bison that successfully complete quarantine will be made available for release on appropriate public lands, subject to mutual agreement with appropriate land management agencies. This part of the agreement effectively "banks" wild bison and their gene pool until they can be restored to public lands. Bison that fail the quarantine are to be slaughtered and distributed to Indian tribes.

The ITBC is dedicated to the restoration of bison in a manner compatible with the spiritual and cultural beliefs and practices of Native American people. At the core of this belief is the knowledge that bison have provided for the needs of Native American peoples since the beginning of time and are to be treated with the utmost respect in all actions. Bison populations must be maintained and enhanced as wild and free-roaming and should be repopulated wherever possible to ensure the continued viability of the species as well as continued and expanded access by Native Americans for cultural and subsistence use.

The NWF and the tribes support ethical hunting of bison on public lands outside of national parks. These hunts must be conducted according to the accepted rules of fair chase and sportsmanship and be consistent with standard wildlife management techniques and goals of state or tribal wildlife management agencies. These hunts must not conflict with, or impinge upon in any manner, existing tribal subsistence bison-hunting activities, and the general hunts will be designed taking this into account.

Popper Plan

Proposals to bring back the bison are not new. The so-called Popper Plan, created by Frank Popper, called for "a buffalo commons" in the Great Plains area that has been losing people for many years. Popper realized that simple economics coupled with strong spiritual values would inevitably lead to the reestablishment of the great bison herds. Bison, it should come as no surprise, cost about half as much to raise as cattle. Because bison co-evolved with native grasses and were well adapted to them, and because they could locate water with their acute sense of smell, bison proliferated across North America where grasslands once covered as much as 60% of the surface. The greatest herds were concentrated on the treeless grasslands stretching from the Mississippi River to the Rockies and from northern Mexico to the Canadian Yukon.

The Great Plains was once an enormously productive region for not only bison in huge herds but also deer, elk, and pronghorn antelope. Waterfowl also congregated in enormous numbers in the once numerous wetlands that peppered the Midwest and Great Plains. Moreover, wildlife survived and thrived without human inventions such as windmills to supply water, antibiotics, predator elimination, artificial insemination, fences to separate herds, and artificial feedlots. Species evolution and co-evolution is indeed a powerful force. Today, 70% of the grain produced in the United States goes to feeding livestock, of which about 45 million are cattle. In other words, since humans took over management of one of nature's most productive regions on earth, productivity has declined. Indeed, the generation that lived through the Great Depression and the Dust Bowl still vividly remembers when all too many farms were turned to wastelands.

How often have we seen on the news pictures of airlifts of hay to feed cattle standing in knee-deep snow? Even in the deepest snows of a Yellowstone winter, bison are able to forage. The bison's powerful neck and shoulders enable the large head to be rocked from side to side to remove snow; hence, bison require no artificial winter feeding. The bison we see today on the Great Plains are the survivors of the great evolutionary test of nature. Of the four species that inhabited North America, *Bison bison* alone survived the trials by fire, blizzard, tornado, and drought and the relentless pursuit of numerous predators. Within 100 years, humans have nearly wiped from the face of the earth a successful evolutionary history.

Bison need only native grasses and the intact ecosystem that produces those grasses. They have the speed to outrun wildfires, the bulk to withstand days without food and water, and the ability to follow storms that can turn parched grasslands green. Domestic cattle are pitiful creatures by comparison. Sadly, when the bison were displaced from the grasslands, our culture lost more than just bison. The elk that once dotted the plains were pushed into the Rocky Mountains. Pronghorns were nearly extirpated. Even small mammals such as prairie dogs were extirpated over large areas.

Fortunately, as a result of evolution, the economics favor raising bison over cattle because bison are better adapted to the harsh plains environment and require less human intervention. Bison produce about the same amount of meat, but the meat is higher in protein and lower in cholesterol than conventional beef. Live bison sell for about twice as much as cattle, and bison meat sells for four times the price of beef. A bison hide tanned to make a buffalo robe can bring in more than $1200. While tourists are willing to pay to see bison, cattle are not a tourist attraction.

As of 1996, the ITBC member tribes (see Appendix 8.1) owned some 7000 bison. The ITBC's campaign to bring back the bison is both wise and timely. Not only does it unite tribes in contributing to a common effort, the goal of which not only makes economic sense but also raises awareness of shared cultural and spiritual values, but it also awakens in all Americans the need for a spirit of adventure that has been too long suppressed by our modern world.

Furthermore, the same economics, if not the same value system, applies to other ranches throughout the Midwest and West. Bison meat is scare because bison are worth more alive than dead. The same cannot be said of cattle. Virtually every bison is kept as breeding stock. Bison ranchers are buying whatever they can to start herds.

As of 1997, about 20,000 bison were managed as wildlife in national, state, or provincial parks and refuges in the United States and Canada, as well as by various conservation groups. The number in zoos is not worth counting. However, there are more than 190,000 bison on private ranches in North America. Several such ranches belong to media magnate Ted Turner.

Mr. Turner bought the Flying D, a 107,000-acre ranch near Bozeman, Montana, in 1990. The ranch now has a herd of some 3350 free-ranging bison. Instead of marketing his surplus animals, he has purchased more ranches in Montana, two in New Mexico, and one in Nebraska's Sand Hills. His five ranches of 657,000 deeded acres have more than 7500 bison. It is well known the Mr. Turner knows how to make money.

Bison Are Coming Back

In June 1998, the National Park Service issued seven different bison management options. Public comment was solicited until October 1998. You can help decide the future of bison in the United States. Visit the first Web site listed below and make your comments known.

The ITBC, the NWF, the National Bison Association, and ranchers like Mr. Turner are determined to bring back the bison to the American landscape. As a result of increased grazing by bison, the grasslands will return as well. Favorable economics and common sense will make possible the bison's comeback. In the not too distance future, Popper's Plan will be revisited and put into action. If you drive across the Great Plains, someday you will see a large dust cloud rising into the air on a perfectly still day. Listening more closely, you will hear distant thunder, yet not a cloud will be in the sky. Small black specks will appear to be moving on the distant horizon. As you drive on, you will see more and bigger black spots and the thunder will become louder. When you stop your car on the top of a small hill, you will witness something denied to too many generations of Americans—thousands of bison on the move, raising dust and flowing across the land like a black river. As you step out of your car to join others who stopped before you, you will behold an age-old ritual played out long before cars and roads were created. After several minutes, the herd will move on and out of sight. When the dust settles, the landscape will look the same, but something will have changed.

Related Web Sites

For more on bison, try the following Web pages:

- http://www.mps.gov/planning/current/htm
- America's West: Development and History: http://www.americanwest. com:80/bison/index.htm
- The Center for Bison Studies at Montana State University–Bozeman: http://www.montana.edu/~wwwcbs/index.html
- "The Bison Is Coming Back" from *Defenders* magazine: http://www. defenders.org/magf-196.html
- "Bison Roundup" from PBS's "Newton's Apple" home page: http:// www.ktca.org/newtons/11/bison.html

To learn more about Native American environmentalism, consider:

- Native Americans and the Environment: http://conbio.bio.uci.edu/nae
- NativeWeb: http://web.maxwell.syr.edu/nativeweb

HORSFIELD'S TORTOISES

In late 1996, a frantic e-mail message was broadcast to many conservation organizations around the world. A thousand endangered Horsfield's tortoises (*Testudo horsfieldi*) had been seized at Arlanda Airport, Stockholm, Sweden, from a Russian dealer. The tortoises had not been fed for seven days, and their health was described as poor. Several had died. The Worldwide Fund for Nature and The Tortoise Trust, a U.K. organization, had officially offered to fund their repatriation to Tadzjikstan and also offered the Swedish authorities specialist veterinary and animal management assistance. A Swedish airline company, SAS Airlines, was prepared to assist with any repatriation. The Swedish Department of Agriculture and the customs agency apparently planned to destroy the tortoises. Indeed, that is what occurred.

The importer, a Russian citizen, had not obtained an import permit from the Swedish Department of Agriculture that was required to import animals to Sweden, and the tortoises were not packaged as required by Swedish law to protect their health. Swedish law also forbids the import of wildlife captured from the wild, as these tortoises had been. The decision to destroy the animals was based on the determination that their condition was too poor to release them back into the wild.

A herpetologist who examined the tortoises suggested that they showed no evidence of any contagious or infectious diseases. There was, however, evidence of poor condition due to inadequate handling and neglect. A salmonella paranoia perpetrated by animal rights activists in Sweden may have backfired. Overblown statistics (e.g., hundreds of thousands of cases of salmonella caused by pet turtles yearly) that were used to exaggerate the threat of turtles and tortoises as potential sources of contagious diseases might have prompted law enforcement officials to take unnecessary action against innocent wildlife.

The drama was given adequate coverage on a Swedish morning news program that aired on an independent television station. Jonas Wahlstrom, Sweden's best-known reptile expert, was critical of the way the tortoises were handled and the fact that authorities failed to call in outside expertise in the case. Normally, experts such as Wahlstrom provide advice and help when extraordinary wildlife cases arise, but he was apparently not contacted in this case. Television news programs showed the tortoises being packed into compartmentalized boxes to kill them.

A spokesman for the Swedish Department of Agriculture defended the decision to destroy the tortoises. He suggested that the Department of Agriculture had applied to the European Commission for dispensation to allow the tortoises to enter Sweden, but permission had been denied. The only recourse, in order to fill the requirements of the CITES convention, was to kill the endangered tortoises. The spokesman also stated the importance of stopping the flow of illegal wildlife and indicated that if the tortoises were allowed to be imported, more would surely follow.

The Tortoise Trust demanded the resignation of the Swedish minister for agriculture and the veterinarian responsible for killing the tortoises. In addition, the trust wanted a guarantee from the Swedish government that nothing of this sort would ever happen again. The trust called for a worldwide boycott of Swedish goods, including Volvos, and vowed to continue the campaign until the authorities responsible left their government posts.

Ultimately, the Swedish authorities euthanized the rare tortoises because of their apparent poor condition. However, agricultural and customs agents in Sweden allowed the tortoises to remain in their crates for a number of days before they were taken out and given water. They were kept at 15°C, far too cold for tortoises to eat or drink. The Tortoise Trust was told that the tortoises were in bad condition after the long trip, but they appeared healthy in photographs. Indeed, the tortoises seemed quite lively for having been on a cold storage room floor and flushed with cold water from a fire hose.

The Swedish border veterinarians and other experts on these animals at Arlanda Airport worked for several days with the tortoises. They examined them; tried to treat, feed, and water them; and sadly came to the conclusion that the tortoises were in such bad condition that, from the *animals' viewpoint*, they had to be destroyed. Denmark, which is also under the CITES convention, places confiscated animals with animal breeders and zoos. The Swedish authorities claimed that no zoo would take them, especially because the importer lacked CITES documentation. A private Swedish citizen filed criminal charges against the Swedish customs agency at Arlanda Airport for violation of the Swedish Animal Protection Law.

Several questions naturally emerged as a result of the publicity surrounding the tortoises. The tortoises arrived on a Friday and were not unpacked until the following Tuesday. They were kept at 15°C, so it is hardly surprising that they "would not feed" because tortoises do not feed when cold. Authorities who saw the tortoises confirmed that their condition was by no means as bad as was reported. The end result for the tortoises was the same. From the animals' point of view, no animal would choose euthanasia over release under any condition. Animals, including tortoises, rarely commit suicide.

APPENDIX 8.1: NATIVE AMERICAN TRIBAL SUPPORTERS OF THE EFFORT TO BRING BACK THE BISON

Native Village of Mekoryuk (Mekoryuk, Alaska)
Elk Valley Rancheria (Crescent City, California)
Round Valley Indian Reservation (Covelo, California)
Southern Ute Indian Tribe (Ignacio, Colorado)
Nez Perce Tribe (Lapwai, Idaho)
Shoshone–Bannock Tribes (Fort Hall, Idaho)
Prairie Band Potawatomi (Mayetta, Kansas)
Sault Sainte Marie Tribe–Chippewa Indians (Sault Sainte Marie, Michigan)
Prairie Island Dakota Community (Welch, Minnesota)
Blackfeet Tribe (Browning, Montana)
Confederated Salish and Kootenai (Pablo, Montana)
Crow Tribe (Crow Agency, Montana)
Gros Ventre and Assiniboine Tribes (Harlem, Montana)
Northern Cheyenne Tribe (Lame Deer, Montana)
Ponca Tribe of Nebraska (Lincoln, Nebraska)
Santee Sioux Tribe of Nebraska (Niobara, Nebraska)
Winnebago Tribe of Nebraska (Winnebago, Nebraska)
Nambe Pueblo Oweenge (Nambe, New Mexico)
Picuris Pueblo (Penasco, New Mexico)

Pueblo of Pojoaque (Santa Fe, New Mexico)
Pueblo of San Juan (San Juan Pueblo, New Mexico)
Taos Pueblo (Taos, New Mexico)
Spirit Lake Sioux Nation (Fort Totten, North Dakota)
Standing Rock Sioux Tribe (Fort Yates, North Dakota)
Choctaw Nation of Oklahoma (Durant, Oklahoma)
Loyal Shawnee Tribe (Bluejacket, Oklahoma)
Modoc Tribe of Oklahoma (Miami, Oklahoma)
Cheyenne River Sioux Tribe (Gettysburg, South Dakota)
Crow Creek Sioux Tribe (Fort Thompson, South Dakota)
Flandreau Santee Sioux Tribe (Flandreau, South Dakota)
Lower Brule Sioux Tribe (Lower Brule, South Dakota)
Rosebud Sioux Tribe (Rosebud, South Dakota)
Sisseton Wahpeton Sioux Tribe (Agency Village, South Dakota)
Yankton Sioux Tribe (Wagner, South Dakota)
Ute Indian Tribe (Fort Duchesne, Utah)
Kalispel Tribe (Usk, Washington)
Spokane Tribe of Indians (Wellpinit, Washington)
Ho-Chunk Nation (Black River Falls, Wisconsin)
Oneida Tribe of Wisconsin (Seymour, Wisconsin)
Northern Arapahoe (Fort Washakie, Wyoming)

EXERCISES

8.1 How can the scientific method be used to test if cattle can contact brucellosis from bison or elk? Design a series of treatments with replicates that can settle this issue.

8.2 Some scientists suggest that bison saliva stimulates native grasses to grow faster. The evolutionary importance of such an adaptation is obvious. How can this hypothesis be tested scientifically?

8.3 What would happen if people began demanding bison burgers at fast-food restaurants? Suppose those health-conscious customers were willing to pay more for this naturally lean meat?

8.4 Cattle ranching on public lands has proved to be a disaster in the West because lands have been badly overgrazed. Do bison overgraze? How was overgrazing prevented?

8.5 Two inventions in the 1850s changed ranching on the Great Plains: barbed wire and the windmill. How did these inventions change the face of ranching west of the Mississippi River? What happened in the American

Southwest, and especially around Tucson, Arizona, as a result of these changes?

8.6 What other animals, birds, and plants have been overexploited to supply the pet trade?

8.7 Of the issues presented in Chapter 3, which species have been threatened by overexploitation?

LITERATURE CITED

Barsness, L. 1985. *Heads, Hides, and Horns*. Texas Christian University Press. Fort Worth.

Callenbach, E. 1996. *Bring Back the Buffalo!* Island Press, Washington, D.C.

Dary, D. 1974. *The Buffalo Book*. Swallow Press, Chicago.

Diamond, J.E. 1989. Overview of recent extinctions. pp. 37–41 *in* Western, D. and M. Pearl (Eds.). *Conservation for the Twenty-First Century*. Oxford University Press, Oxford, U.K.

Guthrie, R.D. 1970. Bison evolution and zoogeography in North America during the Pleistocene. *The Quarterly Review of Biology* **45**(1):1–15.

Haines, F. 1970. *The Buffalo*. Thomas Y. Crowell, New York.

Hodgson, B. 1994. Buffalo back home on the range. *National Geographic* **11**:64–89.

Larson, F. 1940. The role of bison in maintaining the short grass plains. *Ecology* **21**:113–121.

Lott, D.F. 1991. *American Bison Socioecology*. Elsevier Science Publishers, Amsterdam.

Luoma, J.R. 1993. Back home on the range? *Audubon* March/April:46–52.

McDonald, J.N. 1981. *North American Bison*. University of California Press, Berkeley.

McHugh, T. 1972. *The Time of the Buffalo*. University of Nebraska Press. Lincoln.

Mitchell, J. 1993. The way we shuffled off the buffalo. *Wildlife Conservation* January/February:44–50.

Park, E. 1969. *The World of the Bison*. J.B. Lippincott, Philadelphia.

Pielou, E.C. 1991. *After the Ice Age*. University of Chicago Press, Chicago.

Rifkin, J. 1992. *Beyond Beef: The Rise and Fall of the Cattle Culture*. Penguin Books, New York.

Robbins, J. 1997. Indian tribes seek halt to killing of wild bison. *The New York Times* January 21.

Vaughan, T.A. 1986. *Mammalogy*. Saunders College Publishing, Philadelphia.

EVIL QUARTET 2: HABITAT FRAGMENTATION AND DESTRUCTION

9

Alteration of the land is the greatest threat to wildlife today. The most obvious form of alteration is complete destruction of natural areas, such as when forests are clear-cut, land is cleared for housing development and shopping mall construction, fields are cleared for agriculture, wetlands are drained, and streams are dammed. Habitat fragmentation occurs when natural areas become isolated from larger tracts to which they were once connected. In many areas, natural landscapes that were once continuous are a mere patchwork. Studies in many regions have documented local extirpations, extinctions, changes in species composition, and the invasion of "weedy" species like raccoons and opossums.

When a natural landscape begins to undergo fragmentation, generally no species are lost and hence the erosion of biodiversity is not obvious. Certainly, a few individuals of a species and the genetic material they contain are lost, but because no species are lost, attention is not drawn to small amounts of fragmentation. Because landscapes are naturally heterogeneous, altered parts of the landscape seem simply to add to the naturally occurring mosaic. After all, naturally occurring fires "alter" forests, and, in fact, many forests depend on recurrent fires to enable new growth.

As earlier as 1863, Marsh (1907) understood the consequences of landscape fragmentation. While recognizing the importance of human agricultural systems, he expressed concern that the long-term effects of habitat fragmentation and destruction were not understood. In his 1907 book, Marsh wrote:

> Hence, a certain measure of transformation of terrestrial surface, of suppression of natural, and stimulation of artificially modified produc-

199

tivity becomes necessary. This measure man has unfortunately exceeded. He has felled the forests whose network of fibrous roots bound the mould to the rocky skeleton of the earth; but had he allowed here and there a belt of woodland to reproduce itself by spontaneous propagation, most of the mischiefs which his reckless destruction of the natural protection of the soil has occasioned would have been averted. He has broken up the mountain reservoirs, the percolation of whose waters through unseen channels supplied the fountains that refreshed his cattle and fertilized his fields; but he has neglected to maintain the cisterns and the canals of irrigation which a wise antiquity had constructed to neutralize the consequences of its own imprudence. While he has torn the thin glebe which confined the light earth of extensive plains, and has destroyed the fringe of semi-aquatic plants which skirted the coast and checked the drifting of the sea sand, he has failed to prevent the spreading of the dunes by clothing them with artificially propagated vegetation.

Marsh would be stunned today. National parks are often surrounded by vast areas of highly altered, human-dominated areas. The Everglades was declared a national park in 1949. By then, areas outside the park had already been converted to vast agricultural fields and the natural water flows into the Everglades had been seriously altered for human purposes. No park exists in isolation or can be removed from the greater landscape that surrounds it. Habitat fragmentation and destruction beyond the Everglades had long-term detrimental effects within the Everglades. In 1996, the Clinton administration announced a plan to restore the natural ecology of the Everglades at an estimated cost of $7 billion.

Examples of the negative effects of habitat fragmentation and destruction abound. Later we will discuss how dams on the Rio Grande in New Mexico have led to the loss of fish species and threaten the remaining fragmented populations. Here we give specific examples from around the world of the consequences of the second member of the Evil Quartet.

SIBERIA

On September 4, 1995, the cover of *Time* magazine read, "The Rape of Siberia." The article commented that "the natural wonders of the majestic Bikin Valley capture the imagination but also tempt poachers and lumbermen." Note that poachers and lumbermen received equal billing as plunderers of the earth. Siberia, as Linden (1995) wrote, reminds one of a wasteland where subversives of the former Soviet Union were sent to work until they died. This image would be correct. More than 20 million people died in Siberia's gulag. Now, however, portions of Siberia are dying.

Parts of Siberia resemble Alaska in many ways. Glaciers, brown bears, and volcanoes remind one of the Kenai Peninsula of Alaska, although equally rugged Siberia is far more remote and less densely populated. Yet even remote parts of Siberia have been under attack by humans. Plutonium wastes from nuclear reactors have contaminated rivers. Linden suggests that the world's single largest source of air pollution comes from a complex of smelters in central Siberia. The haze, he claims, troubles people as far away as Canada and contributes to global warming. "Perhaps never has so vast an area been so despoiled so rapidly" at the hand of man. Of course, not all of Siberia has been ruined, but then Siberia is about five million square miles, about eight times as big as Alaska. The Sakha Republic is seven times the size of California and is but one of Siberia's 12 political regions.

Siberia is mostly taiga (boreal forest), a vast area of timber. Being a newly independent state, Siberia's economy needs hard cash. These timberlands are already attracting extraction companies. But Siberia has more than forests to offer foreign companies. Oil and gas deposits are believed to rival Saudi Arabia's enormous wealth. Diamonds and rich mineral deposits have been found. Linden claims that much of the gold comes from riverbeds, a clear indication that humans have not explored here before. Already a plethora of companies are exploring for mineral deposits.

With a population of 30 million, a vast area, few environmental laws, and even fewer resources to enforce the laws, poachers have hunted freely. The Siberian tiger once roamed a vast area; now the tiger is endangered and only 180 remain. The Amur leopard is one of the most elusive creatures on earth; fewer than 50 are believed to exist. Logging concessions are routinely given to friends by the very officials appointed to protect the forests. Oil spills have occurred in Siberia, and the permafrost allows the oil to flood large areas. Even our local news programs have shown oil spills in otherwise pristine wetlands in Siberia.

Once again, the United States is leading the environmental charge, in spite of its history of protecting species of plants and animals. Siberia has 25% of its forests intact, fully five times as much as the United States. While the U.S. Congress debates opening the Arctic National Wildlife Refuge to oil drilling, the United States is urging Siberia to set aside vast areas of protected lands that probably contain rich deposits of oil and gas. However, the former Soviet republics had access to an excellent educational system. Fortunately, many highly trained ecologists appreciate the problems Siberia faces. Hopefully, their concerns will be addressed before the rush for resources plunders Siberia.

ARAL SEA

Without doubt, the drying of the Aral Sea is one of the most rapidly occurring environmental disasters ever recorded. The creation of a huge agricultural scheme based on cotton and the opening of the Kara Kum Canal in 1956 precipitated the current problem. The Aral Sea of central Asia, once the world's fourth largest inland body of water, has turned into a mere puddle compared to its former size. This is one of only a few examples where an entire ecosystem has been destroyed so quickly and whose impact is so broad. The sea has contracted by some two-thirds of its surface area. Being an inland sea, as sea level declined, salinity increased. By 1977, the large fishing industry was in ruins. Fish catch had declined by over 75%. Within seven years, the commercial fishery had been eliminated. An industry that employed 60,000 is gone, and once prosperous seaports are now 50 km from the water. Where fruits and vegetables were once grown by local people and fish provided necessary protein, there is now only a wasteland and no fish.

Because the need for hard currency was so great, a large cotton project was begun around the Aral Sea. The hope was that water from feeder rivers could be used to irrigate the cotton crop which could be sold on the world market. The diversion scheme was successful—at first. In 1991, cotton made up 33.6% of Uzbekistan's exports. Cotton continues to be the country's leading hard-currency commodity. Some 17% of the work force was employed in agriculture. In neighboring Kazakhstan, the figures are similar. A substantial proportion of the economies of both nations depends on the canal diversion and the climate moderation provided by the Aral Sea. However, the Aral Sea is the same latitude as North Dakota, and its climate is less than ideal for cotton. Unfortunately, water was not all that the Aral Sea provided for the entire region. Like Easter Island, Mesa Verde, Mesopotamia, and so many places before it, the region has been laid to waste, a victim of ecological arrogance and misuse of man's technology.

The two major freshwater rivers, the Amu Darya (which flows into the Aral Sea from the south) and the Syr Darya (which empties into the sea in the north), were harnessed to irrigate the vast cotton crops. Irrigation canals were built of soil without even plastic liners. As a result, only a small fraction of the water reached its destination. The bulk of the water was sequestered in the walls of the irrigation canals. Indeed, so much water was lost that land nearby the canals began turning to swamp (Ellis 1990).

The Kara Kum Canal diverted large amounts of water from the Amu Darya into the Turkmenistan desert, irrigating millions of hectares of land. As more fresh water was diverted, the sea began to lose surface area and shrink. The sea had received about 50 km^3 water per year in 1965. By the early 1980s, this had been reduced to zero. Everything seemed fine for a while, and it was thought that as the sea dried, bottomlands could be put into production. However, the ecosystem services performed by the sea were never considered important. The Aral Sea, as it turned out, was the great moderator of the local climate. The sea's existence provided a mass of water-vapor-laden air that protected the region from the fierce dry winds from the north. The seasonal dry winds would blow, but the sea's presence would eventually assert itself, providing a buffer that prevented the surface soil from drying.

As the sea surface area was reduced through feeder river diversion, the regional climate began to change. Summers became hotter and drier. Winters, once moderate, became longer and much colder. The climate was hard on people, but it was much harder on cotton. Drier soils fell prey to wind erosion, and windblown salt precipitated from the shrinking sea added to human misery and also blew across the fields, coating the soil and sucking up what little moisture was left. As Mathews (1994) put it, "Thousands of square miles of farmland cannot now grow anything, and on thousands more productivity is dropping, propped up by fertilizer and pesticides and more and more water to rinse the fields of salts after each harvest." Nearly 30,000 km^2 of lake bed, poisoned with salts and chemicals, is toxic to plants. The dry lake bed sediments became windblown, poisoned dust clouds which damaged crops and were hazardous to humans.

Naturally, more fertilizer makes for more salts, which further exacerbates the fundamental problem. As the sea dried and became saltier, the local fish died. Now, toxic water combined with an inadequate healthcare system threaten people's lives. According to Mathews, infant mortality may be among the highest in the world. Groundwater supplies were contaminated. And now the Aral Sea is shared not by one country but by two independent countries. Any solution must be mutually agreed upon before there is any hope of rescuing an already desperate situation.

TROPICAL FORESTS

Myers (1988), and many other ecologists, claimed the world was experiencing an extinction crisis. Although species have gone extinct in the past, extinction rates today are profoundly higher and are directly attributable to human causes. Less than 30 years ago, tropical forests covered 7% (15 million km^2) of the earth's surface and contained 50% of all species. Most disturbing was that these forests were being destroyed faster than any other ecological zone. Today these same tropical forests cover less than 9 million km². Approximately 80,000 km² was being destroyed annually according to 1990 estimates. By now, the rate of deforestation has increased.

While some areas are seemingly under attack, other areas remain undisturbed. Myers (1988) suggested that Papua New Guinea, the Guyana Shield in South America, the Zaire Basin, and the western half of the Brazilian Amazon were being spared. Rapid population growth among communities of small-scale farmers and agriculturists was occurring in areas laid bare by numerous logging companies. According to Myers (1988), broad-scale clearing and degradation of tropical forests were "far and away the main cause of species extinctions."

Myers used three examples to support his conclusion. First, western Ecuador once contained about 9000 plant species, about half of which were endemic. An unknown number of insects and smaller animals, many unknown to science, inhabited the region. Since 1960, at least 95% of the forest has been destroyed to make way for banana plantations, oil wells, and human settlements. In a few year, the forest will be completely gone. The Atlantic Forests of Brazil are known for their uniqueness and high endemism. Originally more than 1 million km², these forests have been steadily reduced to 50,000 km². What is left is highly fragmented. Erosion on the island nation of Madagascar can be seen from orbiting spacecraft. Giant plumes of orange mud can be seen in the otherwise green-blue ocean coasts. Not only is the land being destroyed, but nearby coral reefs and the rich fisheries they support are being buried in mud and silt. Because tropical forest soils are known to be poor, agricultural systems in these forests are not sustainable. Thus, a constantly shifting agricultural system leads to increasing forest destruction.

CITRUS AND SCRUB JAYS

A citrus grower located along the Lakes Wales Ridge, east of Lake Okeechobee in Florida, wanted to plant more orange groves on his 1000-acre farm. The ridge is a unique scrub habitat in Florida and home to the federally threatened Florida scrub jay. The U.S. Fish and Wildlife Service warned the citrus grower that he would be fined under the Endangered Species Act for destroying habitat of a federally threatened species. The grower warned the U.S. Fish and Wildlife Service that he owned the land, and his livelihood depended on his orange grove being productive.

When the citrus grower destroyed 60 acres in July 1996, the U.S. Fish and Wildlife Service served him a temporary restraining order to prevent further destruction until he attended a hearing in federal court. The Endangered Species Act of 1973 allows people to apply for a "take" permit to develop private land. Mitigation might also spare part of the scrub jay's habitat while permitting development. The grower chose to ignore the Endangered Species Act and the permitting process altogether. After the grower had destroyed 350 acres of scrub jay habitat, the U.S. Fish and Wildlife Service invoked federal law to halt further destruction.

Again we must delve deeper into this quandary to truly appreciate what is going on here. Scrub jays live along the ancient sand dunes more than a million years old and today called Lake Wales Ridge, the central spine of Florida. The jay's habitat requirements are specific. Its population in 1989 was estimated at 11,000 birds. Shopping centers, towns, homes, and intensive monoculture agriculture such as citrus groves have caused massive habitat destruction on Lake Wales Ridge and surrounding areas of sand scrub, directly leading to scrub jay population declines. The ridge is 100 miles long and 10 miles wide. In 1987, the Florida scrub jay was added to the endangered species list as a threatened species.

The land in question was recently sold to the citrus grower. The seller probably understood that the land was suitable habitat for the scrub jay and decided to dispose of the property rather than attempt to develop it, although development provisions exist under the Endangered Species Act. The buyer chose, for whatever reason, to ignore an act passed by the U.S. Congress and signed into law by the president of the United States. The executive director for the Highlands County Citrus Grower's Association stated that "the law is confusing to most farmers who just want to plant their crops" (*Tampa Tribune,* August 2, 1996). If the grower chooses to disregard federal officials again, he can be fined up to $100,000 and sent to jail for up to one year. This alternative should not be difficult to understand.

However, the farmer, it must be said, is merely attempting to respond to consumer demands for a luxury food such as oranges and make money through so-called "Florida land succession." Using agricultural subsidies in the form of tax breaks and persuasive advertising to dupe consumers, raw land is first converted to cattle pastures ("real food for real people"), then to citrus groves ("drink more orange juice"), and finally to housing developments or shopping malls ("if you build it, they will come"). With each succession, profits increase. Without the dependent consumer and government subsidies, habitat would not be destroyed.

CORAL REEFS

"Wrecking the reefs," read the title of the science section of *Time* magazine on September 30, 1996. The article by J. Madeleine Nash reported that "across the globe, from the Gulf of Mexico to the South China Sea, people are killing coral reefs." Such ludicrous activities as cyanide fishing, harbor dredging, coral mining, terrestrial deforestation, coastal development, agricultural runoff, shipwrecks, and oils spills, as well as unthinking divers, are hammering reefs around the globe. Although there are thousands of reefs around the world, and most have not been studied, marine scientists suggest that "10% of the earth's reefs have been mortally wounded." There are approximately 400,000 square miles of reef on earth.

Coral reefs provide critical ecosystem functions such as protecting beaches from the ravages of storms. In fact, reefs are adapted to storms and bounce back from natural disturbances. Reefs also harbor unusually high biodiversity. Their rich environment provides a large portion of the earth's population with food.

Coral reefs are tropical and cannot exist in water colder than 60°F, nor can they tolerate prolonged high-temperature water. In the Philippines, reefs have been under siege by fishermen and loggers. A whopping 90% of some 13,000 square miles of reefs is dead or deteriorating. A principal cause has been terrestrial deforestation, which has allowed millions of tons of soil to erode into the nearby shallow seas. Vast deforested tracts have been left unvegetated, and erosion from the tropical rains has laid waste to the land and the rich reefs surrounding many Philippine archipelagos. The history of Philippine reef fishing is a classic lesson in environmental arrogance, a shortsighted profit motive, and unmitigated stupidity on the part of the central government to halt increasingly destructive practices.

Coastal villagers began by harvesting giant clams and large reef fish. When those were gone, fisherman got lazy and just cruised out over the reef, dropped in a stick of dynamite, and harvested the dead fish. The demand by Asians for fresh fish, however, changed their fishing tactics. Fishermen must now take their prey alive. To do this, they first stun the fish with a highly poisonous cyanide compound and then haul their flopping prey to the surface, leaving behind a veritable toxic waste dump. Each year more than 330,000 pounds of cyanide is used to harvest fish. The Philippine government recently moved to halt dynamite and cyanide fishing, but there is precious little left to protect.

Other insidious attackers are devouring reefs as well. The world-famous Great Barrier Reef on Australia's east coast is under attack by coral-devouring starfish. On the other side of the globe, Jamaica's reefs have succumbed to seaweed. In both cases, climate change is blamed. In 1983, 95% of the coral reefs surrounding the world-famous Galapagos Islands died. Marine biologists speculate that a seasonal spike in water temperature wiped out the reef.

Coral reefs off the Florida coast have been attacked by two diseases. White pox attacks elkhorn corals and kills them. The disease appears as white blotches up to ten inches in diameter on the living coral. The coral's tissue decomposes down to the coral skeleton. The disease is unique to the Florida reefs. White plague has attacked elliptical star coral from the Dry Tortugas to the upper Florida Keys. The disease appears to attack the coral stem and kills the living tissue from the bottom of the coral. No known cause has been identified for either disease.

FOREST BIRDS AND FOREST FRAGMENTATION

Robinson et al. (1995) studied forest bird reproductive success in the midwestern states, where vast areas of forests are highly fragmented. They found that nest predation and parasitism by brown-headed cowbirds (*Molothrus ater*) increased with forest fragmentation in five midwestern U.S. landscapes that varied from 6 to 95% forest cover. Observed reproductive rates were low enough to suggest that perpetuation of populations of some species in the most fragmented land-scapes depended on immigration from reproductive source populations in land-scapes with more extensive forest cover. Populations that depend on immigra-tion to exist are referred to as sink populations (Pulliam 1988).

Many neotropical migrant birds are suffering population declines from causes that may include the loss of breeding, wintering, and migration stopover habi-tats (Robinson 1993). Habitat fragmentation may allow higher rates of brood parasitism by cowbirds and nest predation (Gates and Gysel 1978, Temple and Cary 1988). Cowbirds lay their eggs in the nests of other "host" species, which then raise the cowbirds at the expense of their own young.

Populations of cowbirds and many nest predators are higher in fragmented landscapes, where there is a mixture of feeding habitats and breeding habitats. In landscapes fragmented by agricultural fields, levels of nest predation and brood parasitism are so high that many populations of forest birds in the frag-mented landscapes are likely to be population "sinks" in which local reproduc-tion is insufficient to compensate for adult mortality. As landscapes become increasingly fragmented, this reproductive dysfunction may cause regional declines of migrant populations of birds.

Robinson et al. (1995) tested the hypothesis that the reproductive success of nine species of forest birds was related to regional patterns of forest fragmen-tation in Illinois, Indiana, Minnesota, Missouri, and Wisconsin. Cowbird para-sitism was negatively correlated with percent forest cover for all species. Most wood thrush (*Hylocichla mustelina*) nests in landscapes with less than 55% forest cover were parasitized. In some landscapes, there were more cowbird eggs than wood thrush eggs per nest. In contrast, cowbird parasitism levels were so low in the heavily forested landscapes that cowbird parasitism was unlikely to be a significant cause of reproductive failure (May and Robinson 1985).

Levels of nest predation also declined with increasing forest cover for all species. Three ground-nesting warblers (the ovenbird [*Seiurus aurocapillus*], worm-eating warbler [*Helmitheros vermivorus*], and Kentucky warbler [*Opornis formosus*]) and two species that nest near the ground in shrubs (the hooded warbler [*Wilsonia citrina*] and indigo bunting [*Passerina cyanea*]) all had extremely high (6% or higher) daily predation rates in the most fragmented landscapes. Of the 13 cases of daily predation rates exceeding 7%, 12 were in the four most fragmented landscapes. Fragmentation at the landscape scale thus affects the levels of parasitism and predation on most migrant forest species in the midwestern United States. In more fragmented landscapes, the cowbird populations might be more limited by the availability of hosts and may saturate the available breeding habitat, which would result in high levels of parasitism even in the interior of the largest tracts in Illinois. Therefore, landscape-level factors such as percent forest cover determine the magnitude of local factors such as tract size and distance from the forest edge, a result consistent with continental analyses of parasitism levels (Hoover and Brittingham 1993).

Nest predators such as mammals, snakes, and blue jays (*Cyanocitta cristata*) probably have smaller home ranges than cowbirds and might therefore be affected more by local than landscape-level habitat conditions. Small woodlots in agricultural landscapes have high populations of raccoons (*Procyon lotor*). Censuses in both Missouri and Wisconsin have shown blue jay and crow (*Corvus brachyrhynchos*) abundances to be higher in fragmented regions.

Parasitism levels of wood thrushes, scarlet tanagers (*Piranga olivacea*), and hooded warblers and predation rates on ovenbirds and Kentucky warblers were so high in the most fragmented forests that they were likely populations sinks.

Robinson et al. (1995) suggested that a good regional conservation strategy for migrant birds in the Midwest is to identify, maintain, and restore the large tracts that are most likely to be population sources of birds. Further loss or fragmentation of habitats may lead to a collapse of regional populations of some forest birds. Increasing fragmentation of landscapes could be contributing to the widespread population declines of several species of forest birds.

EXERCISES

9.1   What other classification schemes are used to categorize ecological regions of the earth? How do these compare with the classification scheme used here?

9.2 Compare life history strategies of mammals in the different ecozones. For instance, compare different species of deer in the different ecozones. Which have the largest antlers? When are the antlers lost? Offer an explanation of your results.

9.3 Everglades National Park once had millions of wading birds. Even as late as the 1930s, a million or more birds existed in the greater Everglades region. Why are there so few now?

9.4 Not too long ago, there were many species similar to elephants. What were their distributions and how do they compare to distributions of *Loxodonta* and *Elephas* today?

9.5 Do a World Wide Web search for "Aral Sea." Summarize the present situation. Compare maps of the sea before and after the irrigation canal was built.

9.6 Find other regional examples of environmental disasters. What ecozone factors were ignored that made the disaster worse?

9.7 The ecozones we see today did not always exist in their present positions. Using the ecozone system explained here, make an ecozone map of the Carboniferous period 350 million years ago.

9.8 Make an ecozone map of the world during the height of the last glaciation, 22,000 years ago. Where was the boreal forest of North America?

9.9 Tropical forests of the humid tropics are being lost at an alarming rate. How will these losses affect precipitation in that ecozone?

9.10 How does the body size of animals of the same taxa differ across ecozones? Compare the body size of different bears across ecozones. How does body size of panthers differ across ecozones?

9.11 China has proposed building dams along the Yangtze River. Speculate on the long-term environmental ramifications of this action using ecozone information.

9.12 What ecosystem services do forest birds provide? Are any of these services important to humans? What is the economic impact to humans?

LITERATURE CITED

Ellis, W.S. 1990. A Soviet sea lies dying. *National Geographic* **177**:72–93.

Gates, J.E. and L.W. Gysel. 1978. *Ecology* **59**:871.

Hoover, J.P. and M.C. Brittingham. 1993. *Wilson Bulletin* **105**:228.

Linden, E. 1995. The rape of Siberia. *Time* September 4:45–49, 51–53.

Marsh, G.P. 1907. *The Earth as Modified by Humans.* Charles Scribner's Sons, New York.

Mathews, J. 1994. Hard lessons from the death of the Aral Sea. *St. Petersburg Times* October 19.

May, R.M. and S.K. Robinson. 1985. *American Naturalist* **125**:475.

Myers, N. 1988. Tropical forests and their species. *in* Wilson, E.O. (Ed.). *Biodiversity.* National Academy Press, Washington, D.C.

Pulliam, H.R. 1988. *American Naturalist* **132**:652.

Robinson, S.K. 1993. *Transactions of the North American Wildlands Natural Resources Conference* **58**:379.

Robinson, S.K., F.R. Thompson III, T.M. Donovan, D.R. Whitehead, and J. Faaborg. 1995. *Science* **267**:1987–1990.

Temple, S.A. and J.R. Cary. 1988. *Conservation Biology* **2**:340.

EVIL QUARTET 3: INTRODUCED SPECIES   10

A ll species that evolved in some region of the world are said to be *indigenous* to that region. When species that are indigenous to one region of the world are transported to another region, they are *nonindigenous* to the new region. Thus, nonindigenous species are those species that exist in parts of the world where they do not naturally occur. For instance, species that evolved in Europe, such as the house sparrow of North America, are considered to be nonindigenous to North America and indigenous to Europe. The term nonindigenous is relative in nature: every species is indigenous to some region; no species is truly distributed worldwide. However, only a subset of the world's species occur somewhere else as nonindigenous species.

A variety of terms have been employed to describe nonindigenous species. One term that is often used in nontechnical publications is *exotic* species. This term can be somewhat misleading because it may give the impression that the species is in some sense "glamorous" or at least from a glamorous place. But "glamorous" is in the eye of the beholder. House sparrows may not seem glamorous to some people, and many do not think of northern Europe as an especially glamorous part of the world. Therefore, even though exotic is a term that may be misleading, it is used so frequently that we may be stuck with it.

A second term that is often used is *introduced* species. This term carries the connotation that the species was actually introduced somewhere by humans. Some species are physically introduced elsewhere, but their descendants, if born in the new place, are themselves native to the new place, and to label them as being introduced may be confusing. One way to avoid confusion is to use the term *introduced* to refer to the species and not the individuals in the species. In any event, there is no doubt that many of the species that are nonindigenous found their way to the new regions through human introductions (e.g., Caum 1933, Long 1981, Lever 1987).

Another term that is occasionally used with nonindigenous species is *alien* (e.g., Mack 1986, Macdonald et al. 1986) species. We object to this term principally because it has a political rather than geographical or biogeographical meaning.

It is logical to discuss *feral* species at this point. Feral refers to species that have escaped domestication. These species do not have to be nonindigenous. In fact, house cats (*Felis catus*) may be feral yet native to parts of northern Europe. Across the world, feral species have caused a number of problems. Throughout the United States, feral hogs may harbor pseudo-rabies and infect domestic swine. Feral cats have had a devastating effect on native birds of small islands, such as those around New Zealand. Feral dogs may kill wildlife (such as deer) or livestock (sheep or cattle) and sometimes even humans.

The phenomenon of nonindigenous species played a role in Darwin's arguments against creationism. Darwin (1859) argued that one could not explain the absence of rabbits from Australian grasslands on the grounds that the grasslands "down under" could not support rabbits, because European rabbits (*Oryctolagus* or *Lepus*) were thriving there. Darwin also made the case that species in New Zealand, although well adapted to each other for co-existence, were being replaced by the advancing legions of nonindigenous species from Europe.

Indeed, Europeans traveling to the South Pacific and Indian Ocean introduced various types of mammals and birds onto small islands in the hope that the individuals released would generate progeny for later harvesting by future travelers to the islands. Captain Cook released cattle and hogs in Hawaii and elsewhere (Tomich 1986). This made good logistic sense to travelers whose journeys lasted for months or even years. By establishing food supplies in the form of harvestable populations of feral livestock on distant islands, these early traders could avoid transporting the feed and other supplies that livestock on a ship would need for voyages that might last years.

Throughout the world, nonindigenous species have become a hazard of immense proportion both for economic as well as ecological reasons. Species from virtually every taxonomic group have been introduced somewhere. A few examples are mentioned here to provide some insight into the scope of the problem. In terms of invertebrates, we can look to the waterways of the Great Lakes in the United States as they are being choked by introduced zebra mussels (*Dreissena polymorpha*). A nematode (pseudocoelomate) species (*Bursaphelenchus xylophilus*) has affected nearly 25% of Japan's 2.6 million ha of pine forests. So far, at least ten million Japanese trees have been killed by this parasite which apparently was introduced from North America on diseased timber.

From the kingdom Protista, we see red tides caused by dinoflagellates. Red tides may have directly or indirectly led to the deaths of marine mammals in the North Atlantic (Anderson 1994) and manatees in the Gulf of Mexico. Many of the organisms that cause red tides have been transported throughout the world in the same way that marine invertebrates have been able to colonize distant shores: in ship ballast.

From the kingdom Fungi, a number of tragic examples are available. An introduced fungus (*Cryphonectria parasitica*) led to the virtual extinction of the American chestnut tree (*Castanea dentata*). Similarly, another fungus (*Ophiostoma ulmi*) was introduced into North America from Europe and led to the loss of the American elm (van Broembsen 1989).

Among plants, several species have also become severe pests where they have been introduced. For simplicity, we present examples of aquatic and terrestrial species. For terrestrial plants, there are major infestations of an Australian tree (*Melaleuca quinquenervia*) and a South American species (*Schinus terebinthifolius*) in south Florida (Ewel 1986). As for aquatic plants, *Hydrilla verticillata* has created severe problems in Florida waterways and the water milfoil has posed problems in the northeastern United States.

A number of vertebrates have also invaded new environments and posed problems of varying degrees. The gray squirrel from North America was introduced into the United Kingdom (Laycock 1966) and may have had a negative impact on the native red squirrel. Riders on the Texas range may today encounter large grazing mammals from three different continents (Africa, Asia, and Europe), and there may be more blackbuck antelope in Texas than in their native India (Mungall and Sheffield 1994). In south Florida, a diverse fauna of introduced reptiles and amphibians has been spawned through human releases (Wilson and Porras 1983, Butterfield et al. 1997). Finally, visitors to Kapiolani Park on the island of Oahu in Hawaii would have no trouble observing birds introduced from five continents (Pratt et al. 1987): northern cardinal and house finch (North America), red-crested cardinal (South America), common myna (Asia), yellow-fronted canaries and warbling silverbills (Africa), and barred doves (Australia).

THE ORIGIN OF INTRODUCED SPECIES

As we have seen, species from virtually every kingdom of living organisms now occur as nonindigenous species somewhere in the world. But if any species can be nonindigenous somewhere, we might ask: Where do nonindigenous species originate? We can ask this from a biogeographic perspective or from a human dimensions perspective. In the first case (biogeography), the question can be rephrased as: What part of the world do nonindigenous species come from? The answer is that nonindigenous species can come from anywhere in the world. There are, as mentioned above, species of birds from five continents (Asia, Australia, North America, South America, and Africa) now co-existing in the state of Hawaii. In Florida, one of the most serious nonindigenous species problems involves *Melaleuca quinquenervia*, which is indigenous to Australia (Schmitz et al. 1997).

From a human dimensions perspective, we might ask how the nonindigenous species get from one place to another. Are species introduced intentionally as a form of biological control (often the case with insects)? Do nonindigenous species sneak in with tropical fruit from Central America? Do pet owners simply release their pet parrots or alligators when they get bored with them? Are individuals of nonindigenous species released from roadside attractions during hurricanes? Do humans sometimes intentionally release individuals of nonindigenous species to establish populations for recreational or commercial hunting?

The answer to all of these questions is yes. We have already discussed the introductions made by early European visitors to distant lands. More recently, an organization in Hawaii called the Hui Manu (Hawaiian for Bird Club) existed solely for the purpose of buying birds to release in Hawaii, as there was a widespread perception that Hawaii had too few native birds (Berger 1981, Simberloff and Boecklen 1991). On Tahiti (Lockwood et al. 1993) and on Saint Helena (Brooke et al. 1995), humans released large numbers of individuals of several species, again because they were under the impression that the islands had too few native birds. (Tahiti has only three native perching birds, and Saint Helena has none.)

Similarly, in New Zealand (Veltman et al. 1996), the Mascarene Islands (Simberloff 1992), and Australia (Long 1981), acclimatization societies existed solely for the purpose of establishing species that were nonindigenous to the region. These societies served the same function as the Hui Manu in Hawaii. In some cases, the intention of these societies was to liberate species that were thought to be potentially beneficial to humans in agriculture or other ways (Simberloff 1992), whereas in other cases the motivation was to modify the landscape of the new land to more nearly approximate the land from which the colonists came, as in New Zealand (Veltman et al. 1996).

SOME TAXA ARE MORE SUCCESSFUL THAN OTHERS

A reasonable question to ask in dealing with nonindigenous species is whether or not some taxa are more likely to become successfully established than others. For instance, we might wonder if mammals are superior (or inferior) as nonindigenous species when compared to lizards or birds.

There has been no comprehensive analysis aimed at addressing this question. Moreover, it may not be possible to ever answer this question at some levels. For example, it might be possible to compare birds from different families, but it may not be possible to compare birds (class Aves) to mammals (class Mammalia). The reason for this would simply be the scope of the comparison: it might be possible to gather adequate data for a sample of bird families, but not for enough families of birds and mammals to make the comparison at the class level. Nevertheless, whether or not this question is answerable may be irrelevant. In any case, it is an interesting question, and it is reasonable to imagine that differences might exist in relative introduction success among various taxa. Faaborg (1977) made the argument that for a given body size, nonpasseriform (i.e., nonperching) birds might have higher resistance to extinction on islands than comparably sized passeriform (i.e., perching) birds for physiological reasons: nonpasseriforms have lower basal metabolic rates than passeriforms, thus reducing their food requirements for subsistence.

ENVIRONMENTAL CONDITIONS AND NONINDIGENOUS SPECIES

A number of factors may favor the success of nonindigenous species in a new land. The Office of Technology Assessment (1993) lists several factors that favor the success of nonindigenous species in Florida and Hawaii. To appreciate the diversity of species that have been introduced in Florida and Hawaii, vertebrate introductions in both states are summarized in Table 10.1 and nonvertebrate introductions in Florida are summarized in Table 10.2. First, there is the subtropical climate. A subtropical (Florida) or tropical (Hawaii) climate insulates the nonindigenous species to some extent from the hard cold weather so typical of much of the temperate zone. A second factor involves the large number of ports of entry. There are many places for nonindigenous species to enter these states. Both Hawaii and Florida have extensive pet trades. This trade may be tightly controlled for the majority of humans in both Hawaii and Florida through restricted entry at airports and seaports. However, Hawaii also has extensive military operations which might facilitate the entry of some species that would not pass customs from transport on a commercial airline. Some have blamed military personnel for the introduction of the two species of bulbul on the island of Oahu in Hawaii (Berger 1981).

Another factor is extensive habitat manipulation. Once native habitats have been altered and native species reduced in number, nonindigenous species may

TABLE 10.1   Established Nonindigenous Vertebrate Species in Hawaii and Florida

	Number of species	
Taxa	Hawaii	Florida
Fish	?	79
Amphibians	4	4
Reptiles	14	32
Birds	>70	23
Mammals	18	22–26

After Office of Technology Assessment 1993, Butterfield et al. 1997, Layne 1997, James 1997, and Courtenay 1997.

have better chances for successful establishment of a new population. In Hawaii, where many introduced species of plants now comprise new nonindigenous lowland plant communities, native species are no longer adapted to these lowland habitats because they have been so modified. Thus, in suburban areas of Oahu, for example, virtually all the birds one sees are nonindigenous. In south Florida, at least in the Miami area, this is not the case; a variety of native species occur alongside the nonindigenous species.

COSTS OF NONINDIGENOUS SPECIES

The diversity of species that may have high economic or ecological diversity is extremely broad. A list of species from different taxa introduced in Florida

TABLE 10.2   Nonvertebrate Nonindigenous Species in Florida

Taxa	Number of species
Plants	>975
Insects	271
Freshwater snails	6
Land snails	40

After Office of Technology Assessment 1993.

TABLE 10.3 Examples of High-Impact Nonindigenous Species in Florida[a]

Plants	Mammals
Melaleuca	Wild hogs
Schinus	Coyotes
Hydrilla	
Fish	*Birds*
Blue tilapia	Starling
Peacock bass	Monk parakeet
Invertebrates	*Insects*
Asiatic clam	Tiger mosquito

[a] Meaning high cost economically or ecologically.

After Office of Technology Assessment 1993.

is presented in Table 10.3. The costs of introduced species may be paid in various types of currency. For instance, the costs could be assessed purely from a perspective of their economic impact on agriculture. Thus, the Weed Science Society of America (Office of Technology Assessment 1993) reported in 1992 that losses of 46 crops in all states but Alaska amounted to $4.1 billion annually, $2 to $3 billion of which could be attributed to nonindigenous species. Introduced weeds can not only lead to crop losses but can also harbor arthropod pests or produce toxins that are harmful to livestock or other plants.

An entirely separate economic concern involves loss of income from tourists. In Florida, it is widely held that introduced melaleuca trees have a negative effect on tourism (Diamond et al. 1991).

In terms of economics, the tremendous financial commitment needed to combat or limit the spread of certain nonindigenous species must be considered. According to an Office of Technology Assessment (1993) report, the cost of removing introduced melaleuca trees in south Florida ranges from $500 to $2000 per acre, and in 1992 as many as 490,000 acres in Florida were infested. These estimates are for mechanical removal which involves physical uprooting. Other methods for removing melaleuca include herbicide applications. Schmitz et al. (1997) reported that there were 703,000 acres of Brazilian pepper (another noxious weed) in Florida, as well as 373,000 acres of Australian pine. The cost of controlling a handful of nonindigenous plants just in highway rights-of-way exceeded $7 million in fiscal year 1993–1994 (Caster 1994).

ECOLOGICAL IMPACTS

Several ecological effects involving nonindigenous species can occur. These effects include competition, predation and parasitism, disease transmission, and habitat alteration.

Competition, as previously discussed, is the interaction between populations where both species are negatively impacted. The negative impact may manifest itself either in reduced population growth rates or reduced population sizes where the competing species co-occur.

Examples of competition between nonindigenous and native species have been reported in Hawaiian forest birds (Mountainspring and Scott 1985) as well as populations of beach mice in Florida (Humphrey 1992). In both of these cases, the authors reported evidence that the native species appeared to be superior competitors in native habitats, but the nonindigenous species were able to continue to invade the native habitat to the eventual detriment of the native species. In the case of the Florida beach, the native habitat on coastal islands is being rapidly developed and hence lost, which creates additional habitat for the nonindigenous house mouse (*Mus musculus*). The interesting point here is that even if the nonindigenous species might normally be outcompeted because of human activities (i.e., loss of native habitat), the balance is tipped in favor of the nonindigenous species because of the enhanced opportunities to continuously reinvade from ever-expanding stretches of manipulated habitat. Mountainspring and Scott (1985) similarly argued that native forest birds in Hawaii may well be able to outcompete nonindigenous species in native habitats but that nonindigenous species at lower elevations were able to continuously reinvade these native (and higher elevation) habitats.

Predation is the interaction between two populations where one population is impacted negatively and the other positively. One particularly stunning example involves trout and grizzly bears in Yellowstone National Park. In the early 1990s, someone caught a lake trout (*Salvelinus namaycush*) in Lake Yellowstone (Kenworthy 1994). Lake trout are not native to Lake Yellowstone, where the only native species of trout is the cutthroat trout (*Onchorhynchus clarki*). Lake trout are much larger than cutthroat trout (lake trout can grow to over 40 pounds), and the problem is that lake trout may increase in number and start preying on the native trout.

There are countless other examples of introduced vertebrates that prey on native wildlife. Atkinson (1977) presented evidence implicating introduced rats in the demise of native Hawaiian forest birds. Similar cases can be found regarding loss of birds and reptiles from islands following the introduction of cats (e.g., Fitzgerald 1985, Kirkpatrick and Rauzon 1986), but perhaps the most devastating case involves the brown tree snake and its impact on the native forest birds of Guam (Savidge 1987).

Guam is a small island in the Marianas archipelago. Jenkins (1983) described the distribution of the 11 native forest birds of Guam, noting that some species had suffered declines, and suggested a connection between *Boiga irregularis,* the brown tree snake, and the declines which began in the 1960s. Savidge (1987) conducted several tests of alternative hypotheses (see Table 10.4) for the decline of Guam's forest birds but concluded that the main factor was *B. irregularis.*

Disease transmission is yet another possible impact of nonindigenous species on native species. Nonindigenous species can harbor diseases they are immune to but to which natives are vulnerable. For example, Warner worked on forest birds in Hawaii and found that native forest birds were absent from lowland forests possibly because of their susceptibility to avian malaria. Warner (1968) actually trapped native birds at high elevations and kept them in outdoor aviaries at lower elevations. These captives soon died from avian malaria. Warner reasoned that native species were able to persist at higher elevations because the introduced mosquitoes that transmit malaria could not survive the low temperatures at higher elevations.

Nonindigenous species also may alter native habitats by destroying native vegetation. Perhaps the best example of this is wild hogs, which destroy vegetation by wallowing in the soil and thereby tearing up delicate root systems.

TABLE 10.4   Possible Forces in the Decline of Guam's Native Forest Birds

Force	Evaluation
Hunting	Populations have also declined on military lands where hunting is prohibited
Pesticides	Levels of pesticides in bodies of sampled animals were below those needed for adverse impact
Competition	Detailed field studies show that the only nonindigenous forest species uses different foods than native species
Habitat loss	Similar habitat loss on nearby islands has not led to comparable losses of forest birds
Disease	Native birds sampled for bacteria, viruses, and parasites were relatively free from any infestations as recently as 1986

ECOLOGICAL AND ECONOMIC IMPACTS ARE NOT INDEPENDENT

It is important to consider that the breakdown of costs of nonindigenous species into economic versus ecological categories may seem convenient, but when any particular situation is thoroughly investigated, such a separation may often become inappropriate. In all likelihood, economics cannot be divorced completely from ecology because economic losses often result from ecological interactions. A nonindigenous plant species may compete with a crop species for water or nutrients, resulting in reduced crop production. Nonindigenous species may also be toxic to livestock or harbor introduced pests.

Thus, the problems caused by *Melaleuca* infestation in south Florida are both ecological and economic. Economically, the presence of *Melaleuca* trees requires increased maintenance costs for drainage conveyances. Ecologically, *Melaleuca* trees can compete with native species, and this may be especially serious in some ecosystems. Consider the case of coastal mangroves in Florida. Mangrove forests once stretched from Tampa Bay south across the Keys. In many areas, mangroves have been removed in the course of development, and today the total areal extent of mangroves in Florida may be as low as approximately 190,000 ha (Odum and McIvor 1992). Are species from some parts of the world more likely to be successful or unsuccessful than species introduced from other regions?

Moulton and Pimm (1986) tested the hypothesis that perching birds (order Passeriformes) introduced into the Hawaiian Islands from different parts of the world might differ in their chances for success. The reasoning behind this lies in the different competitive milieus in which the species evolved. In any case, no significant differences were found, although the numbers of species introduced from some of the zoogeographic regions (i.e., Australasian, Nearctic) were quite small (see Table 10.5).

SOME INTRODUCTIONS SUCCEED AND OTHERS FAIL

Some species seem to be successful wherever they are introduced, whereas others never appear to succeed. This pattern of nearly always succeeding or failing was discussed in detail by Simberloff and Boecklen (1991). If this all-or-none pattern is a real phenomenon, it would represent a powerful tool for studying species introductions: one would be able to predict the fate of any introduction based only on its previous outcomes as an introduced species. Moulton (1993) argued that this pattern might be merely a sampling artifact. Thus, humans might stop releasing individuals of a certain species after just a few failed introductions of that species. If so, then species that appear to "al-

TABLE 10.5 Outcomes of Passeriform Introductions in Hawaii Categorized by Zoogeographic Region

Region	Location	Number of introductions
Palearctic	Europe, northern Asia, Africa north of the Sahara	9
Ethiopian	Africa south of the Sahara	12
Nearctic	Northern Mexico and North America	6
Neotropical	Central Mexico south throughout South America	7
Oriental	Southern Asia	16
Australasian	Australia and New Guinea	2

ways fail" might simply have not been provided sufficient opportunities to succeed somewhere as an introduced species.

Nevertheless, an all-or-none pattern might occur in some taxa but not others. Thus, hypothetically, it might appear in lizard species but not passeriform birds. The idea is very interesting, and hopefully scientists working on species other than birds will test for its presence.

Assuming for the moment that species vary in their chances for success depending upon the communities where they are released, several possible explanations emerge. First, some nonindigenous species may be released into what is essentially "enemy-free space." These species would have no natural controls (predators or competitors) and hence would be expected to proliferate. One example of this involves European hares and rabbits that were introduced into Australia and Argentina and soon took over huge areas of rangeland.

Finally, we must not overlook the amount of effort invested in any particular introduction. Gamebird introductions in the United States undoubtedly benefited from large numbers of individuals introduced and multiple introductions. Recently, Veltman et al. (1996) tested the hypothesis that species introduced in New Zealand in larger numbers and in greater frequency have had higher rates of success.

SOME REGIONS ARE MORE VULNERABLE TO INTRODUCTIONS THAN OTHERS

The great British ecologist Charles Elton argued in his classic book, *The Ecology of Invasions by Animals and Plants* (1958), that oceanic islands and crop monocultures were more vulnerable to invasions than were more complex ecosystems, such as those seen in forested or tropical areas. Elton's chief argument stemmed from the observation that in simpler ecosystems (i.e., those with fewer species), there may be a greater tendency for population fluctuations, and this may favor local extinctions, thus creating opportunities for invasions.

Birds

Historically, a large number of species introductions (birds and mammals) have occurred for the purpose of establishing populations for either commercial or recreational hunting. In the United States, perhaps the prime example of an introduced gamebird is the ring-necked pheasant (*Phasianus colchicus*). Ring-necked pheasants have been introduced in Hawaii as well as the continental United States (Long 1981). Other gamebird species that have been introduced successfully include the Hungarian or gray partridge (*Perdix perdix*), chukar partridge (*Alectoris chukar*) (Long 1981), eight species of tinamous (South American), and numerous francolin.

In other parts of the world, one example of an introduction for commercial purposes involves the ostrich, which was introduced into south Australia in the 1860s in the hope of establishing a feather trade. This population apparently expanded with influxes from ostrich farmers, who made additional releases following the collapse of the feather industry after the end of World War I (Long 1981).

Mammals

In the United States, one invasive mammal that was introduced to generate a commercial trade is the nutria (*Myocastor coypu*) (Bolen and Robinson 1995). The nutria has been released in the vicinity of Chesapeake Bay and now occurs throughout the eastern United States. Elton (1958) described the introduction of reindeer into Alaska, and volumes have been published on introduced and feral hogs in the United States.

Perhaps the most bizarre example of mammal introductions in the United States occurred in Texas, where a large number of species of ungulates (grazing herbivorous mammals) were introduced for hunting in confined areas. However, seven species have escaped and established wild populations. The species and their origins are listed in Table 10.6.

Notice that the table does not include introduced hogs, which are possibly the most widespread large introduced mammal in Texas today. In 1988, there were an estimated 39,000 free-ranging axis deer in Texas. Despite the success

TABLE 10.6   Texas Nonindigenous Ungulates

Species	Native range
Axis deer	India
Nilgai antelope	India
Blackbuck	India
Aoudad	Africa
Mouflon	Europe
Sika deer	Asia
Fallow deer	Europe

From Mungall and Sheffield 1994.

of these few introductions, several other species that have the potential to establish wild free-ranging populations are still hunted in Texas on ranches.

Other examples of mammal introductions include wild pigs across the United States (Mayer and Brisbin 1991). As many as ten species of deer have been introduced into New Zealand (Daniel 1962), in part to establish herds for recreational hunting. Several of these species became very abundant and spawned a commercial trade, as noted earlier.

Reindeer were introduced into South Georgia Island in the South Atlantic by Norwegian whalers to establish an alternative meat supply that would not require any husbandry and also possibly for recreational hunting (Leader-Williams 1988).

INTRODUCTIONS AND MANAGEMENT OF ENDANGERED SPECIES

Following our discussion of introduced nonindigenous species, it is now appropriate to look at species introduction as a potential management tool. Intentional introductions have been termed "translocations" by Griffith et al. (1989). Across the original ranges of many species, there have been local extinctions or population declines of the proportion where the populations are thought to be unable to recover. Under these circumstances, the forces that led to the declines may now have vanished and the habitat may still be adequate to support populations of the now missing or reduced species. In Florida, a glowing example involves the Florida panther. This subspecies of a once wide-ranging cat now occurs in numbers only west of the Mississippi River. A brewing controversy in Florida is whether or not cougars (as they are called in Texas) should be translocated to Florida to shore up the genetic diversity of the remnant population of 30 to 50 individuals (Alvarez 1993).

Elsewhere in the southeastern United States, the red wolf is being translocated to North Carolina (USFWS 1992). Red wolves (*Canis lupus*) are thought to have suffered from genetic swamping due to interbreeding with coyotes. Other examples of translocations for the sake of establishing companion or satellite populations for endangered species include several species of birds of prey (Cade 1984).

EXERCISES

10.1 Formulate a null hypothesis to test the notion that species from different families vary in their relative chances for introduction success.

10.2 Imagine how you could design a grand-scale experiment to test Elton's idea that communities with fewer species are more vulnerable to invasions than communities with many species.

10.3 Surf the World Wide Web for an example of an introduced wildlife population that is commercially harvested.

10.4 Surf the Web to find an example of techniques for controlling the spread of introduced species.

10.5 Find an example where wildlife biologists have used species introductions to restore populations of endangered or threatened species.

10.6 Find some examples of introduced insects from foreign countries. How do you suppose these species invaded North America?

10.7 Discuss the possibility that species from different taxa (phyla or kingdoms) differ in the severity of their potential impacts when introduced to new environments.

LITERATURE CITED

Alvarez, H. 1993. *Twilight of the Panther: Biology, Bureaucracy and Failure in an Endangered Species Program.* Myakka Publishing, Sarasota, Florida.

Anderson, D.M. 1994. Red tides. *Scientific American* August:62–68.

Atkinson, I.A.E. 1977. A reassessment of factors, particularly *Rattus rattus* (L.), that influenced the decline of endemic forest birds in Hawaii. *Pacific Science* **31**:109–133.

Berger, A.J. 1981. *Hawaiian Birdlife,* 2nd edition. University Press of Hawaii, Honolulu, 260 pp.

Bolen, E.G. and W.L. Robinson. 1995. *Wildlife Ecology and Management,* 3rd edition. Prentice-Hall, Englewood Cliffs, New Jersey.

Brooke, R.K., J.L. Lockwood, and M.P. Moulton. 1995. Patterns of success in passeriform bird introductions on Saint Helena. *Oecologia* **103**:337–342.

Butterfield, B.P. W.E. Meshaka, Jr., and C. Guyer. 1997. Nonindigenous amphibians and reptiles. pp. 123–138 *in* Simberloff, D., D.C. Schmitz, and T.C. Brown (Eds.). *Strangers in Paradise.* Island Press, Washington, D.C.

Cade, T.J. 1984. Reintroduction as a method of conservation. pp. 72–84 *in* Senner, S.E., C.M. White, and J.R. Parrish (Eds.). *Raptor Research Report 5: Raptor Conservation in the Next 50 Years.* Proceedings of a conference held at Hawk Mountain Sanctuary, Kempton, Pennsylvania, October 14, 1984. Press Publishing, Provo, Utah.

Caster, J. 1994. Invasive alien species on DOT right-of-way managed land. pp. 197–205 *in* Schmitz, D.C. and T.C. Brown (Project Directors). An Assessment of Invasive Non-Indigenous Species in Florida's Public Lands. Florida Department of Environmental Protection, Tallahassee.

Caum, E.L. 1933. The exotic birds of Hawaii. *Occasional Papers of the Bernice P. Bishop Museum* **10**:1–55.

Courtenay, W.R., Jr. 1997. Nonindigenous fishes. pp. 109–122 *in* Simberloff, D., D.C. Schmitz, and T.C. Brown (Eds.). *Strangers in Paradise.* Island Press, Washington, D.C.

Daniel, M.J. 1962. Control of introduced deer in New Zealand. *Nature* **194**:527–528.

Darwin, C. 1859. *The Origin of Species by Means of Natural Selection,* Modern Library Reprint. Random House, New York.

Diamond, C., D. Davis, and D.C. Schmitz. 1991. Economic impact statement: the addition of *Melaleuca quinquiveria* to the Florida prohibited plant list. *in* Center, T.D. et al. (Eds.). Proceedings of the Symposium on Exotic Pest Plants, Technical Report NPS/NREVER/NRTR-9106. U.S. Department of the Interior/National Park Service, Washington, D.C.

Elton, C.S. 1958. *The Ecology of Invasions by Animals and Plants.* Methuen, London.

Ewel, J.J. 1986. Invasibility: lessons from South Florida. pp. 214–230 *in* Mooney, H.A. and J.A. Drake (Eds.). *Ecology of Biological Invasions of North America and Hawaii,* Ecological Studies 58. Springer-Verlag, New York.

Faaborg, J. 1977. Metabolic rates, resources, and the occurrence of nonpasserines in terrestrial avian communities. *The American Naturalist* **111**:903–916.

Fitzgerald, B.M. 1985. The cats of Herekopare Island, New Zealand: their history, ecology and effects of birdlife. *New Zealand Journal of Zoology* **12**:319–330.

Griffith, B., J.M. Scott, J.W. Carpenter, and C. Reed. 1989. Translocation as a species conservation tool: status and strategy. *Science* **245**:477–480.

Humphrey, S.R. 1992. Anastasia Island beach mouse: *Peromyscus poilionotus phasma*. pp. 94–102 *in* Humphrey, S.R. (Ed.). *Rare and Endangered Biota of Florida,* Vol. 1: Mammals. University Press of Florida, Gainesville.

James, F.C. 1997. Nonindigenous birds. pp. 139–156 *in* Simberloff, D., D.C. Schmitz, and T.C. Brown (Eds.). *Strangers in Paradise.* Island Press, Washington, D.C.

Jenkins, J.M. 1983. The native forest birds of Guam. *Ornithological Monographs* **31**: 61 pp.

Kenworthy, T. 1994. It's a trout-eat-trout world out there. *The Washington Post National Weekly Edition* October 17–23.

Kirkpatrick, R.D. and M.J. Rauzon. 1986. Foods of feral cats (*Felis catus*) on Jarvis and Howland Islands, central Pacific Ocean. *Biotropica* **18**:72–75.

Laycock, G. 1966. *The Alien Animals.* Natural History Press, Garden City, New York, 240 pp.

Layne, J.N. 1997. Nonindigenous mammals. pp. 157–186 *in* Simberloff, D., D.C. Schmitz, and T.C. Brown (Eds.). *Strangers in Paradise.* Island Press, Washington, D.C.

Leader-Williams, N. 1988. *Reindeer on South Georgia: The Ecology of an Introduced Population.* Cambridge University Press, Cambridge, U.K.

Lever, C. 1987. *Naturalized Birds of the World.* Longman Scientific & Technical, Harlow, U.K.

Lockwood, J.L., M.P. Moulton, and S.K. Anderson. 1993. Morphological assortment and the assembly of communities of introduced passeriforms on oceanic islands: Oahu versus Tahiti. *American Naturalist* **141**:398–408.

Long, J.L. 1981. *Introduced Birds of the World.* David & Charles, London.

Macdonald, I.A.W., F.J. Kruger, and A.A. Ferrar. 1986. *The Ecology and Management of Biological Invasions in Southern Africa.* Oxford University Press, Cape Town, South Africa.

Mack, R.N. 1986. Alien plant invasion into the intermountain west: a case history. pp. 192–213 *in* Mooney, H.A. and J.A. Drake (Eds.). *Ecology of Biological Invasions of North America and Hawaii,* Ecological Studies 58. Springer-Verlag, New York.

Mayer, J.J. and I.L. Brisbin, Jr. 1991. *Wild Pigs in the United States: Their History, Comparative Morphology, and Current Status.* University of Georgia Press, Athens.

Moulton, M.P. 1993. The all-or-none pattern in introduced Hawaiian passeriforms: the role of competition sustained. *The American Naturalist* **141**:105–119.

Moulton, M.P. and S.L. Pimm. 1986. Species introductions to Hawaii. pp. 231–249 *in* Mooney, H.A. and J.A. Drake (Eds.). *Ecology of Biological Invasions of North America and Hawaii,* Ecological Studies 58. Springer-Verlag, New York.

Mountainspring, S. and J.M. Scott. 1985. Interspecific competition among Hawaiian forest birds. *Ecological Monographs* **55**:219–239.

Mungall, E.C. and W.J. Sheffield. 1994. *Exotics on the Range: The Texas Example.* Texas A&M University Press, College Station, 320 pp.

Odum, W.E. and C.C. McIvor. 1992. Mangroves. pp. 517–548 *in* Myers, R.L. and J.J. Ewel (Eds.). *Ecosystems of Florida.* University of Central Florida Press, Orlando.

Office of Technology Assessment. 1993. Harmful Non-Indigenous Species in the United States, OTA-F-565. Government Printing Office, Washington, D.C.

Pratt, H.D., P.L. Bruner, and D.G. Berrett. 1987. *A Field Guide to the Birds of Hawaii and the Tropical Pacific.* Princeton University Press, Princeton, New Jersey.

Savidge, J.A. 1987. Extinction of an island forest avifauna by an introduced snake. *Ecology* **68**:660–668.

Schmitz, D.C., D. Simberloff, R.H. Hofstetter, W. Haller, and D. Sutton. 1974. The ecological impact of nonindigenous plants. pp. 39–61 *in* Simberloff, D., D.C. Schmitz, and T.C. Brown (Eds.). *Strangers in Paradise.* Island Press, Washington, D.C.

Simberloff, D. 1992. Extinction, survival, and effects of birds introduced to the Mascarenes. *Acta Ecologica* **13**:663–678.

Simberloff, D. and W. Boecklen. 1991. Patterns of extinction in the introduced Hawaiian avifauna: a reexamination of the role of competition. *The American Naturalist* **138**:300–327.

Tomich, P.Q. 1986. *Mammals in Hawaii,* 2nd edition. Bishop Museum Press, Honolulu.

USFWS. 1992. Report to Congress: Endangered and Threatened Species Recovery Program. Fish and Wildlife Service, U.S. Department of Agriculture, Washington, D.C.

van Broembsen, S.L. 1989. Invasions of natural ecosystems by plant pathogens. pp. 77–83 *in* Drake, J.A., H.A. Mooney, F. diCastri, R.H. Groves, F.J. Kruger, M. Rejmanek, and M. Williamson (Eds.). *Biological Invasions: A Global Perspective.* John Wiley & Sons, Chichester.

Veltman, C.J., S. Nee, and M.J. Crawley. 1996. Correlates of introduction success in exotic New Zealand birds. *The American Naturalist* **147**:542–557.

Warner, R.E. 1968. The role of introduced diseases in the extinction of the endemic Hawaiian avifauna. *Condor* **70**:101–120.

Wilson, L.D. and L. Porras. 1983. The ecological impact of man on the south Florida herpetofauna. *University of Kansas Museum of Natural History Special Publication* **9**:1–89.

EVIL QUARTET 4: CHAINS OF EXTINCTION

11

U nderstanding the complexity of the living world is perhaps the most difficult problem facing science today. While some space scientists rationalize the need to establish human colonies on Mars, so that humans can migrate when the earth is unfit for human habitation, perhaps money would be better spent on understanding life on our planet, which is still a magnificent sight for space shuttle astronauts. Diamond's last member of the Evil Quartet, chains of extinction, is the most insidious of the four. Because species interact with each other, changes in one species invariably ripple through the greater community of species in ways not yet well understood and certainly not easily predicted. For example, in Chapter 19 we will discuss the impact on songbirds as a result of an overabundance of deer in our eastern forests. Without going into detail here, deer are well known to be browsers that eat understory and mid-canopy forest plants. When there are too many deer, the mid-level of the forest disappears, eaten away by deer. Forest birds that use the mid-canopy for foraging or nesting are left with nothing and soon disappear. Deer contributed to the local extirpation of the birds through a chain reaction by first eliminating the plants whose absence was detrimental to the birds that eat insects that use those plants—the beginning of a possible chain of extinction.

In Chapter 7, we discussed the effect of rabbits and guinea pigs on Laysan Island in the Pacific. These exotic species consumed all vegetation on the island. Once the vegetation was gone, several species of endemic birds passed into extinction, victims of a chain of extinction caused indirectly by exotic herbivores. In fact, as Pimm (1996) has pointed out, extinction mechanisms are inherently synergistic. Once species losses begin, it often becomes impossible to separate the mechanism responsible. For instance, forest fragmentation may inevitably increase the opportunities for exotic species to invade.

In Cameroon, West Africa, the rate of deforestation is increasing. Roads have been cut into remote forest areas, allowing easy access to local bush meat hunters. Armed with shotguns, these hunters are now harvesting three gorillas and chimpanzees a day, in addition to forest birds and antelopes. Although overexploitation is occurring, so too are forest fragmentation and destruction.

HAWAIIAN BIRDS

Species have evolved adaptations so specialized that one species may depend entirely on another for its existence. Such was the case for three species of birds in Hawaii. Introduced goats and pigs have damaged endemic Hawaiian plants, especially lobeliods such as those in the genera *Trematolobelia* and *Clermontia*. These plants co-evolved with several nectar-feeding honeycreepers (Drepanidids) whose long, de-curved bills neatly fit the long, de-curved flowers of the plants (Smith et al. 1995). Pigs and goats ate the plants that the mamo, black mamo, and i'iwi depended upon for their existence. The former two birds are now extinct, and the third has become extinct on two islands and is very rare on a third (Pimm 1996). Hawaii has already lost 90% of its native bird species.

POLLINATORS AND FLOWERING PLANTS

A beneficial relationship between different species is referred to as mutualism. Plant–pollinator mutualism dates back to the Cretaceous period, when insects began to acquire food from the flowers of plants (Kearns and Inouye 1997). At least two-thirds of all plant species today depend on insects for reproduction. Some plants depend on animals and birds to disperse their seeds. Bees are active and reliable pollinators for many plants. Bee-pollinated plants account for 30% of the food humans consume, and many plants depend solely on bees for pollination.

Habitat changes caused, for example, by monocultural agricultural crops have reduced the amount of land to support native vegetation. These areas of reduced plant diversity have led to decreased biodiversity of pollinators (Williams 1986). Grazing, habitat fragmentation, and the use of pesticides can also cause the loss of native pollinators. Moreover, introduced honeybees can have detrimental effects on native vegetation. For example, Paton (1985) enclosed plants in a mesh that excluded bird pollinators but allowed honeybee visits. Mesh-enclosed plants showed significantly lower fruit settings. With an over-abundance of nonnative pollinators, the vegetation community may well change composition, with some species doomed to vanish completely.

Some plants in South Africa depend on native ants for seed dispersal. Argentinean ants (*Tridomyrmex humilis*) have been introduced in South Africa, and these ants do not disperse seeds. Those plants that depend solely on a single pollinator will risk extinction if the pollinator disappears. Worldwide there are about 800 species of figs, and each depends on a unique species of wasp for pollination. Figs flower asynchronously throughout the year, and wasps are dependent on figs. If no flowers are found within a few days of hatching, the young wasps die. Thus, any negative effects on either wasps or figs ultimately affects both.

Janzen (1974) claims that the fates of native bees, orchids, and woody plants are intimately linked. Male euglossine bees pollinate many neotropical orchids that require outcrossing (the exchange of pollen with other members of their species) to reproduce (Roubik 1992). Male bees visit plants and collect scents apparently necessary for mating success. Much of the euglossine–orchid mutualism involves a single bee species. Female euglossine bees travel long distances between woody plants that typically occur at low densities. Because the density of woody trees has decreased due to logging, grazing, and development, bee nesting sites have decreased, leading to declines in the bee population. Female bees shifted their foraging to weedy shrubs in disturbed areas and thus are not aiding woody plant outcrossing. Male bees may have altered their behavior in response to the females' foraging activity, thus adversely affecting the orchids.

Even when systems have been studied for many years, predicting the outcomes of alterations of complex ecological systems is difficult. The sad fact is that most ecological systems have not been studied in great detail.

EVIL QUARTET SYNERGISM

Pimm (1996) explained the connection between habitat loss and fragmentation, the introduction of exotic species, and the inevitable loss of native species. For example, forest fragmentation created highly modified and often species-poor communities that were readily invaded by exotic species. Moreover, modified habitats can harbor species detrimental to native species. More than one conservationist has suggested a "scorched-earth policy," whereby parks and reserves would be surrounded by bare ground to prevent exotic species from invading the protected areas.

Raccoons and opossums are often seen in and around human-occupied areas. Populations of these species invariably increase in the presence of humans, while populations of other more secretive species decline. Because raccoons are able to feed on garbage and pet food, their densities increase, enabling more predation of bird nests and thus increasing the risk to native birds. Indirectly, forest fragmentation and human development lead to a twofold attack on native birds.

Pimm (1996) claims that once any one of the Evil Quartet gets started, the synergism among the other members of the quartet causes extinctions fast and furiously. The consequences of this claim are profound and are supported by a large and accumulating body of evidence. As more parks and protected areas are surrounded by increasingly human-dominated landscapes, extinctions will increase within the protected areas for a variety of reasons all too well understood with the benefit of hindsight.

EXERCISES

11.1 Hawaii has suffered the loss of many native bird species. *Freycinetia arborea* is a Hawaiian native vine that depends on several species of birds for outcrossing. As the birds that pollinated the vine became rare, the vine began to disappear. Today, however, the vine is doing fine. Hypothesize what might have happened to aid the vine.

11.2 Many tropical trees produce large seeds surrounded by fleshy fruit. Nearly all the fruit falls from the tree to the ground and sits locally. A variety of small organisms consume the fruit, but no long-distance dispersal of the seeds takes place. Clearly, the long-term viability of such a strategy must be questioned, yet the tress are living examples of their own reproductive success. Suggest possible mechanisms for seed dispersal.

11.3 Suggest mechanisms by which trees propagate up hillsides.

11.4 The brown tree snake of Guam is infamous. Its introduction on the island has led to the extermination of many bird species on Guam. Describe possible cascading effects of this disastrous introduction.

11.5 Fruit bats are effective pollinators of plants. The loss of suitable roosting sites for the bats could ultimately lead to the loss of plant species. What possible effects might forest fragmentation have on fruit bats?

LITERATURE CITED

Janzen, D.H. 1974. The deflowering of Central America. *Natural History* **83**:49–53.

Kearns, C.A. and D.W. Inouye. 1997. Pollinators, flowering plants, and conservation biology. *BioScience* **47**(5):297–307.

Paton, D.C. 1985. Food supply, population structure, and behavior of New Holland honeyeaters *Phylidonyris novaehollandiae* in woodlands near Horsham, Victoria. pp. 222–230 *in* Keast, A., H.F. Recher, H. Ford, and D. Saunders (Eds.). *Birds of Eucalypt Forests and Woodlands: Ecology, Conservation, and Management.* Royal Australian Ornithologists Union and Surrey Beatty & Sons, Sydney.

Pimm, S.L. 1996. Lessons from the kill. *Biodiversity and Conservation* 5:1059–1067.

Roubik, D.W. 1992. Loose niches in tropical communities: why are there so few bees and so many trees? pp. 327–354 *in* Hunter, M.D., T. Ohgushi, and P. Price (Eds.). *Effects of Resource Distribution on Animal–Plant Interactions.* Academic Press, New York.

Smith, T.B., L.A. Freed, J.K. Lepson, and J.H. Carothers. 1995. Evolutionary consequences of extinctions in populations of a Hawaiian honeycreeper. *Conservation Biology* 9:107–113.

Williams, P.H. 1986. Environmental change and the distributions of British bumble bees (*Bombus* Latr.) *Bee World* **67**:50–61.

HARVESTING
OF WILDLIFE

12

I n this chapter, the term "harvest" is used very loosely and is applied to any type of removal of individuals from the wild state. Thus, we do not differentiate between live capture of parrots for the pet trade, for example, and actual shooting of kangaroos for meat and skins. Throughout evolution, humans have harvested wildlife species. Across the world, we can see a broad spectrum of legal and illegal wildlife uses today. In undeveloped countries or countries with undeveloped areas, we see subsistence hunting. At the opposite end of the spectrum, we see wealthy hunters who pay large sums of money to shoot at what may be little more than a captive animal on a game farm. Somewhere in between these two worlds, we see the commercial hunters—some legal, some illegal. In this chapter, a variety of aspects related to human harvest of wildlife are explored, beginning with subsistence hunting and then proceeding to commercial and recreational hunting. The wildlife trade that is illegal is also examined and reveals that human culture can have a devastating effect on wildlife populations.

Today, there are three central reasons for harvesting wildlife: the first is simply for subsistence, the second is recreation, and the third is commercial/financial. We differentiate between these three because their goals differ, and this in turn may influence the population dynamics of the species that are harvested. In principle, it may seem obvious enough that there is a fundamental difference between hunting to live (subsistence hunters) and living to hunt (recreational hunters). In reality, however, the distinction may be difficult to discern. For the purpose of this discussion, subsistence hunters are defined as hunters who must obtain wildlife meat in order to survive. Hunters in this category would include, for example, tribal and nontribal humans who dwell in Amazonia or tribal humans in Africa or Southeast Asia.

SUBSISTENCE HUNTING: HUNTING TO LIVE

Imagine you are faced with the following problem. You live on the Great Plains of North America in the 1800s. You live in a village and you must live by hunting. Every day you must find food, and that means you must kill an animal. Moreover, it is bitterly cold six to eight months of every year, and animal skins are your only protection from the weather. You use these skins for clothing and to build your dwelling, since there is little structural vegetation for construction.

Native Americans in this region relied heavily on harvesting the American bison (Roe 1951). This was done via cooperative hunting. Cooperation gives humans a tremendous advantage in hunting. A group of humans could, for example, position themselves so as to drive a group of bison over a cliff. Some of the bison would die in the fall and others would suffer broken legs. The injured bison could be easily killed by slitting their throats with knives made from the bones of their ancestors.

When meat was available, Native Americans found ways to preserve it through a drying process. Thus, although the problem of finding food began anew every day, there were intermittent reprieves.

Weapons available to Native American were primitive before Europeans arrived. Nevertheless, the idea has been advanced that humans caused the extinction of a number of species of Pleistocene megafauna.

Now let's fast-forward to a village in the Amazon Basin in the 1990s. It is dawn and you are hungry. You emerge from your hut, made of vegetative material from the surrounding area, and it is pouring rain. In the Amazon Basin of South America, it rains frequently in some seasons. Your hut is in a small village. You have a "new" shotgun which you obtained by trading some artifacts to a passing miner. The shotgun is new only to you. It is in fact very old and partly rusted. You also have ten shotgun shells. Shotgun shells in the Amazon are very expensive and therefore very precious. Each one represents a meal—as long as you don't miss when you shoot. Your hunt for bush meat begins as soon as you load your canoe and leave the village.

Wildlife populations are scarce near the village, as those unwary animals were consumed long ago. Today, you know you must travel a great distance (perhaps several kilometers) to find an area where populations are large enough that your chance for success makes the effort worthwhile.

And what do you hope to find? The farther you travel, the farther you must transport the kill. Nevertheless, you hope for a large animal—one large enough to feed you for more than a day and still small enough that you can kill it with a single shot. (Remember, those shells are worth a fortune in the Amazon.)

In parts of the world, the daily problem of finding food for oneself and one's family has changed little from earlier times. Each day the problem must be solved anew. One way to partially offset this daily grind is to preserve the meat acquired in salt. Village huts are adequate, but there is no television, let alone a refrigerator. If clothes get too gamey, they must be cleaned so as not to betray one's presence to an animal (= a potential meal) with a sharp sense of smell. Clothes are cleaned by taking them to the river to wash them.

Solving the daily energy problems of oneself and one's family may mean finding something to trade for more shotgun shells or reliance on a more primitive means of killing, such as a bow and arrow, a blowpipe, or simply a sharp stick. Failure to make a kill may be offset by sharing food with a successful hunter from one's extended family in the village. (Hunters are expected to share their success.) Physical labor may also be traded for some food.

Every day is a struggle; life is very hard and is becoming more so every day. Consider the effects of deforestation on these populations. The forest is lost and so are the animals that live there. In place of the rain forest, which is often simply burned, grass is sown and cattle are raised on the grass. But the soil is not especially fertile and a lot of rain forest must be converted to rangeland to make it profitable. In the end, these cattle are slaughtered and their meat used in fast-food restaurants in North America. It is ironic that we pay so little for something that is so difficult to come by in some parts of the world.

Let's turn our attention to what kinds of species are harvested. We focus on Latin America because it is one of the last places on earth where subsistence hunting is an important way of life and because we have tremendous knowledge of the area through studies done by scientists at institutions such as the Center for Latin American Studies at the University of Florida.

In Amazonian Peru, "bush meat" and fish constitute much of the animal protein consumed by colonists. Among the Miskito Indians of Nicaragua, wildlife makes up 98% of the meat consumed (Redford and Robinson 1991). In contrast, bush meat comprises less than 20% of the animal protein in the diets of colonists living along the transamerican highway in Brazil (Smith 1976).

If wildlife species are important to local peoples in Latin America, an interesting question involves which species are taken. In one study involving local people in Suriname, it was found that the people killed (for meat and skins) 27 species of mammals, 24 species of birds, 3 species of turtles, and 2 species of lizards (Redford and Robinson 1991). One might imagine that people take animals in proportion to their abundance, but this is too simplistic a view. Other important considerations involve cultural taboos. Mittermeier (1991) conducted in-depth interviews with local people in Suriname to determine their feelings about primate consumption. He found that some subsistence hunters avoided harvesting red-banded tamarins (*Saguinus midas*) or squirrel monkeys (*Saimiri sciureus*) due to local taboos because (1) the animals have sexual behavior too like that of humans, (2) the people believed that the meat would make them sick, (3) some species cradle their offspring like humans, and (4) the animals smelled bad (Mittermeier 1991).

Subsistence hunters may carefully pursue only certain species of mammals. In Suriname, for example, there are eight species of nonhuman primates that the local people hunt for meat and skins. Indeed, hunting for subsistence by local peoples has been identified as the most important factor in the disappearance of primates from Suriname (Mittermeier 1991).

Hunting habits and diets of local peoples are highly vulnerable to even minor development (Ayers et al. 1991). Thus, deforestation reduces population sizes, and roads bring in less knowledgeable hunters who may harvest previously unharvested species. Roads also may bring in supplies of canned meat from other parts of the world.

WILDLIFE FOR SALE

In Chapter 3, we learned that the caiman is a protected species. In Corumba, in southwestern Brazil, cowboys make a living herding cows and hunting caimans (Misch 1992). Although Brazil passed a law in 1967 protecting much of its wildlife from commercial hunting, the law is not enforced, which is not an uncommon occurrence. The cowboys shoot the caimans and sell them to local tanneries for around five dollars each. From there, the hides make their way to North America, Europe, and Asia. Estimates are that some 500,000 caimans are killed in the Pantanal region of Brazil each year. Scientists do not know if such harvests are sustainable.

Caimans are a small part of the illegal wildlife trade, however. Misch estimated the total value of wildlife sold to be from $5 to $8 billion annually. Most of the poaching is done by local people trying to make a better living. Although habitat loss is the primary threat to most of the world's species, trade in wildlife products can also wipe out a species. The passenger pigeon and the Carolina parakeet were two common North American species 100 years ago.

Both were driven to extinction by the wildlife trade. The Orinoco crocodile, the Javan and Sumatran tigers, and the black rhino are being driven to the same fate as a result of the wildlife trade.

Waste in the animal trade is staggering. Some 15 out of every 100 parrots imported into the United States die in transit. Among macaws, 30% die before they reach their owners. The dead birds are, of course, replaced by live birds from the wild. Although these losses seem high, they do not compare to those incurred by primates. When a young chimpanzee is taken from the wild, most of the adults must be killed first. Thus, for every chimp taken from the wild, some ten are killed. The Convention on International Trade in Endangered Species (CITES) (Chapter 2) is supposed to block import and export of endangered species and regulate trade in threatened or potentially threatened species.

Thus, the treaty requires vigilant monitoring, which is proving to be an impossible task for many countries. One of the most notable successes of CITES has been the African elephant. Populations of elephants across Africa had plummeted to less than 600,000 in 1994. The international attention given elephants has raised consumer awareness and made the purchase of carved ivory, boots made of elephant skin, and wastepaper baskets made from an elephant's foot repugnant to most shoppers. CITES, however, has many loopholes that permit countries such as Zimbabwe, South Africa, Botswana, and Malawi to list "reservations" on Appendix I species such as elephants. Appendix II species can be traded provided wild populations are not compromised. But how are these populations monitored? Who verifies that the species is prospering? Some people even suggest that inclusion in the Appendix II list is an advertisement for poachers to collect now because tomorrow may be too late.

CITES agreements are difficult to trace, and no sanctions are placed on nations that violate the treaty. Vietnam, Myanmar, Cambodia, and Laos had not yet signed the treaty in 1996, and these are biologically rich nations. Once a product such as a tiger penis reaches Laos, anyone can purchase it and then attempt to ship it home.

As Misch points out, rangers in most parks are poorly paid and poorly equipped, and most parks are short-handed. It is far easier and safer to take a bribe than to protect an animal when the poachers are heavily armed and determined. Is it reasonable to expect a guard who is paid $5.00 a month to protect a rhinoceros worth $1000 in horn alone. Most often, guards are chosen from the local population and are resented because they have the highest paying jobs in the area. Moreover, the guards are charged with keeping their neighbors from making a living off the park's natural resources. Frequently, the guards are ostracized in their own villages.

COMMERCIAL HARVEST OF WILDLIFE IN NORTH AMERICA*

On Earth Day, April 22, 1996, the *Tampa Tribune* ran a story on a Waldo, Florida, wildlife gift shop. The store sells snake skins to major boot manufacturers, but shoppers can buy many interesting and unusual products. For instance, python cummerbunds or sports jackets and even alligator back scratchers can be purchased. Road-killed stuffed bobcats, opossums, armadillos, and snakes are also available. Among the more exotic gifts are cobra-head belt buckles; rattlesnake belts, ties, and hatbands; and alligator boots. Certainly, animal rights activists actively campaign against such shops. However, although many of us are repulsed by these macabre gift shops, they serve as a constant reminder that humans continue to exploit wildlife and some people make a living doing it.

A large number of species are currently exploited commercially. The taxa involved range from sponges to mammals. In the United States, shellfish, sponges, jellyfish, and several species of fur-bearing mammals are commercially harvested. Historically, a number of species of waterfowl, upland gamebirds, and big game mammals once were commercially harvested in the United States. Some of these species are discussed below, but only after examining some of the tragedies of market hunting in the United States. Hopefully, a pattern will quickly become apparent: there is no safety in numbers.

Fur Industry

There are still six or seven chinchilla farms left in North America and only four buyers of the skins. As of 1993, there were 556 mink farms in the United States, down from some 2000 a few years ago; 57 of these farms also raise foxes.

State	Number of mink farms
Utah	150
Wisconsin	114
Minnesota	72
Oregon	29
Washington	28
Idaho	27

* *International Wildlife Trade: Whose Business Is It?* by Sarah Fitzgerald (World Wildlife Fund, 1989) is recommended as supplemental reading.

Iowa	25
Pennsylvania	21
Illinois	15
New York	12
Ohio	11
South Dakota	7
Other	45

The Passenger Pigeon

When the first Europeans landed in what is now called North America, the passenger pigeon may have been the most abundant bird species (Schorger 1973). Early reports described flocks that blackened the sky and extended for miles. We will never know how many passenger pigeons there once were, but we do know how many there are today—none. Passenger pigeons were victimized by a combination of habitat destruction and overhunting. The last surviving passenger pigeon was a female named Martha; she died in the Cincinnati Zoo on September 1, 1914 (Greenway 1967).

One habit of the passenger pigeon no doubt left the species especially vulnerable to overharvesting: they were very gregarious. Passenger pigeons nested in great concentrations and roosted in enormous flocks. Schorger (1973) described the sad history of the species once Europeans arrived. The vast numbers were reduced by the mid-1800s, and only remnant populations persisted by the late 1800s. Market hunters could kill large numbers by setting out poles for the birds to roost on and then firing their shotguns down the poles. Hunters could also kill vast numbers by smoking their roosts with burning sulfur. The hunters would place little pots of sulfur around the roosts after nightfall. They would then light the sulfur and wait for the birds to drop to the ground. Some individuals were killed outright by the poisonous fumes, and the others were bludgeoned to death. Other techniques involved the use of "stool pigeons." A calling bird would be placed on a perch (i.e., a stool) in a large trap. Wild birds would be attracted to the trap and easily ambushed. Market hunters sold passenger pigeons by the thousands for only pennies. Now the remnants of this species can be seen only as study skins in trays in the bird collections at various museums of natural history.

Waterfowl

Humans interested in making a buck through killing wildlife are often far too ingenious for their quarry, as exemplified by the market hunters who discovered that smoking sulfur would bring roosting passenger pigeons to the ground. Waterfowlers were also rather ingenious. They invented the so-called "punt gun." A punt gun was a large-bore shotgun mounted on the back of a small boat (i.e., a punt). The bore of the gun was much larger than what waterfowl hunters are legally restricted to today, and punt guns were, in fact, little more than cannons on the backs of these little boats (Baldassarre and Bolen 1994). The hunters could row up to a flock of unsuspecting waterfowl and fire away. Because the gun was mounted on the boat, the tremendous recoil would be absorbed by the waterway and not the hunter's shoulder. Waterfowl were slaughtered by the millions in this way. Some examples of the slaughter include more than 400 ducks and 450 geese sent to market in a single day by a market hunter in Louisiana, 122 wood ducks killed before 9:00 A.M. on the Mississippi River, and a group of market hunters who sent 1000 ducks and shorebirds to market per week (Baldassarre and Bolen 1994 and references therein).

Nevertheless, there is no example of a waterfowl species that was forced to extinction by the use of punt guns. Is it possible that market hunters could have continued to use punt guns and kill waterfowl without human-imposed legal limits? Probably not. The slaughter of waterfowl in North America likely was a motivating factor in establishing laws to protect the species. Today, the ducks, geese, and swans remain, but many populations are reduced. Without laws and law enforcement, many of these species no doubt would be gone.

There are examples of waterfowl extinctions that were likely caused by commercial hunting. The common eider (*Somateria mollissima*) population of Maine was reduced to a single nesting pair by 1907 (Greenway 1967), but the species recovered in Maine to some extent after it was protected. The Labrador duck (*Camptorhynchos labrodorius*), which apparently was never very common, nevertheless appeared in markets between 1850 and 1870 but was not cherished for its palatability (Greenway 1967).

COMMERCIAL HARVESTING OF WILDLIFE IN OTHER NATIONS

Kangaroos

Five species of kangaroos currently are harvested in Australia. The three most common species are red kangaroos (*Macropus rufus*), eastern grey kangaroos (*M. giganteous*), and western grey kangaroos (*M. fuligonosus*). It is generally believed that there are more kangaroos today than when Captain James Cook landed in Australia in 1770 (Fitzgerald 1989). When European colonists came to Australia, they cleared forests for agriculture. These activities apparently were not good for smaller species of kangaroos but seemed to be beneficial for larger species. Larger species were seen to be pests as early as the 1850s. Large species were not protected as recently as 50 years ago, but soon it became clear that the commercial harvest of these species had to be regulated. The skins (1 to 1.7 million skins per year between 1981 and 1986) are exported for leather, mostly to Southeast Asia and Europe, and the meat is made into pet food, although at one time it was sold for human consumption. In the early 1980s, more than 1000 tons of kangaroo meat was exported to Europe as "sausage." When undercover agents found that the meat could not pass health inspections, exports dropped to less than 50 tons in 1984. The meat is commonly contaminated with salmonella, in addition to dirt, vegetation, and the parasite *Dirofilaria roemeri* (Shepherd and Caughley 1987).

Commercial harvest of kangaroos is an important management tool because the five species included are often serious pests. State-sponsored shooters are too expensive and recreational shooters do not kill enough to control the populations (Shepherd and Caughley 1987).

Interestingly, the United States once was the largest importer of kangaroo products. However, following a public outcry involving certain endangered species in Australia, the market for kangaroo products dried up (Fitzgerald 1989).

Introduced Deer in New Zealand

Several species of deer (about ten) have been introduced in New Zealand (Daniel 1962). There are no native deer there. Some of the introduced species have become quite common, which has sparked a commercial trade in meat and hides. For the most part, the deer have been hunted on foot, but recently helicopters have been used. Some argue that the populations have since fallen by 75%. This hurt the commercial trade and represents another example of how technology can influence harvesting success.

European Hares in Argentina

European hares (*Lepus europeus*) were introduced in Argentina in the 1800s and have been quite successful. Now they are shot by licensed hunters, and the processed meat is exported to Europe. Between 1976 and 1980, an average of 11,800 tons of hare meat was exported from Argentina to Europe (Fujita and Calvo 1982).

Crocodilians and Sea Turtles

There are 23 species of crocodilians worldwide. In 1971, all 23 species were either declining in number or endangered. Today, seven species are still endangered, but the others have benefited from management. Because the skins of these species have high commercial value, private companies have invested in establishing "crocodile farms" around the world. These farms help by hatching eggs and also work to preserve the species in the wild. Other methods of conservation include carefully regulated hunting (as in Florida) and habitat preservation. This is an example of how commercial value can serve as an incentive to citizens to preserve species.

Management programs for sea turtles include the harvest of eggs of olive ridley sea turtles in Central America. Lagueux (1991) studied human harvest of sea turtle eggs in Honduras from 1982 through 1987. She found that nearly 100% of the olive ridley eggs layed in some areas were harvested, a figure that no population can sustain.

Sponges

Sponges were once commercially harvested along the west coast of Florida and in the Keys. In the 1890s, sponges were harvested and exported to Europe and were used domestically for surgeries. The industry in Key West began in the 1830s, and by the 1890s Key West was one of the four most important centers in the world. The other centers were the Bahamas (Nassau), Cuba, and Calaimo, Greece (Munroe 1987).

Sponges are still commercially harvested along the west coast of Florida, in the vicinity of Tarpon Springs, but there is no more commercial sponge harvesting in the Keys. In the summer of 1995, sponges near Tarpon Springs began to suffer from an affliction of unknown origin that left their bodies brittle and unusable. This may be yet another impact of a recent red tide outbreak.

ILLEGAL TRADE

When discussing commercial uses of wildlife, it is important to keep in mind that there are both legal and illegal uses. It is sometimes difficult to say that a certain type of trade is illegal, since there may not be any legal basis for saying it is. One example is fishing on the high seas. Is France or Taiwan breaking any laws by overharvesting fish on the open oceans? Illegal uses include trafficking endangered species or their body parts. The trade becomes illegal when wildlife traffic is imported into a country against its laws. For the purpose of our discussion, we refer to any trade as being illegal if it violates known laws in the United States, which has some very strict laws.

What Drives the Illegal Wildlife Trade?

In many cases, people involved in illegal harvesting (poaching) can double their annual income with just a few kills (in some cases, one tusk of ivory). Bear gallbladders, tiger bones and penises, spotted cat skins, etc. are all highly valuable. The same is true for wild species captured for the pet trade.

One important point to consider is that the more abundant and common a species is, the less likely it is to be involved in illegal activities. Sadly, it is the rare species, those that can least afford it, that are victimized by human greed. As a species become rare, it becomes more valuable—a simple economic principle.

Examples of Illegal Commercial Use of Wildlife

Rhinoceros

There are five species worldwide, and a variety of products are used, including meat, horn, penis, hide, blood, and urine (Fitzgerald 1989). In India, one zoo sells rhinoceros urine for 44 cents per liter as a cure for asthma or sore throats. The penis of a rhino may sell for $600 and is used as an aphrodisiac in some cultures. In North Yemen, the horns are used for knife handles. Fitzgerald claims that there are now fewer than 11,000 rhinos of all species in the wild.

Elephants

The main product is ivory from the tusks. There are two species: African and Asian. Some isolated populations of Asian elephants (i.e., in Sri Lanka) do not have tusks. One major conservation concern is that there has been elevated inbreeding of Asian elephants in some areas because males have been disproportionately harvested for their ivory. Asian elephants are also used as beasts of burden. The biggest concern for Asian elephants is loss of habitat.

African elephants are seldom used as beasts of burden, but they have high-quality ivory. Between 1979 and 1989, the world population dropped from 1.3 million to 625,000, mostly due to illegal hunting. Other factors responsible for the decline in population are habitat loss, drought, and disease.

RECREATIONAL OR SPORT HUNTING: LIVING TO HUNT

Eisler and Buckley asked an interesting question in *USA Today,* April 25, 1996: "Should the public dictate hunting policy that once was left to the state?" Their question, however, does not go back far enough. We learned in Chapter 2 that the federal government once dictated hunting policy and then granted the states the right to dictate that policy. The question that should be asked is, "Should the public dictate hunting policy that once was left to the federal government?"

SNOW LEOPARD ♦♦ WILD ASS

SPECIAL MONGOLIA HUNT

16 DAY ITINERARY US $7,000; OBSERVER US $3,000 INCLUDES EVERYTHING IN MONGOLIA, EXCEPT TROPHY FEES AS LISTED BELOW.

SPECIE	TROPHY FEE, IF TAKEN
SNOW LEOPARD* (PANTHERA UNCIA)	$13,000
WILD ASS* (EQUUS HEMIONUS)	1,970

ALSO AVAILABLE ON SAME HUNT:

BLACKTAIL GAZELLE (GAZELLA SUBGUTTOROSA SUBGUTTOROSA)	750
LYNX	660
WOLF	200

*BOTH THE SNOW LEOPARD AND WILD ASS ARE LISTED ON THE U.S.A. AND CITES LISTS, AS ENDANGERED (EVEN THOUGH NOT ENDANGERED IN MONGOLIA). THEREFORE, THEY CANNOT BE BROUGHT INTO THE U.S.A. WITHOUT SPECIAL PERMITS. PROBABLY ONLY MUSEUM PERMITS WILL BE CONSIDERED.

PERIOD OF HUNTS WILL TAKE PLACE MARCH AND APRIL OR 01 OCTOBER TO 15 NOVEMBER.

AREA TO BE HUNTED IS SOUTH GOBI. THE MONGOLIAN WILDLIFE OFFICIALS HAVE NOW DETERMINED A SURPLUS OF THE SNOW LEOPARD AND WILD ASS AND ARE ALLOWING VERY FEW PERMITS EACH YEAR.

CONFIRMATION WILL BE TAKEN ON A FIRST COME FIRST SERVED BASIS UPON RECEIPT OF 50% DEPOSIT.

FIGURE 12.1 This ad appeared in 1990. Mongolia has since signed CITES, and hunts like this are now a thing of the past.

A record number of states are putting hunting referenda before the voters, according to Eisler and Buckley. As always, the two ends of the continuum are obvious. On the far right is the National Rifle Association, and on the far left are the animal rights activists. The ballot questions vary widely with region.

In most of the modern world, hunting is no longer really a major means of feeding oneself, although fishing might play a major role in supplementing people's diets in parts of the United States (according to an article in *Florida Sportsman*).

Traditionally, species involved in recreational hunting have been divided into categories. One category includes big game species such as white-tailed and mule deer, elk, mountain sheep, mountain goats, javelinas, and the host of exotics in Texas (discussed in Chapter 10). Big game hunting outside of the United States is also appealing to many recreational hunters (see Figure 12.1). A second category includes small game species, both mammals (such as squirrels and rabbits) and birds. The birds may be either upland gamebirds or waterfowl. Upland gamebirds include species such as ring-necked pheasants, bobwhites, various species of quail and grouse, mourning doves and other doves and pigeons, and turkeys. Waterfowl includes species such as ducks, geese, and swans. There is a very important reason for this distinction between upland gamebirds and waterfowl. Upland gamebirds are typically residents in a state. Moreover, they also include exotic species such as the chukar, grey partridge, and ring-necked pheasant. Waterfowl, on the other hand, are largely migratory, and they cross several state and national borders in any given year. Management of upland species may be handled by individual states, but waterfowl require an interstate and international program.

One measure aimed at producing funds for waterfowl management is the "Duck Stamp" which all waterfowl hunters in the United States must acquire in addition to a small game license.

SPECIAL HUNTS

Alligators in Florida

Alligators were once hunted nearly to extinction, mostly for their skins but also for meat. Even today, the meat may sell for as much as $4 to $5 per pound. The skins are used for leather products and sell for $25 to $35 per linear foot.

Alligators are not only hunted in a special season (in Florida it has been the month of September since 1988) but are also raised on alligator farms. Louisiana has a large alligator farming industry and yields the greatest number of skins of any state.

In Florida, there is a lottery for permits to trap alligators. Florida residents pay $250 for a trapping permit, and nonresidents pay $1000. Each trapper is then assigned an area and has two weeks to remove as many as six alligators. In 1995, there were 573 hunters (selected by computer from 9000 applicants), and the harvest quota for 1995 was 3500. The Florida Game and Fresh Water Fish Commission claims that there are as many as 16,000 alligators in Lake Okeechobee alone; thus, the removal of 3500 from the entire state does not represent a high percentage.

Commercial use of alligators extends to the so-called nuisance alligators. They are taken by licensed trappers. See Figure 3.3 for an idea of the number taken.

Rattlesnake Roundups

Across the southern states, rattlesnake roundups take place each year. One of the largest hunts is in Sweetwater, Texas. People collect snakes and bring them to a large barn on a specified day. Awards are given for the most snakes caught, the largest snake, and so on. The snakes are milked for their venom and then slaughtered for the meat and skins. The harvest is uncontrolled. Because the Sweetwater Round-up occurs in early March, when it is often still too cold for much rattlesnake activity, the hunters flush the snakes from their burrows mechanically with hooks or chemically by pouring gasoline down the burrows. Individuals of many other species are victims of the gasoline. In the southeastern United States, these other victims include indigo snakes (*Drymarchon corais*), gopher tortoises (*Gopherus polyhemus*), and possibly gopher frogs (*Rana capito*) and Florida mice (*Podomys floridanus*).

SUMMARY

Wildlife is harvested for subsistence in many parts of the world. As technology advances, the ability to overharvest species increases. Commercial harvesting typically is not sustainable. Indeed, there are few examples, like the kangaroos in Australia, of species commercially harvested under tight controls where we see any inkling of sustainability.

Recreational hunting is practiced in many parts of world. A number of species are harvested, and this may be economically beneficial to those guiding the hunts. Recreational hunting is probably most popular in states with lower population densities. In states such as Florida, fewer than 2% of the population participates in recreational hunting.

EXERCISES

12.1 Identify three examples of bird species that are harvested for subsistence.

12.2 How would you characterize these species?

12.3 What is the most common reason for use of animal body parts in Asia?

12.4 Pick some states and/or countries and find out what it costs to hunt big game species there.

12.5 Give one example of an introduced species that is harvested commercially somewhere in the world (other than the ones listed in this chapter).

12.6 How many African elephants are there now? How many were there 20 years ago? Why has their population crashed?

12.7 Are hunting reserves the answer to preserving wildlife? If a local village can get $20,000 from a hunter willing to shoot a bull elephant, wouldn't that be a powerful incentive to raise more bull elephants?

LITERATURE CITED

Ayres, J.M., D. dM. Lima, E. dS. Martins, and J.L.K. Barreiros. 1991. On the track of the road: changes in subsistence hunting in a Brazilian Amazonian village. pp. 82–92 *in* Robinson, J.G. and K.H. Redford (Eds.). *Neotropical Wildlife Use and Conservation.* University of Chicago Press, Chicago.

Baldassarre, G.A. and E.G. Bolen. 1994. *Waterfowl Ecology and Management.* John Wiley & Sons, New York.

Daniel, M.J. 1962. Control of introduced deer in New Zealand. *Nature* **194**:527–528.

Fitzgerald, S. 1989. *International Wildlife Trade: Whose Business Is It?* World Wildlife Fund, Washington, D.C.

Fujita, H.O. and J.O. Calvo. 1982. Las exportaciones de productos y subproductos de la fauna silvestre en el quinquenio 1976/1980. *IDIA* **397/400**:1–26.

Greenway, J.C. 1967. *Extinct and Vanishing Birds of the World.* Dover Publications, New York.

Lagueux, C.J. 1991. Economic analysis of sea turtle eggs in a coastal community on the Pacific coast of Honduras. pp. 136–144 *in* Robinson, J.G. and K.H. Redford (Eds.). *Neotropical Wildlife Use and Conservation.* University of Chicago Press, Chicago.

Misch, A. 1992. Can wildlife traffic be stopped? *World Watch* **5**(5):26–33.

Mittermeier, R.A. 1991. Hunting and its effect on wild primate populations in Suriname. pp. 93–107 *in* Robinson, J.G. and K.H. Redford (Eds.). *Neotropical Wildlife Use and Conservation.* University of Chicago Press, Chicago.

Munroe, K. 1987. Sponge and spongers of the Florida reef. pp. 29–39 *in* Oppel, F. and T. Meisel (Eds.). *Tales of Old Florida.* Castle, Secaucus, New Jersey.

Redford, K.H. and J.G. Robinson. 1991. Subsistence and commercial uses of wildlife in Latin America. pp. 6–23 *in* Robinson, J.G. and K.H. Redford (Eds.). *Neotropical Wildlife Use and Conservation.* University of Chicago Press, Chicago.

Roe, F.G. 1951. *The North American Buffalo: A Critical Study of the Species in Its Wild State.* University of Toronto Press, Toronto.

Schorger, A.W. 1973. *The Passenger Pigeon: Its Natural History and Extinction.* University of Oklahoma Press, Norman.

Shepherd, N. and G. Caughley. 1987. Options for management of kangaroos. pp. 188–219 *in* Caughley, G., N. Shepherd, and J. Short (Eds.). *Kangaroos: Their Ecology and Management in the Sheep Rangelands of Australia.* Cambridge University Press, Cambridge, U.K.

Smith, N.J.H. 1976. Utilization of game along Brazil's transamazon highway. *Acta Amazonica* **6**:455–466.

REMEMBERING HISTORY: THE EASTERN U.S. EXPERIENCE

13

A s we saw earlier, wildlife across the United States was disappearing even before the beginning of the 1900s. Imagine hearing old-timers speak fondly of the days when deer and turkeys were plentiful—way back in 1860! Public recognition of this disappearance was slow in coming but deceptively powerful. While some conservation gains were made, those gains were reversed during the depression years of the 1930s as poverty set in. Because ammunition was inexpensive, wild game provided food for the table. While wildlife conservation leaders publicly voiced their concerns, waterfowl numbers were at their lowest in more than 30 years. Gone from the scene were ardent conservationists such as Theodore Roosevelt, George Grinnell, Charles Sheldon, and John Burnham, who had led the charge at the turn of the century. People like Jay "Ding" Darling, M. Hartley Dodge, Charles Horn, Carl Shoemaker, Aldo Leopold, Thomas Beck, Ira Gabrielson, and Frederick Walcott formed a new core dedicated to carrying the banner of wildlife conservation. And carry it high they did. We owe our rising wildlife populations of today to these people.

As the United States was emerging from the global depression, these conservation leaders took advantage of the innovative mood in the U.S. Congress. In a flurry of activity, conservation legislation was passed by Congress and signed into law by the president. By the end of the 1930s, the foundations of many modern and effective U.S. wildlife conservation programs were in place. We have chosen to review and highlight the wildlife conservation programs in eight states in Chapters 13 and 14. Many share several common successes and problems. Each state has its own unique problems, which fostered a degree of independence to address them. Thus, many states have developed thoughtful and innovative solutions. Certainly, one of the common strengths of state wildlife programs is the extraordinary enthusiasm for and dedication to the task. The Federal Aid in Wildlife Rehabilitation Act of 1937 (Pittman–Robertson Act) and the Endangered Species Act of 1973 stand as the most important wildlife conservation laws ever passed anywhere in the world.

Recall from Chapter 2 that the Federal Aid in Wildlife Rehabilitation Act of 1937 (referred to as the Pittman–Robertson Act) empowered states with a continued source of funding to carry out wildlife research programs. This act had two principal purposes: (1) reestablishment of wildlife populations to natural habitats and (2) making wildlife research the basis for science-based wildlife management programs.

Initial progress of wildlife programs was slow, however. World War II saw federal tax receipts from hunters decrease because many would-be hunters entered the armed services. After the war, the Pittman–Robertson Act was the catalyst for state wildlife programs. The Pittman–Robertson Act was quite remarkable in that the tax today on ammunition, firearms, and archery equipment amounts to 11% and is enthusiastically supported by the hunting industry—a sure sign of the success of the act in reestablishing wildlife populations. For example, there were about 100,000 elk in the United States in the 1920s. In 1995, the population of elk surpassed half a million. Wild turkeys were scarce back in the 1930s. Today, they are found in nearly every state and are locally abundant in many states. Their population is estimated to be greater than four million. Wood duck were feared to be nearly extinct in the 1920s. Today, the wood duck is the most common breeding waterfowl in the East. In the 1920s, there were fewer than half a million deer in the entire United States. Today, more than 20 million roam the country and in many areas there are far too many. Pronghorn antelope numbered fewer than 20,000 in 1920. Now they number more than a million strong.

Sadly, many exotic species have also been introduced. Ring-necked pheasants have done extremely well and are a popular gamebird but are not native to the United States. Other exotic hoof stock have been introduced for the pleasure of hunters, but the damage to native plants has been substantial, especially in semi-arid and arid regions of the United States. Nongame species also benefited under the Pittman–Robertson Act. By 1987, more than $2 billion in federal excise taxes had been matched by more than $500 million in state funds (mostly from hunting license fees) for wildlife restoration.

In early 1996, state wildlife agencies received their share of fees collected from federal excise taxes in 1995 paid by anglers, hunters, and recreational shooters to support fish and wildlife restoration and recreation projects. A total of $202.4 million was apportioned to states under the Pittman–Robertson Act, compared to $211 million in 1995. The decrease was due primarily to larger than normal collections available in 1994 from tighter enforcement and collection of back taxes. The money was derived from an 11% excise tax on sporting arms and ammunition, a 10% tax on pistols and revolvers, and an 11% tax on certain archery equipment. One-half of the tax on handguns and archery equipment is made available for state hunter-education programs. An important point to remember is that wildlife restoration funds are made available based on land area and the number of hunting license holders in each state. Distribution of hunter-education funds is based on the relative population of each state.

Another $197.1 million was passed along in 1996 under the Federal Aid in Sport Fish Restoration Act, often called the Dingell–Johnson/Wallop–Breaux Act. This compares to $199.9 million in fiscal year 1995. The slight decrease is attributable to an increase in deductions based on the proposed passage of the Boating Safety Improvement Act. This funding resulted from a 10% excise tax on fishing equipment and a 3% tax on electric trolling motors and sonar fish finders. The Wallop–Breaux legislation of 1984 increased the tax base for sport fish restoration to include a portion of the federal motorboat fuel tax and import duties on fishing tackle and pleasure boats. Distribution of sport fish restoration funds to the states is based on the land and water area and the number of fishing license holders in each state.

Wildlife populations have generally increased and states have reintroduced species that were extirpated many years ago. Indeed, in many states, some wildlife species have increased beyond environmental and social carrying capacity. By studying how far state wildlife programs have come since 1937 and by considering recent referenda concerning wildlife in several states, we can make predictions about the future of wildlife in the United States. We consider here and in the next chapter several states and review their progress since enactment of the Pittman–Robertson Act of 1937 and the Endangered Species Act of 1973. Naturally, many of the early programs emphasized species that could be hunted. Now, however, endangered species programs also play a key role in states' efforts to bring back wildlife.

MAINE

In 1937, the Pittman–Robertson Act provided Maine with the resources to initiate scientific wildlife research programs. Maine's Wildlife Research Division was soon established to undertake waterfowl restoration and research (Kallman 1987). The project lasted nine years, identified the ecological requirements for many migratory species, and was followed by a habitat restoration program. Wood duck populations were rebuilt, and nesting populations of Canada geese, whose populations had been dramatically reduced, were also established. Wildlife management practices were put into place on publicly owned lands, including over 200 seabird nesting islands.

Soon forest management practices that enhanced wildlife populations were integrated into public and private land stewardship programs. Black bear populations were more closely managed (Figure 13.1), and funds became available to rebuild low and, in some places, nonexistent beaver populations. Native wild turkeys were also reintroduced. As a scientifically conducted program, Maine's efforts included not only investigations into the basic biology of target species but also hunter surveys, habitat inventories, habitat acquisition, and environmental impact assessments.

Moose have historically occupied the Northeast. In the 1800s, unregulated market hunting, a commercial activity, had a severe negative impact on Maine's moose population. At the same time, forests were being cleared and replaced with farmland. In 1830, a law was passed by Maine's legislature to restrict

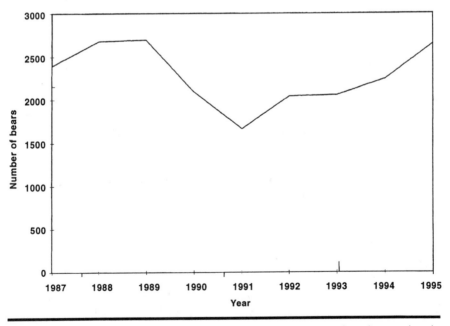

FIGURE 13.1   The number of bears harvested in Maine has been closely monitored.

moose hunting to September through December. By the early 1900s, the southern portion of Maine was farmland, mostly without moose. By 1935, hunting was curtailed indefinitely. Moose continued to occur in low densities until the 1940s, when several hunting laws were introduced in the state legislature but failed. From 1875 to 1979, a period of 105 years, a total of only 1200 moose were legally bagged by hunters in Maine. Not until 1977 did hunting become legal again. In 1980, 700 moose permits were issued. No permits were issued in 1981, but 1000 permits were available in 1982. Opponents of moose hunting successfully collected enough signatures to force a referendum in November 1983 in an attempt to close the hunt. The measure was defeated by a margin of six to four. From 1983 to 1993, 1000 permits per year were issued. In 1994, 1200 moose permits were issued, and 1400 and 1500 permits were issued in 1995 and 1996, respectively. Today, moose are back in Maine. In 1997, Maine hunters enjoyed their 16th consecutive hunting season. Most moose are taken in the more forested northern part of the state. The moose population in Maine was estimated to be between 15,000 and 20,000 in 1985.

Moose are browsers and forest edge species, much like deer. Clear-cutting of forests and Maine's wet ground and remaining forests have created favorable habitat for moose. Lakes in low wet areas are used by moose in summer for thermoregulation. For a $5 fee, residents with a $25 hunting permit can also purchase a big game license for $20 that allows them a limit of one moose. The demand for moose permits, however, far exceeds the supply. Today, some 80,000 to 90,000 hunters apply for the prized license to bag a moose. Moose are now seen in New Hampshire, Vermont, and Massachusetts, and one was even seen in northwestern New Jersey in 1996.

Moose and deer populations fluctuate in Maine, where global climate and land use patterns affect both populations. Moose reach their southern limits in New England, where deer reach their most northern extent. Shifts toward mild climates favor deer, while colder climates favor moose. Thus, management practices must take into account abiotic and biotic factors which influence moose populations that ebb and flow northward and southward. Traffic accidents involving moose have increased from about 160 in 1980 to 700 in 1992 but remain low statewide. Hunter surveys show that moose hunters are selective with regard to moose they shoot, which suggests that moose are commonly seen during the hunting season. Predictably, moose are most often found in commercial forests, and, equally predictably, poaching "for the pot" is heavy with such demand for moose hunting permits. One moose can supply a family with enough meat for a year. Although successful, Maine's moose reintroduction was not a planned program. Habitat changes in Maine and elsewhere combined with healthy moose populations farther north have allowed moose to spill back into areas where they were formerly extirpated. Moose have expanded their range southward and, with adequate protection like that afforded moose in Maine, will probably reach the limit of their physiological constraints (such as an inability to keep cool in summer) somewhere in Pennsylvania and New Jersey.

The history of moose in Maine tracks the history of wildlife in the United States. We can only hope that the recent history of moose in Maine is a harbinger of how wildlife populations will fair across the nation. Indeed, while beavers were once rare in Maine, they have now become a nuisance. One wildlife biologist confessed that there are too many beavers in Maine! Panthers were reported to be rare in Maine at the turn of the century. One panther was taken in 1938, but 80 panther sightings have been reported since 1990. A verified panther report came from a solitary cat in New Brunswick, Canada, in 1993. Also, three panthers were sighted in Vermont in 1993. Eastern panthers are considered federally endangered. Although no wolf sightings occurred in Maine in 1995, wolves occur in nearby provinces in Canada. The politics of reintroducing a top predator can be avoided if the St. Lawrence Seaway freezes for just a short period, which would allow politically neutral wolves to reestablish in Maine.

Deer populations continue to increase in Maine. Approximately 268,000 deer wintered in Maine in 1995–1996, a 23% increase over the 1994–1995 population. Hunters harvested 24,683 deer in 1994 and 27,384 in 1995, a 10.9% increase that fell short of deer population growth. Maine's statewide wintering population target is between 290,000 to 350,000 deer (10 to 12 per square mile). How might this increase in deer numbers affect forest bird species, especially those that feed on the ground, such as the woodcock (*Philohela minor*), whose population in Maine and elsewhere continues to plummet?

The Maine Caribou Project was formed by a private group that wanted to reintroduce into the state woodland caribou (*Rangifer tarandus caribou*), which were extirpated in 1908. Unlike most herbivore reintroduction programs, there was no original intent to establish a population of huntable caribou (McCollough 1990). In 1986, 22 caribou were transported from Canada to Maine. Maine public opinion for the reintroduction program was high, many residents visited the breeding facility on the University of Maine campus, and donations for the "Adopt-a-Caribou" program were consistently high. Within two years, the herd doubled to 45 caribou and soon had to be relocated to a larger breeding facility nearby. Unfortunately, the area chosen for the breeding facility was surrounded by a deer population that acted as vectors for a parasitic disease deadly to caribou.

In the spring of 1989, a dozen caribou were released in Baxter State Park in north central Maine. By fall, ten had died. Three deaths were due to neurologic disease attributed to the parasite. Media coverage was generally negative, perhaps because the media were denied access to the release of the caribou earlier. A local animal rights group objected to the release "because it was cruel to release animals into the wild where they could be killed" (McCollough 1990). Public opinion remained favorable, however, and continued public financial support was apparently secure.

Because disease became a problem in the captive herd, 20 caribou were released in 1990 on private land owned by a paper company. Within seven months, several caribou succumbed to the parasitic disease and many were killed by bears. When Maine's economy declined in the early 1990s, the project was discontinued. Caribou sightings continued for a short time after the project ended, however. McCollough and Connery (1991) concluded that reintroduction failed because of neurologic disease and black bear predation. New calves were killed by bears within one to three days of birth. One calf survived 3.5 months only to be killed by a lynx. Calf predation rates are typically less than 100% in large natural populations, but because the reintroduced herd was small, calf mortality was extraordinary. Predation on adults was also high however, and the parasitic disease acquired in captivity killed excessive numbers.

Although Maine's experimental caribou reintroduction program ended prematurely, the success of the program should be measured not by the number of caribou that were successfully reintroduced but by the number of Maine citizens who were introduced to wildlife conservation programs. That citizen interest remained high in the face of setbacks is evidence enough that the program was successful. Moreover, the simple fact that it is far easier to conserve species *in situ* than to reestablish them once they are gone was not lost on Maine's residents. The wildlife biologists whose involvement and expectations were greater than their experience were disappointed with the results. Species reintroduction is a science in its own right that requires years of training and experience in fields as diverse as public speaking, journalism, group behavior, education, advertising, funding-raising, law, politics, public relations, chaos theory, economics, microbiology, biology, and ecology. Anyone can open a cage door and release animals.

The caribou reintroduction might also have taken place too soon historically. In the past, healthy moose populations probably kept deer densities lower and thus prevented caribou, a higher latitude species, from commingling with deer, a lower latitude species. When moose populations disappeared in Maine, deer invaded northward. With moose populations increasing, deer populations might decrease and caribou might have less chance of contacting a lethal parasite. Also, wild caribou travel in massive herds and females give birth nearly simultaneously. Predators take some young but cannot take them all. When the herd is reduced in size, all the young can be taken.

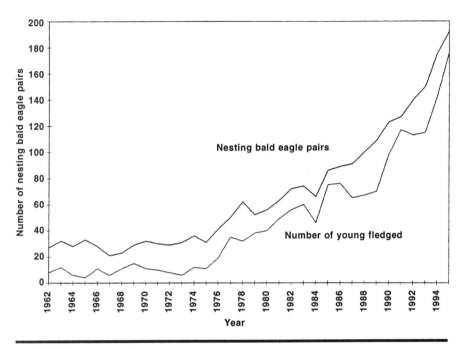

FIGURE 13.2   The number of nesting bald eagles and young fledged in Maine since 1962.

Maine also has an active endangered species program (Hutchinson 1996). A review of wildlife species in the state revealed that only 38 required threatened or endangered status. Thus, funds from purchases of special license plates and income tax checkoffs are not diluted. In 1995, 192 bald eagle pairs produced 176 fledglings, and the eagle continued to recover in Maine (Figure 13.2). Seven peregrine falcon (endangered) pairs produced 12 fledglings. The bald eagle, previously federally endangered, was reclassified as threatened in 1995. A single pair of golden eagles (endangered) was reported in Maine, the sole pair in the Northeast, but has not successfully reproduced in 13 years. Plans to develop a wind power-generating station in the western mountains in the middle of the eagle's range must follow specific guidelines set forth in the Endangered Species Act of 1973. A list of Maine's threatened and endangered species is given in Appendix 13.1

Many species listed in Appendix 13.1 were targeted in Maine's recovery program, such as grasshopper sparrows, piping plovers, least terns, black terns, roseate terns, Blanding's turtles, and spotted turtles. Other species involved in various conservation actions include the Tomah mayfly (recently rediscovered in Maine after it was thought to be extinct), 11 species of freshwater mussels, 3 species of dragonflies, and harlequin ducks. Special habitats were also targeted for monitoring and protection.

The northwestern Atlantic walrus was extirpated by the mid-1700s. Millions of walrus once crowded the rocky shores from Cape Cod, Massachusetts, along the entire extent of Maine's coast, up into the St. Lawrence Seaway and to many areas farther north. The sea mink was also found along Maine's coast. The sea mink has long been extinct, and nothing like it can be brought back. The extinct walrus was a subspecies, however, and populations of another subspecies of walrus in the Bering Sea are thriving under protection. Is it possible that we will, in our lifetimes, once again enjoy walrus basking on Maine's rocky coast? When people in Maine realize that the walrus was as much a part of Maine's coast as moose were a part of Maine's forests, we predict that walrus too will return. Walrus have been gone for nearly 250 years. Only you can bring them back!

NEW JERSEY

New Jersey's geographic location makes the state one of the hemisphere's most important wildlife areas. The organization Defenders of Wildlife ranks New Jersey 11th highest in the country in terms of biodiversity despite the fact that the state's population density (1004 people per square mile in 1997) is higher than that of Japan. The Delaware Bay area is a critically important stopover for migratory waterfowl. The state is also an ecotone where many northern species reach their southern limits of distribution and some southern species reach their northern limits. The state wildlife checklist includes more than 324 birds, 44 reptiles, 35 amphibians, and 89 mammals. The New Jersey Division of Fish, Game and Wildlife initially focused efforts on creating habitat diversity for many wildlife species. In addition, access to wildlife viewing areas was a priority. Using Pittman–Robertson funds, private lands were purchased to increase public land for wildlife. The Endangered and Nongame Species Program receives no state funds and is supported by voluntary contributions, "Conserve Wildlife" license plate sales, and a special line on state income tax forms. At the turn of the century, however, wildlife in New Jersey was disappearing. So were game wardens. In 1897, the first two game wardens, both volunteers, were shot dead on the job.

By 1900, New Jersey's white-tailed deer population was almost nonexistent. Several remnant wild populations hung on and a few small herds were found on private lands. In 1901, hunters shot a total of 20 deer. From that low point

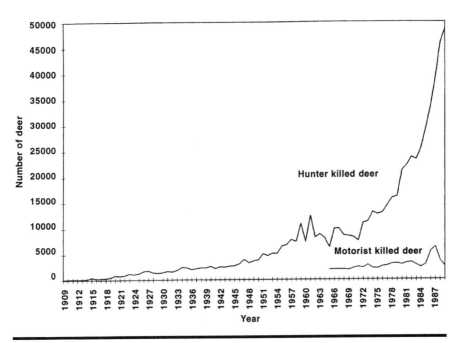

FIGURE 13.3   Deer killed by hunters and motorists per year in New Jersey since 1909.

in the history of wildlife in New Jersey, deer populations rebounded through protection and restocking efforts. By 1940, agricultural damage by deer was extensive enough to use Pittman–Robertson Act funding to investigate the problem. More than 25,000 deer were harvested by sport hunters during the 1984–1985 hunting season. Hunters had a 90-day hunting season and could take eight deer. Today, New Jersey has too many deer despite the number of hunters. Figure 13.3 shows the number of deer killed by hunters and motorists since 1909. A record 51,442 deer were harvested during the six-day hunt in the 1994–1995 deer hunting season. Both bucks and does were hunted. In 1996, the deer population in the state was estimated to be 33 deer per square mile.

Wild turkeys were restocked in New Jersey in 1977 after nearly a century of absence. By 1981, turkeys were legally hunted. In 1987, wild turkeys occupied nearly all available habitat. State estimates show that about 14,000 turkeys occupied 2200 square mile in 1995. The 1995 spring season hunt yielded 1582 gobblers, 411 more than the 1994 season.

As for other hunting successes, 11,650 woodcock, 171,754 squirrels, 215,059 rabbits, 308,332 pheasants, 103,475 quail, 9385 grouse, 56,737 muskrats, 4679

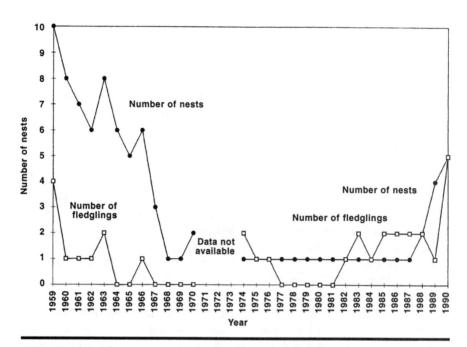

FIGURE 13.4   The number of bald eagle nests and fledglings in New Jersey since 1959.

raccoons, 2209 red fox, 1027 grey fox, 1099 opossum, 398 skunks, 572 mink, 5 coyotes, 172 beavers, and 30 river otters were shot or trapped in the 1994–1995 season (New Jersey Department of Environmental Protection 1995).

The New Jersey Endangered and Nongame Species Conservation Act of 1973 mandated that all forms of wildlife, particularly endangered wildlife, be conserved so as to promote and enhance biodiversity. Bald eagles had declined in New Jersey but are now increasing (Figure 13.4). The state's Landscape Project is an especially exciting program to manage large areas holistically. New Jersey also created a Comprehensive Management Plan for Shorebirds on Delaware Bay, which is utilized each spring by nearly a million migratory shorebirds. Surveys of breeding reptiles and amphibians, American burying beetles (*Nicrophorus americanus*), Mitchell's satyr butterfly (*Neonympha mitchellii*), dwarf wedgemussel (*Alasmidonta heterodon*), woodcocks, wading birds, and migratory songbirds take place regularly. Appendix 13.2 lists endangered and threatened species in the Garden State.

For the purposes of this book. an ecological landscape is defined as a collection of two or more ecosystems. An ecosystem is an area (although it can be the size of a putting green) that has a fairly uniform ecology, such as a pineland, swamp, or field. The Landscape Project in New Jersey attempts to manage large areas that consist of many ecosystems "to insure the long-term protection of rare species populations in landscapes distributed throughout the state" (Landscape Project 1993). Today, we often hear the phrase "ecosystem management" touted by such prestigious organizations as The Nature Conservancy. New Jersey's Landscape Project is two steps ahead of ecosystem management for several reasons.

The Landscape Project recognizes that simply establishing a park or protected area will not guarantee that species will persist in those areas. Phase 1 of the project involves mapping of rare species ranges. Phase 2 provides land managers with species ranges and habitat requirements, and Phase 3 involves coordination of land use regulation so that agencies can incorporate long-term rare species protection methods into current plans. Because land ownership invariably falls between several land management agencies with different land use oversights, habitat protection and conservation efforts are usually hindered. The Landscape Project addresses this problem directly.

The Comprehensive Management Plan for Shorebirds on Delaware Bay recognizes the international importance of the bay to migratory shorebirds. The bay is one of the top ten shorebird stopover areas in the western hemisphere. Because thousands of ships pass through the bay each year, the area is as critical to humans as it is to other species. From May 13 to June 6 each year, about a million shorebirds time their arrival to take advantage of a feast provided by tiny crab eggs produced by spawning horseshoe crabs (*Limulus polyphemus*). Although the crabs (actually they are more closely related to spiders or scorpions) are found from Maine to Florida, Delaware Bay is the species' largest breeding ground. Recognizing the value of this region, New Jersey has adopted a comprehensive approach to its conservation. Public education, including viewing areas, is also part of the plan.

By the early 1900s, bobcats were thought to be extinct in New Jersey, although favorable habitat was plentiful. Occasional sightings occurred, and a bobcat was killed by a car in 1961. Tracks were seen in 1972, but the bobcat was clearly rare in the state. In 1977, a state plan to restore bobcats to New Jersey was created and then sent to the U.S. Fish and Wildlife Service for approval. Upon federal approval of the plan, the state was reimbursed for 75% of the cost of the reintroduction program by the Federal Aid to Wildlife Restoration Act (the Pittman-Robertson Act). In 1978, state biologists released bobcats into the wild as part of a reintroduction program. In 1997, bobcats inhabited most of the counties in the state. Small predators were coming back.

Sadly, voluntary income tax checkoffs to help wildlife slid to their lowest levels in 20 years in one of the country's most well-to-do states. However, volunteers have stepped forward to shoulder the burden and help wildlife. For instance, the New Jersey Department of Environmental Protection claims to have more than 1800 volunteers, some of whom have contributed more than 20,000 hours of unpaid service (Peet 1997). New Jersey's wildlife protection laws still date back to 1939, however. Trafficking in wildlife and poaching remain punishable by a mere $300 fine. An enterprising Maine resident spent six weeks off of New Jersey's coastal waters netting young eels and made about $400,000 by selling them to an Asian marketer. With bear parts increasing in value, how will the last 400 bears in the state survive? When will new laws with increased financial penalties and mandatory jail terms for wildlife trafficking be put into place? When the voters of New Jersey want them!

PENNSYLVANIA

Pennsylvania receives more funding from the Pittman–Robertson Act than any other state except Alaska and Texas because the state has the third largest hunting population. From 1937 to 1987, research projects on eastern cottontail rabbits, ring-necked pheasants (an exotic), snowshoe hares, white-tailed deer, wild turkeys, and woodcock were supported in part by Pittman–Robertson funds. In 1987, Pennsylvania's portion of federal Pittman–Robertson funding was used for habitat development on state game lands and on private property enrolled in the Pennsylvania Game Commission's public access program. As of 1987, 4.4 million acres of private land was enrolled in the program. In 1996, 4.5 million acres of private land was enrolled.

At the time of European colonization, eastern elk (*Cervus elaphus canadensis*) were found in the Northeast, including Pennsylvania, but these large herbivores were extirpated from the state by 1867. Their demise was caused by unregulated hunting and habitat loss. From 1913 through 1926, the Pennsylvania Game Commission attempted to establish a huntable population of elk by reintroducing Rocky Mountain elk (*C. elaphus nelsoni*) into several counties, including Cameron, Elk, Blair, Carbon, Centre, Clearfield, Clinton, Forest, Monroe, and Potter counties; 145 elk were brought in from Yellowstone National Park, 10 from Wind Cave Game Reserve in South Dakota, and 22 from a private reserve in Pike County, Pennsylvania. The program was partially successful, and the first sanctioned elk hunt occurred in 1923. Hunting ceased in 1932 because elk numbers were declining. In the 1970s, a survey by the U.S. Fish and Wildlife Cooperative Unit in the state revealed a population of 30 elk roaming Elk County, Pennsylvania. Elk County is about 30% public land. Since the 1970s, habitat has improved for elk in the state. Elk are grazers and require grasslands to feed. Elk also seek shelter in the forest at night. Reclaimed strip mines were first replanted with grasses, and this habitat proved favorable for elk. Now these lands are kept open for elk. Nearby forests provide cover. The elk population is growing at a rate of about 12 to 14% per year. A public education program about the elk reduced poaching as well.

In 1996, there were more than 300 elk roaming Elk and Cameron counties in north central Pennsylvania, and a hunting season was planned for 1998. The elk population was expanding, and high-quality foraging areas were attracting elk beyond their present range. Road and train kills of elk were extremely rare. Elk viewing was becoming a popular sport in western Pennsylvania, and local economies were benefiting from an increase in tourism. Public education programs in the area were continuing to educate people about wildlife. By 1998, some 67 years after the hunting season on elk ended and 120 years after the eastern elk was extirpated, Pennsylvania might soon have another elk hunt. With public interest in wildlife on the increase, Pennsylvania's elk program received the attention it had earned. Will these elk spread into Ohio, New Jersey, New York, and points farther north? If favorable habitat is available, surely they will.

Pennsylvania's elk reintroduction program was but one of many reintroduction programs the state had successfully executed. Crop damage by elk was low because the state pays for the entire cost of a fence in addition to half the cost of installation. The landowner in turn must sign a ten-year agreement that allows public access to the land. The program had 30,000 participants in 1996, and more than four million acres was involved in the program

Pennsylvania's deer population, like neighboring New Jersey's, has expanded. The state has 12 million residents, about 1.1 million hunters, and an estimated 1 million deer. In 1995, 40,000 deer were road killed and 400,000 deer were taken by hunters (Figure 13.5). One hunter near Philadelphia reported killing 25 deer.

From 1906 until the early 1920s, the Pennsylvania Game Commission stocked deer to increase severely depressed populations. Early in the program, does and juveniles were protected. Does give birth every year and often produce twins when resources are plentiful. Deer populations can therefore double every year. By 1923, agricultural crop damage was becoming a problem. In 1928, the buck season was closed and an antlerless deer season was held. Many people objected to the doe hunt, however.

According to a Pennsylvania deer management brochure, many people, including hunters, disagreed with the female deer hunt. Licenses were purchased and burned. Newspaper ads ridiculed doe hunters. Private landowners put up "No Doe Hunters" signs. In 1928, about 25,000 antlerless deer were harvested by hunters. In 1929, nearly 23,000 bucks were shot. However, by the

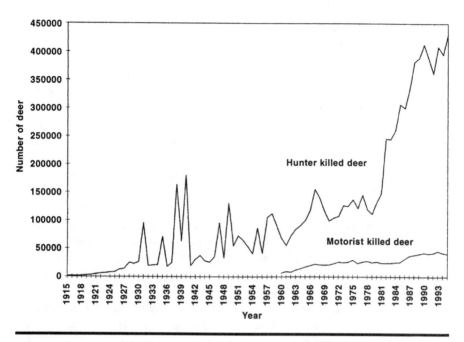

FIGURE 13.5 Number of deer killed by hunters and motorists per year in Pennsylvania since 1915.

1930s, estimates showed that between half a million and a million deer roamed eight million rural acres of Pennsylvania. Estimates now place the number of deer in the state at a million living on 16 million acres.

During the 1980s, the deer population in the state grew rapidly. Harvests were insufficient to control the expanding population. In 1983, the Pennsylvania Game Commission allocated 536,650 antlerless deer licenses, but only 519,000 were sold and the total deer harvest was about 250,000. In the following years, license sales leveled off to slightly more than 500,000 permits per year. Although hunters were taking more deer (Figure 13.5), the deer population continued to increase. The Pennsylvania Game Commission then implemented a statewide "bonus deer program" that allowed hunters to take more than one deer per year. In 1988, 679,300 antlerless permits were allocated and unsold permits were issued as bonus tags. The deer population seemed under control at 1 million deer at a density of 29 deer per forested square mile.

Hunters are an integral part of Pennsylvania's wildlife management program. Thus, deer management in the state depends upon the number and gender of deer taken by hunters. As with most living systems, the human population is dynamic as well. Human demographics are also changing. Pennsylvania's human population has been stable (unlike Florida's, for instance), but the average age is increasing. As people age, they hunt less. The number of people under 25 is forecasted to drop by 5% by the turn of the century. Those age 20 to 24 are forecasted to drop by 20% in the state. The number of preschool children is also forecasted to decrease. These numbers indicate that there will be less recruitment of new hunters. The Anglo population is expected to remain stable, while racial and ethnic minority populations are expected to increase. Hunting license applications show that a smaller fraction of minorities than Anglos hunt. In addition, some 21% of Pennsylvania's families are headed by a single parent, and this parent is usually a female. Generally, fathers, not mothers, introduce their children to hunting. Pennsylvania is becoming more urbanized as well. Urban dwellers are less apt to hunt than people in rural areas. Wildlife viewing, bow hunting, and flintlock hunting are increasing in popularity faster than traditional hunting and hence placing special demands on wildlife management strategies. Clearly, there are many facets to management of wildlife, not the least of which is the human species.

Bear populations are also increasing. Scientists studying the bear population estimate that there are about 9600 bear in the state. In 1995, 2190 bears were taken by hunters. A three-day hunting season in 1996 yielded a total of 1793. Bear baiting and hunting by dogs are not permitted in Pennsylvania.

A record number of wild turkeys has also been reported. Turkeys are forest-edge species, and populations are increasing in the state. River otters and fishers have been successfully reintroduced as well. Otters now occur in all 50 counties in the state. A reintroduction of pine martens might occur soon. Bald eagles, peregrine falcons, and osprey populations are also increasing.

Wolf and panther reintroduction programs have not been discussed. State officials suggest that these top predators require enormous home ranges that would inevitably bring them into contact with people, a situation which is apparently politically unacceptable at this time. What determines the size of the home range of a panther? Could it be determined by the number and size of prey? If so, it could be suggested that panthers would have unusually small home ranges in Pennsylvania because deer densities are so high.

FLORIDA

Florida's geographic location, its favorable climate (the Sunshine State), and its geologic history set the stage for some very interesting wildlife. Add to this unique setting 13.7 million people—with 892 new people per day (1692 entering and 800 leaving) and a projected population of 19 million by 2020—and very few public land and wildlife issues will arise. There are 23 nonindigenous birds, and 27% of the state's plants are nonindigenous. During the ice age 18,000 years ago, the Florida peninsula was twice as wide as it is today. Northern North America was covered by the massive Laurentide Ice Sheet, and the Cordilleran Ice Sheet extended up the northern Pacific coast. By about 7000 years ago, the ice sheets had retreated (Pielou 1991). With all that fresh water locked up in great ice sheets, the sea level 18,000 years ago was at least 85 m lower than today and perhaps as much as 130 m lower. Huge tracts of land along the continental shelf were exposed. Instead of being a long and narrow finger, Florida was a broad but still very flat platform of land (Hoffmeister 1974). As the ice retreated, the sea level rose, and sea water encroached and inundated the land. Peninsular Florida, lacking topographic relief, lost half of its peninsular land to the sea.

At the full extent of the glaciation, Florida, in some cases, had a complex array of large mammals that are no longer found there or anywhere else for that matter. Between 12,000 and 10,000 years ago, mammoths and mastodonts disappeared from the earth. Mastodont fossils have been found in Florida. Florida lost 50 species of mammals (listed in Table 13.1).

The rise in sea level and the loss of so much biodiversity raise an interesting issue. Suppose you were alive 18,000 years ago. Your tribe was comparatively well off because you were living in Florida along what was then its west coast. At the annual gathering of tribes, it was decided that in order to protect the rich hunting grounds, a large reserve should be established so that the animals could breed and roam freely within the reserve. Excess animals leaving the reserve could be hunted without doing harm to the breeding population within the sanctuary. With these goals in mind, a large sanctuary was established and its boundary agreed upon. You were put in charge of the sanctuary. Your assigned task was to do nothing and to make certain no one else did anything either.

The sanctuary functioned properly for generations. The hunting grounds beyond the protected area remained rich, and people colonized these areas, although they remained somewhat nomadic. People began to depend on the sanctuary for excess animals. You were proud of the fact that people, for the most part, respected the boundary of the sanctuary and your authority. You became a legend, and you passed your position and your duties to your offspring.

Slowly the climate began to change. Whereas the western park was once a dry savanna, it was now holding more water. The plant community began to change. Wetland plants began moving in, and dry savanna-adapted plants no

TABLE 13.1   Pleistocene Mammals Now Extinct in Florida[a]

Species	Status	Species	Status
Vampire bat (*Desmodus stocki*)	P	Mustelid (*Trigonictis macrodon*)	E
Chipmunk (*Tamais striatus*)	P		
Armadillo (*Dasypus bellus*)	E	Cheetah (*Felis inexpectata*)	E
Armadillo (*Kraglievichia paranenese*)	E	Jaguar (*Pantera onca*)	P
Armadillo (*Holmesina septentriohalis*)	E	Ocelot (*Felis pardalis*)	P
Glyptodont (*Glyptotherium arizonae*)	E	Jaguarundi (*Felis yagouroundi*)	P
Ground sloth (*Megalonyx leptostomus*)	E	Lion (*Pantera atrox*)	E
Ground sloth (*Megalonyx wheatleyi*)	E	Sabertooth cat (*Smilodon gracilus*)	E
Ground sloth (*Eremotherium mirabile*)	E		
Ground sloth (*Glossotherium chapadmalensis*)	E	Sabertooth cat (*Smilodon floridanus*)	E
Ground sloth (*Glossotherium harlani*)	E	Scrimitat (*Homotherium serum*)	E
Ground sloth (*Nothrotheriops* sp.)	E	Hyena (*Chasmaporthetes ossifragus*)	E
Pocket gopher (*Geomys propinetis*)	E		
Pocket gopher (*Geomys orientalis*)	E	Mammoth (*Mammuthus imperator*)	E
Porcupine (*Erithizon kleini*)	E		
Porcupine (*Erithizon dorsatum*)	P	Elephant (*Cuvieronius* sp.)	E
Capybara (*Hydrochoerus holmesi*)	E	Tapir (*Tapirus veroensis*)	E
Vole (*Pitymysearatai* sp.)	E	Horse (*Equus* sp.)	E
Vole (*Microtus pennsylvanicus*)	P	Peccary (*Platygonus* sp.)	E
Muskrat (*Ondarta idahoensis*)	E	Peccary (*Mylohylus nasutus*)	E
Muskrat (*Ondarta zibethicus*)	E	Llama (*Hemiauchenia macrocephala*)	E
Lemming (*Synaptomys australis*)	P		
Cotton rat (*Sigmodon curtisi*)	E	Llama (*Palaeolama mirifica*)	E
Cotton rat (*Sigmodon bakeri*)	E	Deer (*Blastocerus* sp.)	E
Wolf coyote (*Canis edwardi*)	E	Pronghorn (*Capromeryx arizonenesis*)	E
Dire wolf (*Canis dirus*)	E		
Bear (*Tremarctos floridanus*)	E	Ox (*Euceratherium* sp.)	E
Skunk (*Conepatus leuconotus*)	P	Bison (*Bison antiquus*)	E

[a] Each species listed is either extinct (E) or exists elsewhere outside of Florida (P).

longer were common. The animal community also responded to these changes. Stories of the sanctuary, a verbal history of the area handed down from you to your offspring and in turn to their offspring, told of the western boundary being a dry savanna. Now the area was more like a wetland. The eastern boundary of the sanctuary had not changed, but because of the changes on the western boundary, animals within the park were utilizing the eastern part of the protected area more frequently. When the animals ventured beyond the agreed upon boundary that was now clearly marked by human settlements, they were quickly dispatched and eaten. Indeed, so many people had moved in on the eastern border that poaching along the full extent was becoming a problem.

Year after year, decade after decade, the western wetland invaded eastward. What was once the far western part of the sanctuary became a beach where sea turtles nested. The sanctuary was becoming one dimensional, extending north and south. The east-to-west extent was shrinking. From human generation to human generation, the sea advanced westward and the sanctuary began losing its wildlife. Mastodonts were the first to disappear. The large ground sloths were next. Other animals were only rarely seen. The plant community extended well beyond the eastern boundary of the sanctuary but had been altered by humans. The animals within the sanctuary were trapped between the advancing sea and humans. They were doomed. Your offspring concluded that no sanctuary would be big enough given that the climate was changing and the human population was growing.

Although the above scenario is completely fictitious, the conclusions your offspring reached are valid. Boundaries that keep humans out and keep wildlife in won't work over the long run. The national parks and preserves in place today are cages that must be artificially maintained to survive. The laws that established and maintained these parks are not going to change this human generation. Your task will be to undo these laws and establish new laws that allow people and parks to co-exist across the landscape—all across North America.

Sandhill Cranes

On April 1, 1996, the front page of the *Tampa Tribune* reported that Florida sandhill cranes (*Grus canadensis pratensis*) were showing up in people's backyards. It was no April Fool's joke. The Florida Game and Freshwater Fish Commission estimates that there are between 4000 and 6000 Florida sandhill cranes left. The Florida sandhill crane is a recognized subspecies of sandhill crane; unlike other sandhill cranes, it does not migrate out of Florida to breed. That fact that cranes are showing up in residents' backyards might be a joy for local people, but it sounded an alarm for some ornithologists. More and more sandhill cranes are showing up in urban setting such as golf courses, highway medians, and backyards. There could be several reasons for their appearance.

One reason cranes are more commonly seen is that their population is increasing. Sandhill cranes typically lay two eggs and raise two chicks each year. Another reason might be native habitat destruction. In Pasco County near

Tampa, one ornithologist estimated that up to 100 crane nesting sites remain today. There were thousands of nesting sites a hundred years ago. Sandhill cranes require natural wetlands with short grass to nest and feed. The South Florida Water Management District reported that overpumping of groundwater sources has led to a decrease in the number of local wetlands in Hillsborough, Pasco, and Pinellas (near Tampa) counties. The wellfield drawdown is the number one concern of the Tampa Audubon Society. Entire wetland communities, of which the cranes are a visible part, are disappearing. Wellfield drawdown, however, is a symptom of a more widespread problem.

The reason the sandhill cranes are showing up along highway medians is that water runoff from highways collects in the medians, forming ephemeral wetlands. The cranes then feed in these temporary wetlands. Fortunately, the cranes also nest in other parts of Florida, but this is just one more setback for cranes. Slowly, the crane's habitat is being lost to overdevelopment. Subdivisions, shopping malls, ranchettes, and fast-food restaurants are springing up all over Florida. Most of the land in Florida is privately held. Once development occurs nearby, cashing out becomes the hope of many large tract holders. Taxes increase near developments, making it impossible to make a living from ranching or farming. Habitat destruction due to overdevelopment is the most important problem for the sandhill cranes and for most species living in Florida.

Freshwater Fish

On September 26, 1995, a Florida Game and Fresh Water Fish Commission news release reported than an investigation into the sale of illegal game fish resulted in 27 arrests. Two days earlier, wildlife officers had fanned out across four Florida counties and issued citations to individuals selling bluegill, crappie, channel catfish, largemouth bass, sunfish, striped bass, and sunshine bass. They had exceeded the state-established bag limits, most often deliberately. The individuals arrested had buyers lined up before they ever caught fish, and some even made repeated trips while stashing fish at home or other locations. Officers purchased more than 1100 pounds of fish.

Citizen complaints tipped the officers off. Most complaints of illegal commercialization of freshwater fish are investigated, and steps are usually taken to halt the practice immediately. The charges range from second-degree misdemeanors to infractions. The penalty for a misdemeanor is a fine of $500 and/ or 60 days in jail. An infraction carries a preset fine and court costs. Most people who enjoy recreational fishing do so fully within the law. However, if a few wildlife officers can make 27 arrests in two days, the problem of illegal commercial fishing is probably seriously impacting Florida's freshwater fish.

Florida Panthers

No other animal in Florida symbolizes wildlife issues more than the Florida panther. Coyotes are now invading Florida, and ranchers are shooting them on sight. So-called coy-dogs (a cross between domestic dogs and wild coyotes) are rumored to be domestic dogs by day and evil coyotes by night, striking fear in anyone caught outside after dark! Yet it is the Florida panther that is most hated by some who fear big government and most beloved by others as a symbol of wild places, of what once was in Florida and what could be again.

U.S. Fish and Wildlife Service authorities introduced six Texas panthers into the wilds of northern Florida to test whether panthers could survive there. Several succeeded, but all were eventually removed. The male panthers were all neutered by professionals—well, almost all were neutered. At the time of his capture, Waldo was a free-ranging panther in north Florida near the small town of Waldo. Waldo's genetic makeup indicated that he was of Texas panther descent. His capture was disappointing to many residents and "Free Waldo" T-shirts announced their ire. Many called for Waldo's release, including U.S. Senator Connie Mack (R-Fla.), who questioned the legality of removing an endangered species from Florida. The director of the State Division of Wildlife announced that "its existence near Waldo did not occur naturally, and would have been potentially harmful to Florida wildlife if we had not removed it from the wild."

There is no doubt that panthers are carnivores, and hence some wildlife are harmed by panthers. Let's just admit that panthers eat other wildlife for food. A U.S. Fish and Wildlife Service manager working on the project had earlier been quoted as declaring that "all panthers born in Florida are Florida panthers"—except for Waldo apparently. The contradictions are transparent in so volatile an issue as the Florida panther. And this is just the tip of the iceberg. Six pure-bred, authentic "born in Florida" Florida panthers are caged at a private state-of-the-art captive breeding facility in north Florida. Males are separated from females, and they are prevented from breeding. Many people are asking why. The facility has been prevented from breeding them and is now asking the federal government to find another home for its most controversial occupants.

Some 50 panthers roam freely in Florida, mostly on private land. Habitat destruction is the principal cause of the panther's decline. One hundred years ago, panthers were found from one end of Florida to the other and from coast to coast. So was the eastern timber wolf. Bears were ubiquitous. Bobcats are still everywhere. And there were no coyotes. Some suggest the Florida panther suffers from inbreeding. Some panthers have crooked tails and cowlicks. Some males have only one descended testicle. Without at least one, they would be sterile. Panthers fight for territories, and males kill other males in their territory.

The few that remain are found east of Tampa. Florida has not been kind to panthers.

Cramer (1995) surveyed north Florida residents as part of the Northeast Florida Panther Education Program sponsored by the Florida Advisory Council on Environmental Education. From February 1993 to July 1995, a small population of radio-collared Texas panthers was released into the Pinhook Swamp area near the Georgia–Florida border. Cramer found that fully 73% of the residents favored panther reintroductions into their own county or surrounding counties. Fully 49% felt there should be land development restrictions to preserve natural panther habitat, while 86.7% of the respondents felt that saving wildlife habitat was a good idea. Tax breaks and cooperation from regulatory agencies (state and federal) were incentives favored by 81% of the residents.

Despite the overwhelming apparent support for the panther reintroduction program, Cramer also revealed many interesting problems. Landowners with more than ten acres (31%) were less supportive of panther reintroduction. Education level played a major role in people's decisions. People with less education showed the least support for the program (55%), the highest level of ignorance of the program (30%), and the highest outright opposition (15%) to the reintroduction program. Those people with a high-school degree or more favored the program (85%). More men (62.6%) felt that panthers were probably not a threat to deer populations, while only 42.5% of women felt the same way. More women (53%) than men (41.5%) felt that panthers were a threat to children.

A closer look at the question of preserving habitat for wildlife revealed an interesting trend. When asked whether the responsibility for habitat preservation was the landowner's, 100% of respondents with ten acres or more used for residential purposes agreed that saving habitat was their responsibility, whereas 94.1% of those who used their land for forestry agreed, 86.7% of those who used their land for ranching agreed, and 72.0% of those who used their land for farming agreed. Clearly, the more land was used, the less responsibility for habitat preservation people were willing to accept. Nevertheless, seven out of ten agreed that it was an important responsibility of the landowner.

Demographically, respondents 18 to 49 years old supported the panther reintroduction program in higher percentages than did those over 49 years of age. Those 60 and over saw the panther as a threat. All landowners with ten or more acres were asked if they would allow certain species to live on their land. Again the trend was clear. Fully 86.5% would permit gopher tortoises, 92.2% red-cockaded woodpeckers, 93.8% fox squirrels, 84.4% otters, 64.6% bobcats, 57.8% panthers, and 53.1% black bears. Only 11% of respondents said they presently did not take any actions to benefit wildlife.

Specialty license plates in Florida are carried on 12% of registered vehicles, and at $27 each, total revenues were nearly $4 million for the state coffers in 1995. Since its introduction in 1990, 217,737 Florida panther specialty license plates have been sold, raising $13.2 million. The Challenger license plate, introduced in 1987, has sold 597,685 plates and raised $28.7 million. The manatee plate is second at 323,288 plates and $14.4 million raised. Fees vary from plate to plate, but the popularity of specialty plates is so broad that there are 36 from which the public can choose. Although each new introduction dilutes the effectiveness of each individual plate, the manatee and the panther both continue to be leading license plates, which indicates that these two popular species will continue to be of concern to Florida residents.

The Florida panther issue is extremely complex, and books have been written describing efforts to save the species (Alvarez 1993). Dave Maehr was chief of the state game commission for eight years. He suggested that the commission's strategy to interbreed Texas panthers with Florida panthers was poorly conceived (Maehr personal communication). Data show that the native Florida panthers are reproducing just fine. Who cares if a panther has a crooked tail or a cowlick, as long as the females continue to have an average of 2.3 cubs per year. Maehr claims that habitat destruction, in part caused by poor water management, is the panther's principal enemy. Furthermore, more panthers are located on private rather than state or federal lands, and hence more money should be spent on creating beneficial relationships with local landowners. The director of research of the Florida Game and Fresh Water Fish Commission agrees that spending any additional funds on a population of panthers that is expected to remain at about 50 for the foreseeable future is irresponsible. Yet in 1995 alone, Florida specialty license plates generated about $1.7 million to save the panther. That kind of money is worth fighting over even if the panther cannot be saved. Thus, political problems are bound to occur.

In 1993, a woman jogger in New Mexico was threatened by a panther. Panthers are found throughout New Mexico. Several residents tried to get a glimpse of the cat but failed. Indeed, seeing a panther in the wild is a unique experience, and most people who enjoy the outdoors would welcome the opportunity to see a panther. However, most backpackers in New Mexico will never see a panther, although the chances are excellent that a panther has seen them. As one professor at the University of Florida bluntly put it, if we are worried about people being killed, we should do away with all cars. In fact, you probably have a better chance of finding a winning lottery ticket on the street than of being killed by a panther in Florida. Panthers simply avoid people.

The Arizona State Game and Fish Department estimates that 3000 panthers live in Arizona. Until 1970, the state paid a bounty of $200 for each panther killed. A few people in Arizona consider panthers to be a nuisance. Ranchers routinely hire private hunters to kill panthers.

Frog Gigging

On April 29, 1996, an article on frog gigging appeared in the *Miami Herald.* Two days later, another article reported that frog gigging in Big Cypress National Preserve would be outlawed the following day. The disclosure that poachers (a word not used in the April 29 article) had been hauling tons of frogs a month out of the national preserve had caused a public outcry. Commercial giggers had plundered too much, too often. Also, two endangered birds that used the marshes were a consideration.

Sportsmen and conservationists were outraged by the previous report and demanded the preserve be closed to such activities. The preserve is 2400 square miles and supports hunting, fishing, and off-road vehicles. It took but two days of public pressure for the chief ranger to close the preserve to gigging. Intense, relentless public pressure did what scientists could not do. The South Florida Sportsmen's Association was among the groups calling for an end to the pillaging of resources. The Everglades Coordinating Council, a coalition of 14 sportsmen's groups, had called for statewide regulation of frogging.

The chief ranger suggested that the ban will eventually be lifted when limits are placed on the number of frogs giggers can take. Other areas of the state have already established limits. For instance, no more than 50 frogs a night can be taken from the Miccosukee Indian Reservation.

As is the case in many instances, scientists understand the intricate food webs of aquatic systems and the role each animal and insect plays, but it is the public that tells the park service what to do. When the public gets behind a wildlife issue, one way or another the issue will be resolved. What the public demands, the public gets.

Rabies and Small Mammals

With the loss of Florida's wolves and panthers, small mammals have been able to thrive. Raccoons, opossums, armadillos, bobcats, feral hogs, and foxes are seen often. They have no top predator to control their populations, and some of them are able to profit from a landscape dominated by humans. Raccoons are ubiquitous in Florida. Many people feed raccoons in the mistaken belief that they are helping to save wildlife, when in reality the opposite is true. With the rise of so-called meso-mammals, eggs of ground-nesting birds are more likely to be eaten. Even the eggs of some birds of prey are eaten by raccoons.

Alligators are top predator in Florida wetlands. Although female alligators guard their nests, raccoons often prey on the eggs. There has been a role reversal in Florida. The alligator eggs are eaten by raccoons, but because there are fewer alligators, fewer raccoons fall prey to them. Raccoons in the drier parts of Florida find gopher tortoise eggs a delicacy. Some raccoons have learned that the tortoises carry the eggs around inside them before they are laid. The raccoons attack the tortoises for their eggs and in doing so kill the tortoises. Gopher tortoises are threatened in Florida.

On January 6, 1996, the *Gainesville Sun* reported that cases of rabies were on the rise in 1995. Thirteen cases of rabies were reported in Alachua County in the first seven months of 1995. Aside from two cases in a neighboring county, there were no cases of rabies in Alachua County in the 18 previous months before the first reported case. There were 251 cases of rabies in 1995 and 258 cases in 1994. Of the 1995 cases, 184 were from raccoons, 18 from foxes, and 27 from domestic cats. The domestic cats may have contracted rabies from encounters with raccoons.

With the increasing numbers of raccoons and other meso-mammals, disease outbreaks such as rabies are more common. When populations increase beyond historical limits, aided in part by humans, their density becomes too high. Viral epidemics such as rabies pass through a population quickly, knocking the numbers, and hence the density, back.

Red-Cockaded Woodpeckers

Old-growth longleaf pine forests are the home of the red-cockaded woodpecker (RCW, *Picoides borealis*), a federally endangered species. Of course, longleaf pine forests are themselves rare. Longleaf pine forests once covered extensive areas of the southeastern United States and much of Florida. Trees more than a meter thick were common. Today, the longleaf pine forests are disappearing. RCWs are primary cavity nesters; that is, over a period of months, they excavate their own nest sites only in living trees. They create holes in trees to nest, which other birds can occupy after the RCWs leave.

RCWs have declined throughout their range because of extensive loss of nest trees. They usually nest in trees more than 80 years old. Traditional forestry practices require harvesting trees between 40 and 50 years old. Thus, the trees favored by the birds have disappeared. Furthermore, other less favorable tree species such as slash pine have been planted. Harvesting trees and replanting with other species more favorable to harvesting has led to a steady decline in RCW numbers. There are an estimated 700 nesting pairs of RCWs in the entire state of Florida and 2000 in the southeastern United States, their traditional range.

RCWs are cooperative breeding birds. That is, young birds from the previous brood help to raise the young of their parents. In the morning, the RCWs all fly off together to feed. They return to their cavity nests just before the sun sets. The sight of them returning to the nest site is an experience enjoyed by a decreasing number of people. The squawking birds all fly into the nest stand together, land on the sides of their respective trees, and immediately hop into their cavities. RCWs are about 20 feet off the ground. One of North America's most beautiful birds, the RCW is a symbol of a healthy forest with many old-growth trees and snags. The U.S. Forest Service plans to regulate timber cutting on more than 3000 square miles of land around nesting sites in an attempt to increase RCW numbers. Also, the U.S. Fish and Wildlife Service plans to replace harvested areas with slower growing longleaf pines. The Florida Forestry Association has suggested that such a policy is unfavorably restrictive.

APPENDIX 13.1: MAINE'S THREATENED AND ENDANGERED SPECIES AND THEIR FEDERAL STATUS AS OF SEPTEMBER 1, 1995[a]

Endangered species

Birds
Peregrine falcon (*Falco peregrinus*)*
Golden eagle (*Aquila chrysaetos*)
Piping plover (*Charadrius melodus*)**
Least tern (*Sterna antillarum*)
Roseate tern (*Sterna gougallii*)*
Sedge wren (*Cistothorus platensis*)
Grasshopper sparrow (*Ammodramus savannarum*)
Eskimo curlew (*Numenius borealis*)**?

Mammals
Gray wolf (*Canis lupus*)**?
Eastern panther (*Felis concolor couguar*)*?
Right whale (*Eubalaena glacialis*)**
Humpback whale (*Megaptera novaeangliae*)**
Finback whale (*Balaenoptera physalus*)**
Sperm whale (*Physeter catodon*)**
Sei whale (*Balaenoptera borealis*)**

Reptiles
Leatherback turtle (*Dermochelys coriacea*)**
Atlantic Ridley turtle (*Lepidochelys kempi*)**

Box turtle (*Terrapene carolina*)
Black racer (*Coluber constrictor*)

Fish
Shortnose sturgeon (*Acipenser brevirostrum*)**

Beetles
American burying beetle (*Nicrophorus americanus*)**?

Threatened species

Birds
Bald eagle (*Haliaeetus leucocephalus*)*

Mammals
Northern bog lemming (*Synaptomys borealis*)

Reptiles
Loggerhead turtle (*Caretta caretta*)*
Blanding's turtle (*Emydoidea blandingii*)
Spotted turtle (*Clemmys guttata*)

---

[a] * = federally threatened, ** = federally endangered, ? = presence in Maine uncertain.

Source: Wildlife Division Research and Management Report 1996.

APPENDIX 13.2: NEW JERSEY'S THREATENED AND ENDANGERED SPECIES AND THEIR FEDERAL STATUS AS OF JUNE 3, 1991[a]

Endangered species

Birds
Pied-billed grebe (*Podilymbus podiceps*)
Bald eagle (*Aquila chrysaetos*)*
Northern harrier (*Circus cyaneus*)
Cooper's hawk (*Accipter cooperii*)
Red-shouldered hawk (*Buteo lineatus*)
Peregrine falcon (*Falco peregrinus*)**
Piping plover (*Charadrius melodus*)*
Upland sandpiper (*Bartramia longicuda*)
Least tern (*Sterna antillarum*)
Roseate tern (*Sterna gougallii*)**
Black skimmer (*Rynchops niger*)
Short-eared owl (*Asio flammeus*)

Sedge wren (*Cistothorus platensis*)
Loggerhead shrike (*Lanius ludovicianus*)
Vesper sparrow (*Pooecetes grammineus*)
Henslows's sparrow (*Ammodramus henslowii*)

Mammals
Bobcat (*Lynx rufus*)
Eastern wood rat (*Neotoma floridanar*)
Sperm whale (*Physeter catodon*)**
Finback whale (*Balaenoptera physalus*)**
Sei whale (*Balaenoptera borealis*)**
Blue whale (*Balaenoptera musculus*)**
Right whale (*Eubalaena glacialis*)**
Humpback whale (*Megaptera novaeangliae*)**

Reptiles
Bog turtle (*Clemmys muhlenbergi*)
Atlantic hawksbill (*Eretmochelys imbricata*)
Leatherback turtle (*Dermochelys coriacea*)**
Atlantic Ridley turtle (*Lepidochelys kempi*)**
Loggerhead turtle (*Caretta caretta*)*
Corn snake (*Elaphe guttata*)
Timber rattlesnake (*Crotalus horridus*)

Amphibians
Tremblay's salamander (*Ambystoma tremblayi*)
Blue-spotted salamander (*Ambystoma laterale*)
Eastern tiger salamander (*Ambystoma tigrinum*)
Pine Barrens treefrog (*Hyla andersonii*)
Southern gray treefrog (*Hyla chrysoscelis*)

Fish
Shortnose sturgeon (*Acipenser brevirostrum*)**

Threatened species

Birds
American bittern (*Botaurur lentiginosos*)
Great blue heron (*Ardea herodias*)
Little blue heron (*Egretta caerulea*)
Yellow-crowned night heron (*Nyctanassa violaceus*)
Osprey (*Pandion haliaetus*)
Northern goshawk (*Accipter gentilis*)
Red-shouldered hawk (*Buteo lineatus*)
Black rail (*Laterallus jamaicensis*)
Long-eared owl (*Asio otus*)

Barred owl (*Strix varia*)
Red-headed woodpecker (*Melanerpes erythrocephalus*)
Cliff swallow (*Hirundo pyrrhonota*)
Savanna sparrow (*Passerculus sandwichensis*)
Ipswich sparrow (*Passerculus sandwichensis princeps*)
Grasshopper sparrow (*Ammodramus savannamu*)
Bobolink (*Dolichonyx oryzivrus*)

Reptiles
Wood turtle (*Clemmys insculpta*)
Atlantic green turtle (*Chelonia mydas*)
Northern pine snake (*Pituophis melanoleucus*)

Invertebrates
Mitchell's satyr (butterfly) (*Neonympha mitchellii*)**
Northeastern beach tiger beetle (*Cicindela dorsalis*)
American burying beetle (*Nicrophorus americanus*)**
Dwarf wedge mussel (*Alasmidonta heterodon*)

Amphibians
Long-tailed salamander (*Eurycea longicauda*)
Eastern mud salamander (*Pseudotriton montanus*)

---

[a] * = federally threatened, ** = federally endangered.

Source: New Jersey Department of Environmental Protection 1995.

EXERCISES

13.1 Devise a program to reestablish the walrus on Maine's coastline. What groundwork needs to be done before walrus are released? How will this program be paid for? Would people in Maine donate money to reintroduce the walrus? Create a "Bring Back the Walrus" Web page.

13.2 Find out how Pittman–Robertson Act funds are being used in your state. How much is spent on nongame species? Are insects afforded treatment similar to hunted species? Explain your answer.

13.3 Are there plans in your state to protect salamanders and toads? How about spiders, snails, or moths?

13.4 What other states have too many deer?

13.5 The U.S. Fish and Wildlife Service is the principal agency through which the federal government carries out its responsibilities to conserve, protect, and enhance the nation's fish and wildlife and their habitats for the continuing benefit of people. Explore the possibility of becoming a summer volunteer with the U.S. Fish and Wildlife Service. Volunteers play an important role in aiding the U.S. Fish and Wildlife Service to conserve and protect wildlife in the United States. Individuals 18 or older can volunteer to participate in a variety of programs, including research programs. Activities include bird banding, photographing natural and cultural resources, working as a technician, leading tours, and conducting fish and wildlife population surveys. Many of these skills are necessary for a career in wildlife biology. Furthermore, volunteers help decide which program they will participate in. The service has many field stations around the country that can use your skills.

13.6 Examples of state allocations for fiscal year 1996 in million of dollars are as follows:

State	Sport fish restoration	Wildlife restoration
Arizona	4.1	4.8
New Mexico	3.6	4.2
Oklahoma	3.8	3.8
Texas	10.0	9.8

Every year Texas receives the maximum apportionment allowable for a state under the Federal Aid in Sport Fish and Wildlife Restoration acts. The Federal Aid funds are divided among states using formulas based upon land area, population, and licensed sportsmen. How much did your state receive? What programs can your state invest in to increase its allocation?

13.7 Some states continue to allow bear baiting and the use of dogs to hunt bears. For example, out of 2645 bear taken in Maine in 1995, at least 2020 were taken by baiting, 329 were taken by hunters using dogs, and 25 were trapped. Other states have outlawed bear baiting and the use of dogs to hunt bears. Florida closed bear hunting altogether because hunters and road mortality threatened the state's bear population. Gather information on bear hunting in other states. Are bear baiting and the use of dogs on the way out?

13.8 A native of the Mediterranean, mouflon are wild sheep. Where are they legally hunted in the United States? These sheep may be damaging native vegetation in sensitive areas. What can and should be done with these exotics?

13.9   Ring-necked pheasants of Asian origin were stocked in Oregon in 1881. They have become a popular gamebird in many states. Do these birds compete for food with any native species? What rules govern hunting of pheasants in your state?

13.10   Chukar partridges (*Alectoris chukar*) range from Turkey to throughout Asia. The Indian subspecies found in the mountains of Pakistan and northern India was successfully introduced in the United States in 1893 when five pairs were released. By 1968, 806,000 chukar had been released, some even in Hawaii. Create a distribution map for the chukar. Why has this popular gamebird not spread throughout the United States? Begin with reading about the chukar in Kallman (1987).

13.11   Whooping cranes are being released in Florida by the U.S. Fish and Wildlife Service. Investigate the status of whooping cranes using the World Wide Web and a conventional library, and summarize your findings.

13.12   Find out more information on the Florida Manatee Sanctuary Act of 1978. What does the act provide for? Are fines assessed for violating the act?

13.13   Do a more complete study of manatee introductions and reintroductions. Do reintroductions succeed more often than introductions? Offer an explanation.

13.14   Armadillos invaded eastward from Texas and into Florida. How did they cross the Mississippi River? Investigate this fascinating story of the successful armadillo invasion. What was it about the breeding biology of this diminutive creature that helped it succeed?

13.15   Investigate the coyote invasion into Florida. Why is this happening? Can it be prevented? Why didn't coyotes invade 100 years ago?

13.16   Find out how specialty license plate money is spent in your state. How much money really goes to conservation of the species on the license plate?

13.17   Speculate on the complete loss of a primary cavity maker such as the red-cockaded woodpecker. What effect would its demise have on other birds?

13.18   The Clinton administration wants to restore the Florida Everglades to its former glory and bring back the massive colonies of wading birds that once existed. How many species of wading birds use the Everglades, and what were their populations in 1850? Trace the history of the wading birds of south Florida. Why are there so few today?

LITERATURE CITED

Ackerman, B.B., S.D. Wright, R.K. Bonde, D.K. Odell, and D.J. Banowetz. 1995. Trends and patterns in mortality of manatees in Florida, 1974–1992. *in* O'shea, T.J., B.B. Ackerman, and H.F. Percival (Eds.). *Population Biology of the Florida Manatee.* National Biological Service, Washington, D.C.

Alvarez, K. 1993. *Twilight of the Panther.* Myakka River Publishing, Sarasota, Florida.

Beeman, K. 1996. Manatees dying less frequently. *Tampa Tribune* May 2.

Cramer, P. 1995. The Northeast Florida Panther Education Program. Florida Advisory Council on Environmental Education, Tallahassee.

Hoffmeister, J.E. 1974. *Land from the Sea.* University of Miami Press, Coral Gables.

Hutchinson, A.E. 1996. Annual Report, 1995. Maine's Rare & Endangered Wildlife. Maine Department of Inland Fisheries and Wildlife, Bangor.

Kallman, H. (Ed.). 1987. Restoring America's Wildlife, 1937–1987: The First 50 Years of the Federal Aid in Wildlife Restoration (Pittman–Robertson) Act. Fish and Wildlife Service, U.S. Department of the Interior, Washington, D.C.

Landscape Project. 1993. New Jersey Department of Environmental Protection, Division of Fish, Game, and Wildlife, Trenton.

McCollough, M.A. 1990. Final Report, Maine Caribou Project 1986–1990. Maine Department of Inland Fisheries and Wildlife, Bangor.

McCollough, M.A. and B.A. Connery. 1991. An Attempt to Reintroduce Woodland Caribou to Maine 1986–1990. Maine Department of Inland Fisheries and Wildlife, Bangor.

New Jersey Department of Environmental Protection. 1995. Annual Report 1994–95. Division of Fish, Game, and Wildlife, Trenton.

O'shea, T.J., B.B. Ackerman, and H.F. Percival (Eds.). 1995. *Population Biology of the Florida Manatee.* National Biological Service, Washington, D.C.

Peet, J. 1997. No beast is a burden to them. *The Star Ledger* February 18:1.

Pennsylvania Game Commission. 1996. What You Should Know About the Pennsylvania Game Commission. Pennsylvania Game Commission, Harrisburg.

Pielou, E.C. 1991. *After the Ice Age.* University of Chicago Press, Chicago.

Wildlife Division Research and Management Report. 1996. Maine Department of Inland Fisheries and Wildlife, Bangor.

Wright, D.D., B.B. Ackerman, R.K. Bonde, C.A. Beck, and D.J. Banowetz. 1995. Analysis of watercraft-related mortality of manatees in Florida, 1979–1991. *in* O'shea, T.J., B.B. Ackerman, and H.F. Percival (Eds.). *Population Biology of the Florida Manatee.* National Biological Service, Washington, D.C.

FROM EXPLOITATION TO REINTRODUCTION: WILDLIFE IN THE WESTERN UNITED STATES

14

Wildlife programs in Maine, New Jersey, Pennsylvania, and Florida share many common problems, though each state must deal with unique problems as well. The attitude toward predators in particular is quite diverse. While voters in California have mandated physical protection for panthers and committed to spend $30 million a year on habitat purchase programs, people in Florida feel quite differently. In the Northeast, top carnivores are viewed as politically unacceptable. In Idaho, Montana, Wyoming, Wisconsin, Michigan, and Minnesota, wolves are coming back. In this chapter, we continue our investigation of wildlife programs in California, Arizona, Colorado, and New Mexico.

CALIFORNIA

Pittman–Robertson Act funding helped start six projects in California in 1940. The first projects concentrated on games species such as beaver, mule deer, other furbearers, sage grouse, and quail. The impact of agricultural practices, including the use of fertilizers and pesticides, on game species was a highlight of the state's early wildlife programs. Critical habitat was also purchased with the aid of federal funding. By 1987, Tule elk and bighorn sheep were being reintroduced into their historical ranges and their populations were increasing.

285

In 1968, California established the Nongame Wildlife Investigation Project and work began on the California condor and San Joaquin kit fox. As a precursor to the state's current endangered species program, California led the way in what has become the frontier of nongame species conservation efforts nationally. Since 1979, Pittman–Robertson Act funding has been the sole source for work on such species as golden eagles, sandhill cranes, bobcats, and spotted owls as well as other important species.

California was also the second state to establish a hunter safety program. In many ways, California's visionary programs have now become routine across most of the United States. Indeed, one must consider what is happening in California as a precursor to future wildlife programs in the United States.

California's geographic location and extent have contributed to the large biological diversity the state claims. More than 5200 different species of plants are found in California, 30% of which are endemic. There are 63 freshwater fishes, 46 amphibians, 96 reptiles, 563 birds, 190 mammals, and more than 30,000 species of insects. Since 1850, 23 native vertebrates and 40 native plants have passed into extinction in California. Today, 20% of the state's vertebrate species are listed as threatened or endangered. California, however, has an aggressive program to protect biodiversity (Steinhart 1990).

California's role in bringing back wildlife has expanded significantly since Pittman–Robertson funding began, because protecting biodiversity means protecting more than games species. California believes that biological resources must be managed across landscapes. Recall that landscapes are collections of ecosystems.

In 1984, the state mandated the California Endangered Species Act and expanded the role of the Department of Fish and Game in protecting biodiversity statewide. Among other declarations, the act stated that certain species of wildlife, fish, and plants were threatened with extinction and therefore required special attention for aesthetic, ecological, economic, educational, historical, recreational, and scientific reasons. The habitat of these species required statewide concern.

Three years later, in 1987, Defenders of Wildlife claimed that California spent five times more on nongame than game species and committed to those species a higher portion of its budget than any other state except one. The human population in California is about 32 million and growing. Thus, given California's rich biodiversity and large human population, it is not surprising that the state has more species under consideration as candidates for federal listing as threatened and endangered than any other state. In that same year, a report written for the California Senate Committee on Natural Resources and Wildlife created guidelines that serve as the foundation for an expanded nongame species protection program (Annual Report on the Status of California State Listed Threatened and Endangered Animals and Plants 1992).

In 1991, the California Executive Council on Biological Diversity was founded. The council consists of representatives from federal and state natural resource agencies, the University of California, and county boards of supervisors. The state was divided into ten bioregions established by drainage, topography, habitat type, and climate, that is, landscapes. The council's mission is to ensure that economic development is compatible with the conservation of biodiversity. California's biodiversity is indeed impressive. Appendix 14.1 lists the endangered and threatened species in California.

Many of California's threatened and endangered species are riparian specialists. Since the state has lost 95% of its riparian habitat and at least 50% of the endangered, threatened, and special concern land birds in the state require riparian habitat, clearly something has to be done to reestablish riparian areas throughout the state quickly.

The panther (puma, mountain lion) has a political history in California. Form 1907 to 1963, panthers carried a bounty and were shot as vermin. Some 12,461 were shot during that time (Torres et al. 1996). From 1963 to 1968, panthers were managed as a nongame species. From 1969 to 1972, panthers were classified as game animals and 118 were taken by hunters. In 1972, a moratorium on hunting was put in place which lasted until 1986, when the panther again became a game animal. Successful court challenges in 1987 and 1988 led to passage of Proposition 117, which designated the panther as a specially protected species in California in 1990. Proposition 197, which would have allowed the panther to become a game animal, was defeated in March 1996, and the panther remains a protected species in California today. Accordingly, panther management goals in California include (1) ensuring that viable populations exist, (2) minimizing conflicts with humans and their property, (3) protecting important habitats, (4) recognizing their ecological role and value, (5) monitoring populations and conducting research, and (6) educating the public.

The simple fact is that Californians want panthers in their state and are leaving places for panthers to co-exist with humans. This progressive attitude is typical of Californians, and invariably these values spread eastward over time. Humans have come in contact with panthers in California, and at least one person was attacked and killed by a panther. Many Californians realize, however, that automobiles are far more dangerous than panthers and that the likelihood of serious injury from a motor vehicle far exceeds the same possibility from any wild animal. No one has called for a ban on automobiles. In California, the most populous state in the United States, panthers are there to stay and will probably remain protected. Contrast this attitude with that found in Florida.

ARIZONA

Federally supported programs got under way in Arizona in 1939, when projects were launched to study pronghorn antelope, wild turkey, quail, and beaver. One interesting study began in January 1971, when 11 deer were enclosed in a 600-acre predator-free environment. Comparison studies were begun in similar habitats where bears, coyotes, and panthers were present. By December 1975, there were 37 deer in the enclosure. Predators (or some other means of population control) were obviously necessary to keep herds in check. Other studies demonstrated that deer, not cattle, were responsible for the lack of browse in semi-arid environments in the state. Radio tracking of bears equipped with transmitters was demonstrated in Arizona as well. Appendix 14.2 lists Arizona's species of special concern

Arizona Heritage Fund

On November 6, 1990, by a margin of almost two to one, Arizona voters established the Heritage Initiative, which allocates $20 million each year from state lottery funds for wildlife and recreational purposes. Heritage funds are divided equally between the Arizona Game and Fish Department and Arizona state parks. Other funds are provided by the federal government through Pittman–Robertson and the Nongame Income Tax Checkoff program. In 1994–1995, sportsmen provided about $15 million and federal programs provided $13 million; clearly, the impact of Heritage Fund contributions is substantial. Total spending by the Game and Fish Department was $41.9 million in 1994–1995.

The Heritage Fund allocations are specifically assigned to support different programs. The fund provides $3.6 million for the identification, inventory, acquisition, protection, and management of Arizona's biodiversity; $2.4 million is set aside for habitat acquisition, $1.5 million for urban wildlife, $1.5 million for habitat protection, $500,000 for environmental education, and $500,000 to provide information on public access for recreational use of public lands.

Heritage funds have been used to support a variety of programs, including studies of the desert tortoise, Sonoran topminnow (*Poeciliopsis occidentalis*), desert pupfish, peregrine falcon, Harris hawk, and Sonoran pronghorn antelope; thick-billed parrot reintroduction; bald eagle management; and the black-footed ferret reintroduction program.

California Condor Program

On December 12, 1996, the U.S. Fish and Wildlife Service and the Peregrine Fund released six California condors from the Vermilion Cliffs of northern Arizona. It was the first time the giant birds have been seen in the skies of the American Southwest since 1924. The Arizona release, undertaken under provisions of the Endangered Species Act, placed the birds—part of a population of 120 left in the world—near the Paria Plateau, about 115 miles north of Flagstaff, Arizona, an area that once supported the condor. The rugged Coconino County terrain provided the necessary remoteness, ridges, cliffs, and caves favored by the carrion-eating birds.

California condor populations have declined sharply since the 1800s; by 1940, their number was estimated at 100 birds. By the late 1970s, the estimate had dropped to 25 to 30 birds. In 1979, the U.S. Fish and Wildlife Service, the California Department of Fish and Game, and the National Audubon Society, among other government and private groups, began a joint effort to study and preserve the bird. To increase the condors' egg production, biologists began removing eggs laid in the wild in 1983; eggs were taken to the San Diego Wild Animal Park and the Los Angeles Zoo for hatching.

Until 1985, biologists had planned to leave at least some condors in the wild in the belief that the free-flying birds would provide role models to captive-hatched birds. Then disaster struck. Members of four of the five remaining breeding pairs disappeared over the winter of 1984–1985, and the wild population was reduced from 15 to 9 birds. Biologists decided to capture all remaining wild condors and bring them into a captive breeding program. However, even with condors in such small numbers that they had to be retrieved from the wild, the service and its partners were focused on a release of the birds into the wild. In the fall of 1988, the service began a three-year reintroduction experiment using Andean condors as stand-ins for their endangered cousins.

The released condors were bred in captivity at the Los Angeles Zoo and at the Peregrine Fund's World Center for Birds of Prey in Boise, Idaho. In late October, they were transferred to the Vermilion Cliffs, where they have been kept away from people and allowed to acclimate to their new surroundings. Officially, condors were classified as an "experimental, non-essential" population under the Endangered Species Act, which allows for the birds to be managed with fewer restrictions than those normally covering endangered species. The classification is also designed to ensure that protected, reintroduced species are compatible with current and planned activities in the project area. If the initial release is successful, the service will seek to release more condors over a period of years until there is a self-sustaining population of about 150 in the project area.

The condor recovery plan calls for two self-sustaining, geographically separate wild populations to reclassify the condor from endangered to threatened. Northern Arizona offers the new population a greater degree of security and isolation than it has had at release sites in California, where seven introductions have occurred.

Adult condors weigh up to 20 pounds and have a wingspan of nearly 10 feet. In prehistoric times, the bird ranged from Canada to Mexico, across the southern United States to Florida, and north along the East Coast to New York State. The birds managed to maintain a strong population until settlement of the West, when shooting, poisoning, and egg collecting began to take a heavy toll. By 1987, the population of the birds in the wild had dwindled to seven. In what was then a controversial decision, the service decided to remove the remaining birds from the wild for captive breeding in a last-ditch effort to avert the extinction of the condor.

COLORADO

Colorado became a state in 1876. The Colorado Game, Fish and Parks Division was created in 1897 when the Eleventh General Assembly passed a session law establishing the Department of Forestry, Game and Fish. The new commissioner received an annual salary of $1200 with an additional $500 to cover travel expenses. Concern for natural resources, however, dates back to the 1861 territorial government. One act passed by the Territorial Assembly provided for a fine of $25 to $50 a day for the use of seines, nets, baskets, or traps, with half of the fine going to the county treasury and the other half to the witness. There were no other laws protecting wildlife, and their wholesale slaughter continued.

Market hunters who provided meat to the laborers laying railroad track across the plains and into the Rocky Mountains received as much as two cents per pound. Territorial Governor Edward M. McCook was an early advocate of laws to protect all wildlife. In his opening address to the Territorial Assembly on January 3, 1872, he warned that unless the useless and pitiless destruction of bison, deer, elk, antelope, and trout ceased, these animals would become extinct in Colorado. He pointed out that "the Indian entirely lacks...the wasteful cruelty and unsportsman-like attributes which seem to characterize the professional hunter and the orthodox tourist" and urged the passage of laws to protect wildlife. That same year, an act was passed to protect birds and animals, including orioles, flycatchers, and ravens. In 1866, grasshoppers by the millions swept across parts of Colorado, consuming all crops in their path. Clearly, avian predators of grasshoppers required protection. In 1873, quail hunting was made illegal from January 1 to September 31.

In 1874, quail and bobwhite hunting ceased for four years to allow those populations to recover. In 1876, bounty laws were passed to encourage the hunting of wolves, mountain lions, and coyotes, which were a nuisance to growing livestock interests. By 1885, antlered elk and deer could only be legally hunted from October 1 to November 15. During this time, wildlife resources decreased and rangelands were overgrazed by cattle and sheep. In 1887, laws were passed prohibiting the hunting of bighorn sheep for ten years and bison for eight years. No provisions for hunting licenses of any kind were made. By this time, various species of fish had been introduced into most of Colorado's lakes and streams.

In 1891, Gordon Land became the first game and fish commissioner. Land, an outspoken advocate of wildlife protection laws, called for the complete replacement of existing laws. He further recommended laws to restrict or prevent trafficking in game, hides, heads, and horns, as well as live animals. He also recommended that no bounty laws that placed premiums on mountain lions or bear be passed because there was more game in the state before these laws came into existence. He let it be known that "as far as the killing of stock by mountain lions, I am of the opinion that the worthless beings, I will not call them men, who pursue and hunt these animals with dogs solely for the reward, are far more destructive to the livestock interests than are these solitary beasts of prey." He gave in to bounties on wolves and coyotes, however. By 1891, most professional hunters had left Colorado. Wildlife populations continued to decline however, because enforcement was difficult and stockmen encouraged hunters to take wildlife that competed with their livestock for food.

By 1897, J.S. Swan, then forest, game and fish commissioner, was critical of the enforcement of game laws. He reported to the governor that few if any bison remained in Colorado, that a reduced elk hunt would help rebuild diminished populations, and that bighorn sheep numbers were so small that they should not be hunted. He indicated that bears also deserved protection and that a hunting season should be established for them. This might seem odd today, but there was no hunting season on bears back then—they could be hunted all year! He also reported that the silver-tip bear, or grizzly, was plentiful in western Colorado. By the end of the century, killing a bison meant serving time in a state penitentiary. Nevertheless, four hunters killed four bison in Lost Park, Colorado, in 1897. About this time, the subject of requiring a hunting license was again brought before the governor.

During the early years of the 1900s, hunting seasons were lengthened, so that by 1908 only 2500 deer were harvested. Elk populations were doing poorly despite the fact that hunting on them ceased in 1902, but bighorn sheep and antelope populations were recovering. By 1910, about 700 deer were taken by hunters. Two years later, approximately 400 were taken. Several years later, the hunting season for deer was closed.

In 1920, hunting licenses doubled in cost to two dollars for residents. Waterfowl populations were increasing because of the closure of the spring hunting season. The 23rd General Assembly set aside $25,000 for the eradication of predatory animals such as coyotes, mountain lions, and wolves. In 1922, 100 exotic Hungarian partridges were released. Pheasant were also released. Hay was delivered to help thousands of deer, elk, and bighorn sheep make it through the severe winter of 1921–1922. By 1925, turkeys were scarce in Colorado, with only 1000 birds remaining. In 1929, the first elk season was held since its closing in 1902. By the 1930s, estimates showed that deer and elk populations were increasing through protection efforts. Ranchers and farmers were demanding restitution for damage done by protected furbearers and game animals. Wildlife feeding programs were paying off.

On May 10, 1939, the Pittman–Robertson Enabling Act passed the Colorado legislature, making the state eligible for federal funding. That same year, the Beaver Act made feral cats and dogs predators that could be shot on sight. Women were allowed to purchase small game and fishing licenses at half price. Bows and arrows were finally recognized as hunting weapons. The mule deer population reached 248,000 in 1941, up from a low of 16,000 in 1913. The first doe permits were issued since 1907. Elk were also opened to hunting by both sexes. By 1939, the antelope had recovered to 4000 from a low of 1200 in 1909. In the early 1940s, antelope were transplanted to suitable areas, where they are now found. The first antelope season was held in 1945, after being closed in 1899. In 1939, the state legislature passed a law granting complete protection to all furbearers not classified as predators. A program was initiated to transplant beaver where they had been extirpated. Extensive turkey transplanting programs began, and in 1949 the first turkey hunt was held in almost 30 years. Sadly, by 1941, only five grizzly bears remained in Colorado.

In 1953, bighorn sheep hunting resumed after more than 64 years, and 169 licenses were issued; 58 sheep were harvested. A special bear season was established in 1955 to keep damage claims down. The use of an artificial light to take wildlife was finally made illegal in 1959. By the 1960s, hunters and fishermen were complaining that private landholders were blocking their access to potentially large resources. In 1963, 147,000 deer where taken by hunters, the most ever in Colorado. By 1964, grizzly bear hunting was outlawed, but it was already too late to save the grizzly in the state. A year later, 25 herds of bison were established throughout Colorado.

Since passage of the Endangered Species Act of 1973, Colorado has had an active reintroduction program. In 1976, river otters were reintroduced in the Black Canyon of the Gunnison River. Desert bighorn sheep were reintroduced in Colorado National Monument. The last grizzly bear was killed in self-defense by an outfitter near Platoro Reservoir. However, continuing its ignorance of predators, the U.S. Fish and Wildlife Service spent $500,000 on predator control in Colorado. In 1989, hunters took 41,276 elk, and an auction brought $38,700 for a single bighorn sheep. In 1996, voters successfully stopped all recreational trapping in the state of Colorado. Clearly, voters are demanding and winning a voice in the management and protection of wildlife. Voters, not wildlife managers, are bringing about rapid change. Could this be due to an influx of Californians into Colorado? Appendix 14.3 lists Colorado's threatened and endangered species.

NEW MEXICO

Wildlife issues in New Mexico are just as exceptional as in any other state. A large ten-year program to study free-ranging panthers was one of the most comprehensive of its kind in the United States. Meanwhile, in a nearby wildlife refuge, Mexican wolves await their release onto federal lands in New Mexico against the wishes of Governor Gary Johnson. Funding for the state's nongame program peaked in 1983 and spiraled downward as the number of threatened and endangered species climbed. Fully half of the state's fish were threatened or endangered in 1996. Meanwhile, exotic herbivores such as aoudad (*Ammotargus lervia*), gemsbok (*Oryx gazella*), Persian wild goat (*Capra aegagrus*), and Siberian ibex (*Capra ibex*) have prospered across the state and are now managed as game animals. Native plants have suffered where exotic wildlife exists.

As introduced species are the third member of the Evil Quartet, a brief history of exotic introductions in New Mexico is in order. Gemsbok, Persian wild goat, and Siberian ibex were intentionally introduced by the New Mexico Department of Game and Fish. Release of the aoudad, a native of North Africa, was unauthorized in 1956. In 1950, Elliot Barker, chief game warden for the state of New Mexico, purchased 12 aoudad from a private landowner and released them into the Canadian River Gorge in northern New Mexico. There have also been unintentional escapes of aoudad from private game parks in the state (Morrison 1983). These releases and escapes resulted in aoudads becoming established in huntable populations all across New Mexico except higher mountain elevations.

Gemsbok (also known as oryx) are natives of Africa and the Middle East. Gemsbok were released in 1969 in White Sands Missile Range in south-central New Mexico. Other releases occurred in White Sands as well. This population is also hunted. The Persian wild goat, a native of Iran, Saudi Arabia, Pakistan, and Afghanistan, was introduced between 1970 and 1976. Range expansion has resulted in a huntable population. The Siberian ibex, a native of Pakistan, Afghanistan, India, and South China, was introduced in 1975 and supplemented in 1977. Hunting began in 1981.

These exotics are now hunted using different sustainable harvest strategies. The key word here is, of course, sustainable. Morrison (1983) wrote that the success of the harvest strategies was dependent on an open-minded administration that allowed experimentation with bag limits and harvest dates in the face of opposition from misinformed public organizations and lay groups that opposed hunting. Presumably the lay groups as well as the New Mexico Department of Game and Fish both opposed unlimited hunting to completely eliminate these exotics whose impact on the local flora and fauna has not been thoroughly investigated.

Funding for New Mexico's Share with Wildlife program came from a state income tax checkoff program. In 1981, $240,000 was raised to support a broad range of wildlife research. In 1990, $100,000 was raised and by 1997 the program brought in just $35,000 statewide—an amount equivalent to just 15 cents per year per resident. Fortunately, federal programs also contributed to funding the Share with Wildlife program.

New Mexico has many interesting wildlife issues and a diverse array of wildlife management programs. Somehow, we must gain an understanding of these seemingly diverse problems and integrate them into a single program. One way to achieve such a view is to consider the underlying parameter or set of parameters that is key to each issue. In New Mexico, access to fresh water is a key issue in many disputes that extend well beyond wildlife issues. Appendix 14.4 lists New Mexico's threatened and endangered species.

In 1996, New Mexico was suffering from several problems. Due to a severe drought, the Rio Grande was drying up in stretches and threatening the federally endangered Rio Grande silvery minnow (*Hybognathus amarus*). The Rio Grande bosque was famous for its large cottonwood trees, and residents have long feared that the cottonwoods were not regenerating. Indeed, young cottonwood trees were difficult to find in many areas of the bosque. Residents often accused beavers of killing the young trees. Cattlemen feed their cattle in the bosque, and some residents felt that cattle were destroying wildlife habitat. Exotic trees such as tamarisk had become widespread and were crowding out native vegetation. Cattlemen claimed that cattle were good for the bosque and kept vegetation down, thus preventing runaway fires from destroying the remaining cotton-woods. Albuquerque residents used more than four billion gallons of water a month in 1996. The water was pumped from deep-water wells. New Mexico's forests were on fire in 1996 because of a severe lack of moisture.

Clearly, water is the central theme that threads these issues together. Is it possible to separate wildlife issues from agricultural issues, forest fire issues, and grazing issues? What perspective must be taken to consider these issues? Rarely are issues isolated events. Indeed, the cyclic nature of the global climate is largely responsible for recurring droughts in the Southwest. The bottom of each cycle is defined by severe drought, and each successive bottom in the cycle seems to get deeper. Humans cannot seem to learn from the past and seem destined to repeat each mistake in every cycle. First, let's explore all the seemingly diverse wildlife issues and follow the different threads until they come together.

"A Desert called the Rio Grande" was the caption to a front-page article in the May 12, 1996 *Albuquerque Journal.* The subtitle read, "A 60-mile stretch of parched riverbed attests to an arid year and unrelenting demands for water." A desiccated catfish was pictured on a sandy riverbed. Cottonwood trees on either side of the dry riverbed were green and lush, their roots able to find groundwater. On another page was a photograph of a protester's sign: "Hungry? Out of work? Eat a silvery minnow."

In April 1996, U.S. Fish and Wildlife Services fisheries biologists found more than 10,000 Rio Grande silvery minnows piled up more than six inches deep in a channel on the Rio Grande. The dead fish were located in a dried pool near San Antonio, New Mexico, just south of the San Acacia diversion dam. What was left of the mighty Rio Grande had dried up, leaving the fish stranded. The Rio Grande silvery minnow was classified as endangered in 1994 under the Endangered Species Act. Water from the Rio Grande was crucial for crop irrigation along its route from southern Colorado, through the length of New Mexico, and across the international border between the United States and Mexico.

In May 1996, Forest Guardians of Santa Fe, New Mexico, an environmental organization, joined the Sierra Club and the Southwest Center for Biological Diversity and filed a "notice of intent to sue," which is required before the federal government can be sued. After the waiting period of 60 days, a suit was filed. Forest Guardians filed suit in federal court, charging the Middle Rio Grande Conservancy District with "allowing the minnow's habitat to erode" and "deliberately caus[ing] the extinction of the species." The Middle Rio Grande Conservancy District managed water allocations from Cochiti Dam south of Santa Fe to Bosque del Apache just south of Socorro.

All water had been diverted by the Middle Rio Grande Conservancy District to provide fresh water for crop irrigation, leaving isolated pools to dry up under the intense New Mexico sun. Both the silvery minnow and people needed the water. In spite of a clear and deliberate violation of the Endangered Species Act, all the water was diverted to serve human purposes. The fish lost. In a "normal year," officials claimed, water would have been available to support both irrigation and the fish. The primary minnow population had been squeezed into a 20-mile stretch of the Rio Grande south of Albuquerque and north of Socorro. The entire population was at risk of extinction.

Language in the Endangered Species Act expressly prohibits habitat destruction, and failure to allow sufficient water to maintain existing habitat was a clear and flagrant violation of the act, in spite of state laws that might suggest otherwise. The Forest Guardians had a viable case.

Long stretches of the Rio Grande have dried up not because of a lack of water but because water had been diverted out of the river by humans and for humans. About 60,000 property owners had water rights along the river, including 2500 agricultural users. Others used the water to keep their lawns green. State law reserved water for irrigators and guaranteed them rights to water under the law. The district was not "against the fish," but "thousands of crop growers who make their living along the river's banks" apparently needed the water more than the fish, according to a district spokesperson. State law did not recognize natural water flow needed by nonhumans as a beneficial use of the water. Indeed, water used for nonhuman benefit was apparently a waste of water under New Mexico law.

In mid-May 1996, rescue efforts began to save what was left of the silvery minnow population. Approximately 10,000 minnows were moved 20 to 30 miles upstream. The minnows themselves tried to migrate upstream, but found their route blocked by diversion dams. The small minnow could not leap the impediments and thus had to be moved by humans. Some 70% of the minnow population lived along a 35-mile stretch of the Rio Grande near Socorro. Some 10% were expected to be lost in the move. Many of the minnows were bearing eggs.

Why should we care about a minnow only three inches long? A report in the *Albuquerque Journal* claimed that "the silvery minnow is nothing special biologically" (Linthicum 1996). Vernon Tabor, a fisheries biologist with the U.S. Fish and Wildlife Service, knows and understands the Rio Grande well, or at least he thinks he does. Although he said, "If the last silvery minnow dies, the ecosystem won't grind to a halt," he realized that the tiny fish was as much a part of the New Mexico bosque as the cottonwood tree (Linthicum 1996). Tabor claimed that if the minnow disappeared, some other fish would take its place. The minnow was just one of a number of small fish that lived in the Rio Grande, eating mosquitoes and dragonfly larvae and eaten by songbirds, egrets, and blue herons. The silvery minnow was also the last native minnow left in New Mexico. The blunt-nosed shiner and the phantom shiner both became extinct in the 1960s. So much for some other fish. Jeff Whitney of the U.S. Fish and Wildlife Service's Rio Grande Silvery Minnow Rescue Team realized the true meaning of the loss of a species. He suggested the silvery minnow was an "integral part of the ecosystem" and "is a kind of a caution sign for us, an indication that all is not well in our environment." How many minnow species must be lost before we have lost too many? Or was a local farmer correct in asserting that "Most of us couldn't care less if that fish is here tomorrow." Will he care when the mosquitoes hatch?

There were at least seven major dams along the Rio Grande in New Mexico in 1996. The Cochiti Dam was just south of Santa Fe. Angostura Dam was just north of Albuquerque. San Acacia Diversion Dam was north of Socorro; Elephant Butte Dam, Caballo Dam, Leasburg Diversion Dam, and La Mesilla Diversion Dam all ensured that New Mexico received its fair allotment of river water before it reached Texas. And make no mistake—the water was owed to Texas through legal agreements.

In May 1996, usage of water in Albuquerque was up 25% from the previous year. The city's water was pumped from 94 deep-water wells. Total usage was 3.7 billion gallons of water during May. In April, only three billion gallons was used. Apparently because of the severity of the drought, residents had to water their lawns more often. Residential water customers increased their water usage by 30% over 1995 figures. Institutional users such as parks increased their usage by a mere 21%. Albuquerque's water supply came from an underground aquifer. Experts once thought the aquifer was as big as Lake Superior. Now they believe it is much smaller (Dmorzalski 1996). With usage increasing dramatically, droughts preventing aquifer recharge, and experts revising downward the size of the water supply, residents might think twice about having a nice lawn or draining the Rio Grande for a few acres of onions and opt instead to save water for drinking.

In January 1996, water restrictions went into effect in Ruidoso, New Mexico. By June 1996, residents were allowed to water their lawns only twice monthly. Violators could be fined $100 for the first violation and $500 for the second. Ruidoso residents knew that without water, there would be no town. One of the city's reservoirs was completely dry. Pumping from wells had increased and the water pressure was reduced, halving flow rates. The city was attempting to buy water rights from downstream farmers who also did not have enough. Seven new well sites had been marginal. By June 1996, other cities in the state were suffering. Brian Wilson of the State Engineer's Office summed up the obvious: "We've simply reached the point where our population has exceeded the limits of the environment to support it" (Taugher 1996).

Forest Fires

In May 1996, the U.S. Department of Agriculture declared five southern New Mexico counties primary disaster areas. Seven other adjoining counties were declared disaster areas. Farmers and ranchers in these counties were suffering severe economic losses due to the continued drought. Five "drought-management seminars" were planned for ranchers in the affected areas.

About 43,000 acres of forest in New Mexico went up in smoke in the first five months of 1996. More fires were expected as the drought wore on. Firefighting cost taxpayers more than $10 million in New Mexico alone. Forests were seeded after the fires, but a forest requires more than a century to mature. Just how dry was New Mexico in 1996? Rainfall in the spring was 0.02 inches, compared to 1.62 inches normally. It had not been drier since 1860. Across the state, rainfall was 50 to 5% of normal levels. River flows were 10% of more normal levels. Irrigation projects in the state were seeing their water allotments dry up. The world-famous Bosque del Apache National Wildlife Refuge was also burning out of control during part of 1996.

Bosque Vegetation

Dams along the Rio Grande had completely altered the character of vegetation along the river. The bosque was once filled with young cottonwood trees; now there are few. Some long-time residents blame the beavers. Beavers seem to prefer young cottonwoods, and residents believe the beavers prevented new cottonwoods from establishing. However, cottonwoods require spring floods to germinate. When the bottomlands along the river were flooded in spring, millions of cottonwoods would spring to life. Now that the river is controlled and no longer floods, cottonwoods cannot germinate. Instead, invasive species such as tamarisk move in and crowd out native species. The character of the Rio Grande is changing due principally to flood control and irrigation.

Grazing Along the Rio Grande

Ranchers claimed that damage along the Rio Grande was caused by humans littering the banks of the river with beer bottles, used refrigerators, old mattress springs, abandoned automobiles, and garbage. Cattlemen charged that vandals were destroying the riverbanks. Many ranchers were tired of hearing the complaints. Ranchers insisted that cows helped the bosque. "Cows eat the tall grass, which is fuel for fires. Two years ago a fire here burned for over a half a mile and killed a lot of big trees. Cows keep the grass short; they don't start fires," said one rancher (Sanchez 1996).

Many of these ranchers have grazed their cattle in the bosque, land owned by the Middle Rio Grande Conservancy District, for decades. In exchange for grazing privileges, ranchers installed gates and tried to keep the area clean. One rancher had more than a hundred head in the bosque. Another rancher said his family had run cattle in the bosque for three generations. Grazing in the bosque was not a privilege but a right—a family right.

Hunters, on the other hand, suggested that cows were destroying the Rio Grande bosque, a unique vegetation community of grand old cottonwood trees. Several cattle had been shot dead and tossed into the river by people who apparently disagreed with the ranchers. One duck hunter complained that he was hunting on land that looked like it should have been a corral.

Fisherman also turned on the cattlemen. Free-ranging cattle were trampling the banks and ruining fish habitat. A permit system and gates were installed to control access to some areas. Cattle, however, ignored the gates and were apparently destroying the native vegetation. In 1995, several fishermen and hunters complained about locked access to public land. They found it hard to imagine why hunters and fisherman were excluded when cattle were not.

"The bosque should be big enough to meet the needs of both cattlemen and sportsmen, if all will keep their cool and try to understand the other's needs," suggested an editorial in the May 14, 1996 *Albuquerque Journal*. The real damage comes from "nocturnal citizens," one article asserted. These people "can tear down a fence, shoot at cows (or anything else that moves), or start fires. They go there to drink or just dump trash." The article went on to suggest that barriers ought to be erected to keep people out.

Which of these assertions can be tested using the scientific method? Can we test to see if cattle and their grazing habits are good for the bosque?

Public Lands

New Mexico ranchers take it for granted that their ranching needs should be met at the expense of the public. Public lands in the state were heavily degraded by overgrazing, yet ranchers still demanded access to land they did not own. New Mexico is a dry state, and cattle grazing has changed the face of the land from north to south and east to west. While most cattle ranches in the midwestern and eastern United States are on private lands, most grazing in the West takes place on public lands, lands owned by the federal government with free access to all citizens.

When the Forest Guardians filed suit in federal court against the Bureau of Land Management to prevent overgrazing of public lands, ranchers responded predictably. They accused the Forest Guardians of being elitist, against jobs, against humans, against ranchers and their families, and insatiable and greedy. "The fate of 100 ranchers and their families can be cavalierly dismissed as meaningless and below any reasonable consideration," one rancher said in a local editorial (Jackson 1996).

Should a few people dictate policy for the majority of citizens in the United States? The Bureau of Land Management manages about 163 million acres in the 11 western states for livestock grazing (Jacobs 1991). As Jackson (1996) said, the debate is not about a choice between jobs or the environment; "it is about whether private individuals will be allowed to survive on public lands." Should people be allowed to live off public lands at public expense? Should people be allowed to harvest wildlife or timber off public lands to make a living? Is it fair that those cattle ranchers who own their land must compete with ranchers who do not and are subsidized by the public? Cattle in the West have damaged wildlife habitat over millions of acres for decades. Perhaps the drought (see below) will help solve these age-old problems. Or will the public now be called upon to provide water for privately owned cattle on public lands?

Global Climate

From January 1 through April 30, 1996, Albuquerque received 0.38 inches of rain. The average is normally 2.12 inches. Other cities along the Rio Grande, such as Socorro, received 0.35 inches. Average rainfall is normally about 1.41 inches. The 1930s, 1950s, and 1970s saw drought conditions across much of the United States. The drought was linked to cyclical global climate patterns and the La Niña cold-water phenomenon in the southern Pacific Ocean. La Niña events foster dry conditions across the American Southwest during the winter months. El Niño events occur when ocean waters along the western coast of South America warm. El Niño causes heavy precipitation and often flooding in the Southwest.

Global climate specialists are not certain what causes La Niña or El Niño events. Some scientists have suggested that variations in the sun's energy output cause the events, but no one really knows. Although the drought adversely affected large mammal populations, small mammals such as kangaroo rats and pocket mice experienced population increases, perhaps because the seeds they consume were not rotting from excess moisture. Reduced moisture also caused fewer seeds to germinate, thus ensuring a plentiful, long-lasting food supply.

Global climate influences precipitation worldwide. La Niña and El Niño events affect rainfall in New Mexico. Climate cycles are long term and also affect precipitation patterns in the Southwest. Rio Grande dams have been very damaging to the ecology of the bosque. Dams have prevented fish movement, prevented flooding and hence regeneration of the precious cottonwoods, and directly aided exotic plant invasions such as tamarisk. The Rio Grande beavers are not responsible for the loss of cottonwoods—humans are. Agricultural water diversions have acted to exacerbate drought-year effects. Leaf litter accumulation has been aided by a lack of spring floods, and as a result, fires are more harmful than they would otherwise be. The more humans attempt to manage the bosque, the more harm seems to result. The true cost of decades of mismanagement is cumulative and will be paid sooner or later.

Other species that depend on New Mexico's riparian habitats are also in trouble. The southwestern willow flycatcher (*Empidonax traillii extimus*) is state and federally listed as endangered. About 90% of the flycatcher's habitat has been altered or destroyed by humans. Water diversions and flood control, groundwater pumping, livestock grazing, invasions of exotic plants (Evil Quartet member 3), heavy recreational use of riparian areas, habitat fragmentation (Evil Quartet member 2), and urban encroachment have contributed to the bird's decline. Because it breeds in New Mexico, Arizona, and California and winters in Latin America, riparian areas must be protected in the United States.

New Mexico's aquatic species in general were in trouble in 1997. Fully half of the state's fish were listed as federally threatened or endangered. Overgrazing by cattle for many decades in arid grasslands and woodlands brought about massive habitat destruction. The nongame and endangered species program in the state was funded by a tax checkoff program that Governor Gary Johnson, a conservative Republican, threatened to cancel. He had already vetoed a bill passed by the state congress that would have created a specialty license plate to support endangered species. In early 1997, the governor rejected $250,000 of federal money earmarked for the nongame program. In 1996, the funds raised to support the nongame and endangered species program amounted to just 15 cents per person per year in the state.

SUMMARY AND SYNTHESIS OF STATE WILDLIFE PROGRAMS

Early in America's history, fresh water, prairies, forests, and wildlife were considered almost infinite resources. The human population around 1800 was a mere ten million people, most of whom lived in the Northeast. Scattered Native America tribes existed throughout the continent. Canadians living farther north and also along the coast had wiped out the walrus 40 years earlier, but no one seemed to notice. Fifty years later, the forests in the Northeast were cleared, the elk were all but finished, turkeys were disappearing, and people were pushing west to seek a new start. In the late 1800s, the American bison was nearly wiped off the earth. Elk were gone east of the Rocky Mountains. Wolves and panthers were nearly extirpated east of the Rockies as well. Panther populations held on in Florida a bit longer. Although there was plenty of land, certain areas of the United States lacked fresh water until the windmill was invented in the mid-1800s. Florida was a veritable wildlife paradise, with teaming populations of deer, bears, panthers, wolves, birds, and millions of wading birds, but the paradise had been plundered by the turn of the last century.

Habitat fragmentation, commercial exploitation, and ignorance wiped away much of the nation's wildlife resources before most people knew it was missing. Wildlife was certainly not the treasure it has become today. Commercial harvesting of wild game cannot be sustained by any wildlife population. Passenger pigeons, with their enormous populations and migratory behavior, could not be sustainably harvested. Habitat destruction finished off the populations that illegal and indiscriminate harvesting could not reach. By the turn of the 19th century, wildlife was in trouble virtually nationwide. By the mid-1930s, however, people across the United States were demanding that once extant wildlife be reintroduced, particularly for hunting. From some 500,000 deer at the turn of the 20th century, the deer population has swelled to an estimated 20 million.

There is no doubt that Pittman–Robertson Act funding played the key role in bringing many wildlife species back from the brink of extirpation in most states. Populations of managed game species have increased dramatically. Indeed, in some cases, game species have become a nuisance. However, some species, such as many songbirds, have decreased. The words "to manage populations" have become synonymous with "to increase populations," so much so that the term "nuisance wildlife" was coined. Have we humans managed some populations so well that we have, out of neglect, mismanaged other species nearly out of existence? Can humans really manage wildlife? Humans certainly have demonstrated that they can increase a game species' numbers. However, can humans manage communities of wildlife across a variety of landscapes that evolved over millions of years? Can humans simultaneously juggle all species, from soil microbes to top predators, whose populations are dynamically responding to environmental change? Must humans manage every species? What should the grand management strategy be?

APPENDIX 14.1: CALIFORNIA'S THREATENED AND ENDANGERED SPECIES AND THEIR FEDERAL STATUS AS OF OCTOBER 1, 1996[a]

Endangered species

Gastropods
Trinisty bristle snail (*Monadenia setosa*)
Morro shoulderband (*Helminthoglypta walkeriana*)**

Crustaceans
Riverside fairy shrimp (*Streptocephalus woottooni*)**
Conservancy fairy shrimp (*Branchinecta conservatio*)**
Longhorn fairy shrimp (*Branchinecta longiantenna*)**
Vernal pool tadpole shrimp (*Lepidurus packardi*)**
Shasta crayfish (*Pacifastacus fortis*)**
California freshwater shrimp (*Syncaris pacifica*)**
San Diego fairy shrimp (*Branchinecta sandiegoensis*)***

Insects
Zayante band-winged grasshopper (*Trimerotropis infantilis*)****
Santa Cruz rain beetle (*Pleocoma conjugens conjugens*)****
Mount Herman June beetle (*Polyphylla barbata*)****
Mission blue butterfly (*Icaricia icarioides missionensis*)**
Lotis blue butterfly (*Lycaeides argyrognomon*)**
Palos Verdes blue butterfly (*Glaucopsyche lygdamus lotis*)**
El Segundo blue butterfly (*Euphilotes battoides allyni*)**
Smith's blue butterfly (*Euphilotes enoptes smithi*)**
San Bruno elfin butterfly (*Incisalia mossi bayensis*)**
Lange's metalmark butterfly (*Apodemia mormo langei*)**
Quino checkerspot (*Euphydryas editha quino*)****
Laguna Mountains skipper (*Pyrgus ruralis lagunae*)****
Callippe silverspot butterfly (*Speyeria callippe callippe*)****
Behren's silverspot butterfly (*Speyeria zerene behrensii*)****
Myrtle's silverspot butterfly (*Speyeria zerene myrtleae*)**
Delhi Sands flower-loving fly (*Rhaphiomidas terminatus abdominalis*)**

Fish
Winter-run chinook salmon (*Oncorhynchus tshawytscha*)**
Paiute cutthroat trout (*Oncorhynchus clarki seleniris*)**
Delta smelt (*Hypomesus transpacificus*)
Mohave tui chubb (*Gila bicolor mohavensis*)**
Owens tui chubb (*Gila bicolor snyderi*)**
Bonytail (*Gila elegans*)**

Sacramento splittail (*Pogonichthys macrolepidotus*)**
Colorado squawfish (*Ptychocheilus lucius*)**
Lost River sucker (*Deltistes luxatus*)
Shortnose sucker (*Chasmistes brevirostris*)**
Razorback sucker (*Xyrauchen texanus*)**
Desert pupfish (*Cyprinodon macularius*)**
Cottonball marsh pupfish (*Cyprinodon salinus milleri*)**
Owens pupfish (*Cyprinodon radiosus*)**
Armored threespine stickleback (*Gasterosteus aculeayus williamsoni*)**
Tidewater goby (*Eucyclogobius newberryi*)**
Tecopa pupfish (*Cyprinodon navadensis calidae*) delisted—extinct
Thicktail chubb (*Gila crassicauda*) delisted—extinct

Amphibians
Santa Cruz long-toed salamander (*Ambystoma macrodactylum croceum*)**
Desert slender salamander (*Batrachoseps aridus*)**
Arroyo southwestern toad (*Bufo microscaphus californicus*)**

Reptiles
Leatherback sea turtle (*Dermochelys coriacea*)**
Blunt-nosed leopard lizard (*Gambelia silus*)**
Island night lizard (*Xantusia riversiana*) proposed**
Black legless lizard (*Anniella pulchra nigra*) proposed**
Alameda whipsnake (*Mawsticophis lateralis euryxanthus*)**

Birds
California brown pelican (*Pelecanus occidentalis californicus*)**
California condor (*Gymnogyps californianus*)**
Bald eagle (*Haliaeetus leucocephalus*)**
Peregrine falcon (*Falco peregrinus*)**
California clapper rail (*Rallus longirostris obsoletus*)**
Light-footed clapper rail (*Rallus longirostris levipes*)**
Northern spotted owl (*Strix occidentalis caurina*)
Great gray owl (*Strix nebulosa*)
Gila woodpecker (*Melanerpes uropygialis*)
Gilded northern flicker (*Colaptes aurayus chrysoides*)
Willow flycatcher (*Empidonax traillii*)
Southwestern willow flycatcher (*Empidonax traillii extimus*)**
San Clemente loggerhead shrike (*Lanius ludovicianus mearnsi*)**
Arizona Bell's vireo (*Vireo bellii arizonae*)
Belding's savannah sparrow (*Passerculus sandwichensis beldingi*)

Mammals
Riparian brush rabbit (*Sylvilagus bachmani riparius*)
Point Arena mountain beaver (*Aplodontia rufa nigra*)**

Pacific pocket mouse (*Perognathus longimembris pacificus*)**
Morro Bay kangaroo rat (*Dipodomys heermanni morroensis*)**
Giant kangaroo rat (*Dipodomys ingens*)**
Stephens' kangaroo rat (*Dipodomys stephensi*)**
Tipton kangaroo rat (*Dipodomys nitratoides nitratoides*)**
Salt-marsh harvest mouse (*Reithrodontomys raviventris*)**
Amargosa vole (*Mictotus californicus scirpensis*)**
San Joaquin kit fox (*Vulpes macrotis mutica*)**
Gray whale (*Eschrichtius robustus*)**
Sei whale (*Balaenoptera borealis*)**
Blue whale (*Balaenoptera musculus*)**
Finback whale (*Balaenoptera physalus*)**
Humpback whale (*Magaptera novaeangliae*)**
Right whale (*Balaena glacialis*)**
Sperm whale (*Physeter catodon*)**
Peninsular bighorn sheep (*Ovis canadensis cremnobates*)****

Threatened species

Crustaceans
Vernal pool fairy shrimp (*Branchinecta lynchi*)*

Insects
Delta green ground beetle (*Elaphrus viridis*)*
Valley elderberry longhorn beetle (*Desmocerus californicus dimorphus*)*
Kern primrose sphinx moth (*Euproserpinus euterpe*)*
Bay checkerspot butterfly (*Euphydryas editha bayensis*)*
Oregon silverspot butterfly (*Speyeria zerene hippolyta*)*

Fish
Coho salmon (*Oncorhynchus kisutch*) proposed*
Little Kern golden trout (*Oncorhynchus mykiss whitei*)
Lahotan cutthroat trout (*Oncorhynchus clarki henshawi*)
Klamath Mountains Province steelhead (*Oncorhynchus mykiss*)
Bull throat (*Salvelinus confluentus*)
Modoc sucker (*Catostomus microps*)
Rough sculpin (*Cottus asperrimus*)

Amphibians
Siskiyou Mountains salamander (*Plethodon stormi*)
Tehachapi slender salamander (*Batrachoseps stebbinsi*)
Kern Canyon slender salamander (*Batrachoseps simatus*)
Shasta salamander (*Hydromantes shastae*)
Limestone salamander (*Hydromantes brunus*)
Black toad (*Bufo exsul*)
California red-legged frog (*Rana aurora draytonii*)*

Reptiles
Desert tortoise (*Gopherus agassizii*)*
Green sea turtle (*Chelonia mydas*)*
Olive Ridley sea turtle (*Lepidochelys olivacea*)*
Barefoot banded gecko (*Coleonyx switaki*)
Coachella Valley fringe-toed lizard (*Uma inornata*)
Flat-tailed horned lizard (*Phryosoma mcallii*) proposed*
Southern rubber boa (*Charina bottae umbratica*)
Giant garter snake (*Thamnophis couchi gigas*)*

Birds
Aleutian Canada goose (*Branta canadensis leucopareia*)*
Swainson's hawk (*Buteo swainsoni*)
California black rail (*Laterallus jamaicensis conturniculus*)
Greater sandhill crane (*Grus canadensis tabida*)
Western snowy plover (*Charadrius alexandrinus nivosus*)*
California least tern (*Sterna antillarum browni*)*
Marbled murrelet (*Brachyramphus marmoratus*)*
Western yellow-billed cuckoo (*Coccyzus americanus occidentalis*)
Elf owl (*Micrathene whitneyi*)*
Bank swallow (*Riparia riparia*)
California gnatcatcher (*Polioptila californica*)*
Least Bell's vireo (*Vireo bellii pusillus*)*
Inyo California towhee (*Pipilo crissalis eremophilus*)*
San Clemente sage sparrow (*Amphispiza belli clementeae*)*

Mammals
San Joaquin antelope squirrel (*Ammospermophilus nelsoni*)
Mohave ground squirrel (*Spermophilus mohavensis*)
Fresno kangaroo rat (*Dipodomys nitratoides exilis*) state rare
Sierra Nevada red fox (*Vulpes vulpes necator*)
Island fox (*Urocyon littoralis*)*
Guadalupe fur seal (*Arctocephalus townsendi*)*
Northern sea lion (*Eumetopia jubatus*)
Wolverine (*Gulo gulo*)
Southern sea otter (*Enhydra lutris nereis*)*
California bighorn sheep (*Ovis canadensis californiana*)

---

[a] * = federally threatened, ** = federally endangered, *** = proposed endangered, **** = federally proposed endangered.

Source: California Department of Fish and Game 1996

APPENDIX 14.2: ARIZONA'S SPECIES OF SPECIAL CONCERN

Mammals

Jaguar (*Felis onca*) extirpated
Sonoran pronghorn (*Antilcapra americana sonoriensis*)
Ocelot (*Felis pardalis*) extirpated
Southwestern river otter (*Lutra canadensis sonora*)
Black-footed ferret (*Mustela nigripes*) extirpated
Grizzly bear (*Ursus arctos*) extirpated
Mexican wolf (*Canis lupus baileyi*)
Meadow jumping mouse (*Zapus hudsonius*)
Camp Verde cotton rat (*Sigmodon arizonae arizonae*)
Mesquite mouse (*Peromyscus merriami*)
Navajo Mexican vole (*Microtus maxicanus navaho*)
New Mexican banner-tailed kangaroo rat (*Dipodomys spectablis baileyi*)
Mount Graham red squirrel (*Tamiasciurs hundonicus grahamensis*)
Black-tailed prairie dog (*Cynomys ludovicianus*)
Greater western mastiff bat (*Eumops perotis*)
Towsend's big-eared bat (*Plecotis townsendii*)
Western yellow bat (*Lasiurus xanthinus*)
Western red bat (*Lasiurus blossevillii*)
Lesser long-nosed bat (*Leptonycteris curasoae yerbabuenae*)
Water shrew (*Sorex palustrus*)
Arizona shrew (*Sorex arizonae*)

Birds

American bittern (*Botaurus lentiginosus*)
Least bittern (*Ixobrychus exilis*)
California condor (*Gymnogyps californianus*)
Bald eagle (*Haliaeetus leucocephalus*)
Northern goshawk (*Accipiter gentalis*)
Swainson's hawk (*Buteo swainsoni*)
Ferruginous hawk (*Buteo regalis*)
Northern Aplomado falcon (*Falco femoralis septentrionalis*) extirpated
Masked bobwhite (*Colinus virginianus ridgwayi*)
California black rail (*Laterallus jamaicensis coturniculus*)
Yuma clapper rail (*Rallus longirostris yumanensis*)
Thick-billed parrot (*Rhynchopsitta pachyrhyhcha*) extirpated
Western yellow-billed cuckoo (*Coccyzus americanus occidentalis*)
Cactus ferruginous pygmy-owl (*Glaucidium brasilianum cactorum*)
Mexican spotted owl (*Strix occidentalis lucida*)
Elegant trogon (*Trogon elegans*)

Green kingfisher (*Chloroceryle americana*)
Southwestern willow flycatcher (*Empidonax traillii extrimus*)
Buff-breasted flycatcher (*Empidonax fulvifrons*)
Thick-billed kingbird (*Tyrannus crassirostris*)
Rose-throated becard (*Pachyramphus aglaiae*)
Azure bluebird (*Sialia sialis fulva*)
Veery (*Catharus fuscescens*)
Swainson's thrush (*Catharus ustulatus*)
Gray catbird (*Dumetella carolinensis*)
Sprague's pipet (*Anthus spragueii*)
Baird's sparrow (*Ammodrammus bairdii*)
Arizona grasshopper sparrow (*Ammodrammus savannarum ammolegus*)
Five-striped sparrow (*Amphispiza quinquestriata*)

Snakes
Ridgenose rattlesnake (*Crotalus willardi*)
Massasauga (*Sistrurus catenatus*)
Mexican garter snake (*Thamnophis eques*)
Narrowhead garter snake (*Thamnophis rufipunctatus*)

Lizards
Arizona striped whiptail (*Cnemidophorus inornatus arizonae*)
Flat-tailed horned lizard (*Phrynosoma mcallii*)
Bunchgrass lizard (*Sceloporus scalaris*)
Mohave fringe-toed lizard (*Uma scoparia*)

Turtles
Desert tortoise (*Gopherus agassizii*)
Yellow mud turtle (*Kinosternon flavescens flavescens*)

Frogs
Plains leopard frog (*Rana blairi*)
Chiricahua leopard frog (*Rana chiricahuensis*)
Relicit leopard frog (*Rana onca*)
Northern leopard frog (*Rana pipiens*)
Lowland leopard frog (*Rana yavapaiensis*)
Ramsey Canyon leopard frog (*Rana subaquavocalis*)
Tarahumara frog (*Rana tarahumarae*)

Treefrogs
Mountain treefrog (*Hyla eximia*)

Leptodactylid frogs
Barking frog (*Eleutherodactylus augusti*)

Mole salamanders
Sonoran tiger salamander (*Ambystoma tigrinum stebbinsi*)

Fish

Mexican stoneroller (*Campostoma ornatum*)
Yaqui shiner (*Cyprinella formosa*)
Humpback chub (*Gila cypha*)
Sonora chub (*Gila ditaenia*)
Bonytail chub (*Gila elegans*)
Gila chub (*Gila intermedia*)
Yaqui chub (*Gila purpurea*)
Roundtail chub (*Gila robusta*)
Virgin chub (*Gila seminuda*)
Desert pupfish (*Cyprinodon macularis*)
Quitobaquito pupfish (*Cyprinodon macularis eremus*)
Gila topminnow (*Poeciliopsis occidentalis occidentalis*)
Yaqui topminnow (*Poeciliopsis occidentalis sonoriensis*)
Apache trout (*Oncorhynchus apache*)
Gila trout (*Oncorhynchus gilae*)
Yaqui catfish (*Ictalurus pricei*)
Yaqui sucker (*Catostomus bernardini*) extirpated
Flannelmouth sucker (*Catostomus latipinnis*)
Razorback sucker (*Xyrauchen texanus*)
Woundfin (*Plagopterus argentissimus*)
Colorado squawfish (*Ptychocheilus lucius*)
Loach minnow (*Rhinichthys cobitis*)
Virgin spinedace (*Lepidomeda mollispinis mollispinis*)
Little Colorado spinedace (*Lepidomeda vittata*)
Spikedace (*Meda fulgida*)

Amphipods

Arizona cave amphipod (*Stygobromus arizonensis*)

Mollusks

California floater (*Anodonta californiensis*)

Snails

Pinaleno Mountain snail (*Oreohelix grahamensis*)
Yavapai Mountain snail (*Oreohelix yavapai cummingsi*)
Kanab ambersnail (*Oxyloma haydeni kanabensis*)
Bylas springsnail (*Pyrgulopsis arizonae*)
Grand Wash springsnail (*Pyrgulopsis bacchus*)
San Bernardino springsnail (*Pyrgulopsis bernardino*)
Kingman springsnail (*Pyrgulopsis conica*)
Desert springsnail (*Pyrgulopsis deserta*)
Verde Rim springsnail (*Pyrgulopsis glandulosa*)
Page springsnail (*Pyrgulopsis morrisoni*)
Fossil springsnail (*Pyrgulopsis simplex*)

Brown springsnail (*Pyrgulopsis sola*)
Three Forks springsnail (*Pyrgulopsis trivialis*)
Squaw Peak talussnail (*Sonorella allynsmithi*)
San Xavier talussnail (*Sonorella eremita*)
Pinaleno talussnail (*Sonorella grahamensis*)
Wet Canyon talussnail (*Sonorella macrophallus*)
Gila tryonia (*Tryonia gilae*)
Quitobaquito tryonia (*Tryonia quitobaquitae*)

APPENDIX 14.3: COLORADO'S THREATENED AND ENDANGERED SPECIES

Endangered species

Fish
Razorback sucker (*Xyrauchen texanus*)
Bonytail (*Gila elegans*)
Humpback chub (*Gila cypha*)
Colorado squawfish (*Ptychocheilus lucius*)
Rio Grande sucker (*Catostomus plebeius*)

Birds
Plains sharp-tailed grouse (*Tympanuchus phasianellus*)
Whooping crane (*Grus americana*)
Least tern (*Sterna antillarum*)

Mammals
Gray wolf (*Canis lupus*)
Grizzly bear (*Ursus arctos*)
Black-footed ferret (*Mustela nigripes*)
Wolverine (*Gulo gulo*)
River otter (*Lutra canadensis*)
Lynx (*Felis lynx*)

Amphibians
Western toad (*Bufo boreas boreas*)

Threatened species

Fish
Arkansas darter (*Etheostoma cragini*)
Greenback cutthroat trout (*Onchorynchus clarki stomias*)

Amphibians
Wood frog (*Rana sylvatica*)

Birds
Peregrine falcon (*Falco peregrinus*)
Bald eagle (*Haliaeetus leucocephalus*)
Greater sandhill crane (*Grus canadensis tabida*)
Mexican spotted owl (*Strix occidentalis lucida*)
Greater Prairie-chicken (*Tympanuchus cupido)*
Lesser Prairie-chicken (*Tympanuchus pallidicnctus)*
Piping plover (*Charadrius melodus*)

APPENDIX 14.4: THREATENED AND ENDANGERED SPECIES OF NEW MEXICO

Endangered species

Mammals
Arizona shrew (*Sorex arizonae*)
Mexican long-nosed bat (*Leptonycteris nivalis*)
(Penasco) least chipmunk (*Eutamias minimus atristriatus*)
Gray wolf (*Canis lupus*)
(Arizona) montane vole (*Microtus montanus arizonensis*)
(Desert) bighorn sheep (*Ovis canadensis mexicana*)

Birds
Brown pelican (*Pelecanus occidentalis*)
Aplomado falcon (*Falco femoralis*)
White-tailed ptarmigan (*Lagopus leucurus*)
Whooping crane (*Grus americana*)
Piping plover (*Charadrius melodus*)
Least tern (*Sterna antillarum*)
Common ground dove (*Columbina passerina*)
Buff-colored nightjar (*Caprimulgus ridgwayi*)
Elegant trogon (*Trogon elegans*)
Northern beardless tyrannulet (*Campostoma imberbe*)
(Southwestern) willow flycatcher (*Empidonax traillii extimus*)
Thick-billed kingbird (*Tyrannus crassirostris*)

Reptiles
Gila monster (*Heloderma suspectum*)
Gray-checkered whiptail (*Cnemidophorus dixoni*)
Mexican garter snake (*Thamnophis eques*)
Plainbelly water snake (*Nerodia erythroaster*)
New Mexico ridgenose rattlesnake (*Crotalus willardi obscurus*)

Amphibians
Lowland leopard frog (*Rana yavapaiensis*)
Western boreal toad (*Bufo boreas*)
Great Plains narrowmouth toad (*Gastrophryne olivacea*)
Spotted chorus frog (*Pseudacris clarkii*)

Fish
Gila chub (*Gila intermedia*)
Chihuahua chub (*Gila nigrescens*)
Roundtail chub (*Gila robusta*)
Rio Grande silvery minnow (*Hybognathus amarus*)
Arkansas River shiner (*Notropis girardi*)
Phantom shiner (*Notropis orca*)
(Rio Grande) bluntnose shiner (*Notropis simus simus*)
Southern redbelly dace (*Phoxinus erythrogaster*)
Colorado squawfish (*Ptychocheilus lucius*)
(Zuni) bluehead sucker (*Catostomus discobolus yarrowi*)
Blue sucker (*Cycleptus elongatus*)
Pecos gambusia (*Gambusia nobilis*)

Crustaceans
Noel's amphipod (*Gammarus desperatus*)
Socorro isopod (*Thermosphaeroma thermophilum*)

Mollusks
Paper pondshell (*Anodonta imbecillis*)
Texas hornshell (*Popenaias popei*)
Socorro springsnail (*Pyrgulopsis neomexicana*)
Roswell springsnail (*Pyrgulopsis chupaderae*)
Pecos assiminea (*Assiminea pecos*)
Wrinkled marshsnail (*Stagnicola caperatus*)
Shortneck snaggletooth (*Gastrocopta dalliana dalliana*)
Florida mountainsnail (*Orehelix florida*)

Threatened species

Mammals
Least shrew (*Cryptotis parva*)
Southern long-nosed bat (*Leptonycteris curasoae*)
Spotted bat (*Euderma maculatum*)
Southern yellow bat (*Nycteris ega*)
White-sided jackrabbit (*Lepus callotis*)
(Organ Mountains) Colorado chipmunk (*Eutamias quadrivittatus australis*)
Southern pocket gopher (*Thomomys umbrinus*)
Meadow jumping mouse (*Zapus hudsonius*)
American marten (*Martes americana*)

Birds
Neotropic cormorant (*Phalacrocorax brasilianus*)
Bald eagle (*Haliaeetus leucocephalus*)
Common black hawk (*Buteogallus anthracinus*)
Peregrine falcon (*Falco peregrinus*)
(Gould's) wild turkey (*Meleagris gallopavo mexicana*)
Whiskered screech owl (*Otus trichopsis*)
Boreal owl (*Aeglius funereus*)
Costa's hummingbird (*Calypte costae*)
Lucifer hummingbird (*Calothorax lucifer*)
Violet-crowned hummingbird (*Amazilia violiceps*)
White-eared hummingbird (*Hylocharis leucotis*)
Broad-billed hummingbird (*Cyananthus latirostris*)
Gila woodpecker (*Melanerpes uropygialis*)
Bell's vireo (*Vireo belli*)
Gray vireo (*Vireo vicinior*)
Varied bunting (*Passerina versicolor*)
Abert's towhee (*Pipilo aberti*)
Baird's sparrow (*Ammodramus bairdii*)
(Arizona) grasshopper sparrow (*Ammodramus savannarun ammolegus*)
Yellow-eyed junco (*Junco phaeonotus*)

Reptiles
Western river cooter (*Pseudemys grzugi*)
Sand dune lizard (*Sceloporus arenicolus*)
Bunchgrass lizard (*Sceloporus scalaris*)
Giant spotted whiptail (*Cnemidophorus burti*)
Mountain skink (*Eumeces callicephalus*)
Green ratsnake (*Senticolis triaspis*)
Narrowhead garter snake (*Thamnophis proximus*)
(Mottled) rock rattlesnake (*Crotalus lepidus lepidus*)

Amphibians
Jemez Mountains salamander (*Plethodon neomexicanus*)
Sacramento Mountains salamander (*Aneides hardii*)
Colorado River toad (*Bufo alvarius*)

Fish
Gila trout (*Oncorhynchus gilae*)
Mexican tetra (*Astyanax mexicanus*)
Arkansas River speckled chub (*Macrhybopsis aestivalis tetranemus*)
Spikedace (*Meda fulgida*)
(Pecos) bluntnose shiner (*Notropis simus pecosensis*)
Suckermouth minnow (*Phenacobius morabilis*)
Loach minnow (*Rhinichthys cobitis*)

Gray redhorse (*Moxostoma congestum*)
Pecos pupfish (*Cyprinodon pecosensis*)
White Sands pupfish (*Cyprinodon tularosa*)
Gila topminnow (*Poeciliopsis occidentalis*)
Greenthroat darter (*Etheostoma lepidum*)
Bigscale logperch (*Percina macrolepida*)
Brook stickleback (*Culaea inconstans*)

Mollusks
Swamp fingernailclam (*Musculium partumeium*)
Raymond's fingernailclam (*Musculium raymondi*)
Long fingernailclam (*Musculium trabsversum*)
Lilljeborg's peaclam (*Pisidium lilljeborgi*)
Sangre de Cristo peaclam (*Pisidium sanguinichristi*)
Gila springsnail (*Pyrgulopsis gilae*)
Pecos springsnail (*Pyrgulopsis pecosensis*)
New Mexico hot springsnail (*Pyrgulopsis thermalis*)
Alamosa springsnail (*Tryonia alamosae*)
Koster's springsnail (*Tryonia kosteri*)
Star gyro (*Gyraulus crista*)
Ovate vertigo (*Vertigo ovata*)
Hacheta Grande woodlandsnail (*Ashmunella hebardi*)
Cooke's Peak woodlandsnail (*Ashmunella macromphala*)
Mineral Creek mountainsnail (*Oreohelix pilsbryi*)
Doña Ana talussnail (*Sonorella todseni*)

EXERCISES

14.1 Design an experiment to test whether cattle are good for the bosque. What types of exclosures could be used? What would the control be? What variables would be measured?

14.2 Design an experiment to test whether cottonwood seeds need spring floods to germinate. What is the null hypothesis?

14.3 The New Mexico bosque is a riparian area in an otherwise dry, semi-arid land, and hence is a magnet for wildlife. The bosque is also a fragile environment. Establishing a series of reserves along the Rio Grande might aid migrating birds but would do nothing for the Rio Grande silvery minnow. Ultimately, the issue is water. What can be done for farmers whose livelihoods depend on water from the Rio Grande? Is this a political issue or a wildlife issue?

14.4 How are amphibians, reptiles, crustaceans, and mollusks apparently doing in Colorado? Compare threatened and endangered fish species in New Mexico and Colorado.

14.5 Spiders and insects are important components of biodiversity and are protected by the Endangered Species Act of 1973. These species are often overlooked, however. What can be done to increase awareness of these critical species?

LITERATURE CITED

Annual Report on the Status of California State Listed Threatened and Endangered Animals and Plants. 1992. Department of Fish and Game, State of California, Sacramento.

Dmorzalski, D. 1996. City water conservation efforts dry up. *Albuquerque Tribune* June 11.

Endangered and Threatened Animals of California. 1996. Department of Fish and Game, State of California, Sacramento.

Jackson, J.M. 1996. Elitist environmentalists belittle people's needs. *Albuquerque Journal* June 18.

Jacobs, L. 1991. *Waste of the West: Public Lands Ranching.* Lyn Jacobs, Tucson, Arizona.

Linthicum, L. 1996. A desert called the Rio Grande. *Albuquerque Journal* May 12.

Morrison, B.L. 1983. The Use of Harvest Strategies to Control Exotic Ungulate Populations in New Mexico. New Mexico Department of Game and Fish, Santa Fe.

Sanchez, T. 1996. People hurt the bosque, ranchers contend. *Albuquerque Journal* May 9.

Steinhart, P. 1990. California's Wild Heritage. California Department of Fish and Game, Sacramento.

Taugher, M. 1996. Drought takes new victims. *Albuquerque Journal* June 16.

Torres, S.G., T.M. Mansfield, J.E. Foley, T. Lupo, and A. Brinkhaus. 1996. Mountain lion and human activity in California: testing speculations. *Wildlife Society Bulletin* **24**(3):451–460.

GLOBAL PERSPECTIVES OF WILDLIFE

15

Dewar (1996) makes a compelling point—America's wildlife is dwindling. Meanwhile, 300 million people visited America's national parks in 1995. Populations of bees, birds, frogs, and many other species, including native plants, are declining. Populations of raccoons, armadillos, fire ants, and introduced species of plants and animals are increasing in the United States. Clearly, something is terribly wrong. More than half of the neotropical migrant birds are declining in number. The American goldfinch population has declined 25%. Raccoons are becoming nuisance wildlife. Bees are disappearing. In fact, in California and Florida bees are brought in to pollinate crops. And these bees are usually European bees! America's native bees are checking out. As Dewar points out, bobwhite populations are crashing along the east coast, the common honeybee is becoming rare, frogs are disappearing worldwide, butterflies are vanishing, once common eastern meadowlarks are losing ground, and even prairie dogs are disappearing across the short-grass prairie. What is happening?

In the *Albuquerque Journal* on April 21, 1996, Dewar summarized what people are seeing across the United States—strange invasions of coyotes into Florida, mesopredator populations such as red foxes increasing dramatically, the once common summer tanager of New Mexico now rarely seen. The tanager depends on riparian habitat that is shrinking or damaged by river projects. Even snakes such as the eastern indigo snake are now endangered. And when insect populations decline, something is definitely wrong. A recent study by The Nature Conservancy showed that 17% of America's butterflies are experiencing population declines.

Who should care about insects? Fully 60% of U.S. agricultural crops are pollinated by bees. Some 3000 varieties of native bees are in trouble. Nabhan of the Arizona–Sonoran Desert Museum estimates that 25% of all managed beehives have disappeared in the last five years, having fallen victim to a pair of parasites from Asia and Europe.

Habitat loss and introduced species, the second and third members, respectively, of the Evil Quartet, are partly to blame. Human activities are responsible for both, however. Fortunately, we are entering crisis mode, and people often seem to mobilize around crises. Like a sleeping giant, nothing seems to motivate us until a crisis occurs. However, Americans seem to respond to crises like no other society. Their solutions are bold, unprecedented, and cannot be predicted. The status quo is thrown aside completely. Dramatic new laws are passed without regard to the shackles of tradition or grandfather laws. Bold, sweeping, grand proposals become law. Private citizens step forward with new initiatives. The private sector responds just as it did 220 years ago when the nation was established.

Hardly a day passes without one wildlife issue or another appearing in a local newspaper in every city and town across the United States. Every child learns about wildlife and is attracted to wildlife every day. Parents encourage their children to enjoy wildlife. Nature shows appear daily on television across the United States. Presidential candidates and members of Congress want to be seen supporting wildlife. People are catching on that if you want to see wildlife, now is the time. Something is happening across the United States. If the history of the dusky seaside sparrow is representative, perhaps it is crisis time.

DUSKY SEASIDE SPARROW

At one time, there were ten recognized species and subspecies of seaside sparrows and three species and subspecies of sharp-tailed sparrows from the Texas Gulf coast to the Gulf of St. Lawrence in Canada. Beecher (1955) pointed out that the Texas seaside sparrow (*Ammodramus maritima sennetti*) and the Acadian sharp-tailed sparrow (*A. caudacuta subvirgata*) of the Gulf of Saint Lawrence are closely related species. The seaside sparrows range from the Texas Gulf coast to the coast of Massachusetts. The sharp-tailed sparrows range from Delaware to the Gulf of Saint Lawrence, and so overlap of the seaside and sharp-tailed sparrows occurs on the Atlantic coast.

As the Pleistocene glaciers melted, the sea level rose some 100 m, and North America's coastal crust also rose due to the loss of an immense weight of ice. Beecher suggested that the loss of land along the Gulf and Atlantic coasts brought about the isolation of a once contiguous population of an ancestral sparrow species. As land was inundated by the rising ocean, the ancestral sparrows became isolated populations and therefore speciated in response to differing environmental conditions. The Delaware River, for instance, separates the common sharp-tailed sparrow (*A. caudacuta caudacuta*) from the southern sharp-tailed sparrow (*A. caudata diversa*). Galveston Bay separates the Texas seaside sparrow from the Louisiana seaside sparrow (*A. maritima fisheri*). The Mississippi River divides the Wakulla seaside sparrow (*A. maritima macgillivrayii*) from the Louisiana seaside sparrow. Traveling eastward along the Gulf coast, the bay at Mobile divides Scott's seaside sparrow (*A. maritima peninsulae*), and the Cape Sable seaside sparrow (*A. maritima mirabilis*) is isolated in the Everglades. The dusky seaside sparrow (*A. maritima nigrescens*) was on the east coast of Florida near Merritt Island, which, according to Beecher, "could have been greatly affected by rise in sea level." Continuing farther north up the Atlantic coast, other populations of seaside and sharp-tailed sparrows are similarly isolated.

Complicating Beecher's analysis is the fact that glacial advances have occurred at least 18 times during the Pleistocene and that each advance and recession of the great ice sheets has affected the sparrow populations differently. Any seaside sparrow populations would have been driven southward during glacial growth. Indeed, other populations of seaside and sharp-tailed sparrows probably existed and have become extinct. Environmental processes such as hurricanes acted to maintain the isolation between populations. So did extinction.

The extinction of the dusky seaside sparrow is the story of jobs and growth versus those who would set aside even a small piece of ground to save a bird (Walters 1992). Dusky seaside sparrows inhabited the salt marsh areas of broomgrass and rushes. Before Merritt Island became the site of America's launch pad for the space race, nests were reported to be 10 to 15 m apart. The size of a typical sparrow, duskies lived on insects, caterpillars, and spiders. In the 1960s, NASA bought much of Cape Canaveral and North and South Merritt Island and built Kennedy Space Center. The U.S. Air Force had previously used the area as a launch site. Mosquito spraying eliminated an important food source for the duskies and had reduced their population even before NASA arrived. As early as 1955, Longstreet (1955) warned that mosquito control programs might harm the dusky population. But mosquitoes became a public menace and had to be done away with. In 1960, it was realized that the diminutive dusky was already in trouble.

Indeed, the mosquito problem was apparently so bad that NASA decided to construct dikes on the Indian River to inundate mosquito breeding grounds. Naturally, neither the loss of a food source nor the flooding of critical habitat was good for the dusky, but few cared. The vegetation community also changed.

The avian community responded to the vegetation change, and larger birds such as red-winged blackbirds and boat-tailed grackles moved in. Salt water was replaced by fresh water. Fresh water enabled ducks to take residence, and so, at first appearance, it seemed that the number of birds increased and all was well. Of course, the delicate balance of the salt marsh was broken. The Spartina grass where the duskies nested was almost gone. Without appropriate nesting habitat, the bird would disappear.

In 1967, the Endangered Species Act of 1966 (see Chapter 2) passed through Congress and was signed into law by the president. The original endangered species list included more than 50 animals; among the 10 from Florida were the West Indian manatee, the Florida panther, the Everglades kite (snail kite), the Cape Sable seaside sparrow, and the dusky seaside sparrow. In 1969, the duskies on Merritt Island were disappearing, but another population was found to the west in the coastal marshes of the St. Johns River. By 1973, only three dusky seaside sparrows remained on Merritt Island, their population a victim of over-zealous mosquito control and habitat destruction. Unfortunately for the approximately 2000 remaining duskies in the marshes at St. Johns, the state wanted to build a highway through the middle of their habitat. Today, the Beeline Expressway in Brevard County cuts through what was once prime dusky habitat.

With the new highways came real estate developers. Canaveral Groves and Port St. Johns now abut what was prime habitat for duskies. By 1968, the entire population of duskies had been reduced to only a few. Land purchases by The Nature Conservancy and U.S. Fish and Wildlife Service established the St. Johns National Wildlife Refuge around 1970 and provided the dusky a temporary reprieve. With the borders of the 4200-acre refuge established, the countdown to the dusky's extinction had begun.

As we learned earlier (Chapter 9), even large reserves fail to prevent extinction. Furthermore, a heterogeneous landscape is needed to support many species. If, for instance, the dusky could only nest in areas of vegetation that had been burned three years prior and could only survive in those areas for four years, then a heterogeneous landscape maintained by periodic fires widely spread around the landscape would be essential to the dusky's survival. Such a regime would be impossible to maintain in a small refuge. Six fires, in fact, burned in the refuge from 1970 to 1977, but because the refuge was too small, several fires were damaging. By 1976, only 28 males remained in the refuge. Despite repeated past warnings about the devastating effect fires might have, it was not until 1979 that the U.S. Fish and Wildlife Service cut fire lanes. It was already too late, however. By then, only a few males survived, presumably along with an unknown number of females.

By 1979, reality had set in, and the dusky's slide to extinction caused a bureaucratic panic. By 1980, only four birds remained in the wild, three of which joined others that had been captured earlier. The last wild dusky, a male, eluded capture and presumably died in the wild. All the duskies in captivity were males, but the dusky was not extinct yet. A dusky could be crossed with another of the seaside sparrows; by crossing and then crossing the offspring again, each time with a dusky male, a nearly purebred dusky could be recreated in six generations. Indeed, a female Scott's seaside sparrow from the Florida Gulf coast had been successfully crossed with a dusky in captivity and viable offspring were produced. However, the U.S. Fish and Wildlife Service refused to allow a cross-breeding program. The late Herb Kale, a vice-president of the Florida Audubon Society and a central figure in the story of the dusky, was furious. "The decision by USF&WS was one of the worst decisions made anywhere at any time" (Kale personal communication).

Having searched for more duskies, the U.S. Fish and Wildlife Service insisted that a female could be found. One has to ask, if a female could be found in the wild, couldn't the males do a better job of it than humans? Kale simply never gave up. He hatched a plan to transfer the three (two died earlier) remaining males to Disney World, where a captive breeding program began. But it was too late, and old age had taken its toll. In July 1987, the last pure dusky died. The aviary at the Disney facility where the hybrids were kept was neglected and in disrepair when rats ate them. Even the hybrids had been lost through the carelessness of humans (Walters 1992).

FISHING IN U.S. AND INTERNATIONAL WATERS

Atlantic bluefin tuna can weigh up to three-quarters of a ton and are able to swim 50 miles per hour. On the docks of Halifax, Nova Scotia, bluefin can sell for $20,000 apiece. In Japan, however, a bluefin can sell for $350 a pound. In 1975, the northwest Atlantic population of bluefin tuna was estimated to be 250,000. Twenty years of overfishing reduced the population to about 22,000. In 1995, the bluefin was proposed for listing as an internationally endangered species. Man's greed is solely responsible for the decline.

The Magnuson Fisheries Conservation and Management Act (the Fish Conservation and Management Act, see Chapter 2) set up regional fishery management councils that are today dominated by fishing interests as regulators (Swardson 1994). As a result, the U.S. fishing industry underwent massive growth, in part subsidized by federal loan guarantees. With improving technology and a complete reluctance to regulate itself, the fishing industry is overcapitalized and far too large. Unfortunately, the seas were found to harbor only finite resources. The situation in the United States is not unique however. Worldwide harvests of fish in 1989 brought in a catch worth about $70 billion. The cost to operate the global fishing fleet was about $90 billion. The difference had to be subsidized. The Japanese government extended $19 billion in credit to its troubled fishing industry, and no one seriously expects the money to be paid back. Even the fiscally conservative Norwegians subsidized their fleet at $150 million. In 1994, the Clinton administration issued a $30 million aid package for the sinking New England fishing industry. Canada must, of course, support the Newfoundland economy.

Who consumes all this fish? It is estimated that 6.6% of the animal protein consumption of North Americans comes from fish. Fish accounts for 12% of European consumption of protein. In Africa, the figure is 19% and in Asia 29%. It is thus no surprise that Asians lead in aquaculture technology. It is easy to suggest that the people of Third World countries will ultimately pay for depleted fish stocks worldwide. However, their investment in capital equipment such as processing factories and large fleets is much less. As is usually the case, subsidies will spread the cost throughout the economy.

On March 11, 1995, the *St. Petersburg Times* reported that Canada had seized one of 14 Spanish fishing trawlers operating within Canada's international waters. Canadian gunboats fired across the bow and gained control of the ship after a chase on the high seas. Relationships between the European Economic Community and Canada have since deteriorated sharply over the dispute, the article suggested. Canada continued to protect its dwindling fish stocks after the failure of intense high-level political negotiations. As we saw, Canadian fish stocks, especially in the northwest Atlantic, were in trouble.

One day later, on March 12, 1995, the *Miami Herald* reported that United Nations Food and Agricultural Organization (FAO) experts had concluded that oceans around the world were being "ravaged by overfishing that has decimated natural stocks and jeopardizes the future of the seas as a source of food for the Earth's rapidly growing, ever-hungrier population." The experts concluded that the history of fishing is to deplete the resource until there are no fish. The world's fish catch was a stagnant 101 million tons in 1993, about 83 million tons of which were caught in marine and freshwater fisheries. Aquaculture accounted for the remaining 18 million tons, and this figure has been increasing yearly for some time. Approximately 85% of fish farms are in Asia, and aquaculture is now a $30 billion industry annually.

The world take of Atlantic cod and such prized species as haddock, hake, flounder, and shrimp has declined principally because of overfishing. Less valuable fish such as Alaskan pollock, Peruvian anchoveta, and Chilean jack mackerel are now sought-after species. According to the FAO, over 70% of all conventional species are fully exploited, overexploited, depleted, or are in the process of recovering from overexploitation. In West Africa, Bangladesh, and Southeast Asia, fish that were once taken when they were 1 m in length are now taken at 10 cm in length. The world's fishing fleets have swelled over the past quarter century so that fishing effort has increased to bring in the same tonnage each year. Twenty nations account for over 80% of the total harvest of fish. China leads the world with ten million tons caught offshore. Japan, Peru, Chile, and the United States follow with 5.8 millions tons.

The FAO researchers claim that fishermen discard upwards of 27 million tons of fish yearly, a tremendous loss. Less valuable fish are tossed overboard to make room for more valuable fish. How to manage the world's fisheries is a difficult problem. In developing nations, populations are growing faster than food output, and thus it is nearly impossible to limit the number of fishermen. High unemployment, political upheaval, and pest outbreaks all drive people to fish for themselves. If all fishing were to stop immediately, some fish stocks would require a decade or more to recover. "What we are talking about is a worldwide management crisis of natural resources," said one U.N. official.

The U.S. fisheries also suffer from overharvesting. Fish resources in Puget Sound in the Pacific Northwest have seriously declined, and harvests are at their lowest level in over 55 years. Annual catches of over 26 million pounds were routinely harvested in the past. Now the total harvest is a pitiful 3.6 million pounds. Populations of Pacific cod, hake, and pollock have been severely reduced. One researcher suggested that "essentially the fisheries have pretty much ceased for these three stocks."

Outraged by foreign fishing fleets slightly more than 3 miles off U.S. coastlines, the U.S. Congress expanded the coastal economic zone to 200 miles in the 1970s. Although the foreign fleets backed off, U.S. factory ships quickly filled the void. Driven by greed, these trawlers were said to take everything "their nets could snag." On March 15, 1995, the *Miami Herald* reported that Oregon Indian tribes urged President Clinton to declare a state of emergency in the Pacific Northwest to make good on a 140-year-old treaty that would help bring salmon populations back. The tribes have depended on the salmon fishery for more than 10,000 years. Hydroelectric dams have caused the demise of several species of salmon that have been added to the federal list of threatened and endangered species. The Indian tribes, state fish and wildlife officials, and conservation groups sued the federal government over its failure to protect the salmon.

Dams have been shown to block fish migrations upstream to breed and to slow travel downstream after the fish spawn. The National Marine Fisheries

Service proposed barging fish, but conservationists claim that more water needs to be spilled over dams to speed migration to the sea and also lower reservoir levels. Such measures are opposed by irrigation farmers, shippers, and the aluminum industry, all of which depend on water from the rivers. The Indian tribes claim that if nothing is done, the several salmon species will be extinct by the turn of the century. If the salmon become extinct, the very culture and religion of the tribes will be lost as well. Clearly, there is a wildlife issue here.

Northwest Pacific salmon such as coho salmon are born in freshwater streams, migrate to live in the sea, and then return to freshwater streams to spawn and die. Each year, fewer and fewer fish return to their place of birth to spawn. In 1995, the National Marine Fisheries Service reported that West Coast ocean catches of coho fell to only 292,000 fish in 1993 from 5.3 million fish in 1976, a 95% decline. The Chinook salmon catch declined by 80% from 1988 to 1992. Sockeye salmon populations also declined. Steelhead trout are being considered for the endangered species list.

On Oregon's Rogue River, all migrating salmon must pass through a narrow underwater passage to get around Gold Ray Dam. Each passing salmon is recorded by a video camera. Biologists in the region agree that most species of Pacific salmon are in deep trouble as a result of overfishing at sea and destruction of their freshwater habitat by dams, agricultural schemes, and industrial and urban development. Logging near upstream spawning areas caused massive siltation of the gravel beds where the salmon laid their eggs. Insect populations that fed salmon were decimated when the trees were taken. Streams were straightened and obstructions removed, and as a result, the complex environment so necessary to support salmon was destroyed. Human activities were blamed for the loss of fish. As one logger put it, "I've got to feed my family. It's either us or them fish."

INTRODUCED FISH

Introduced species, the third member of the Evil Quartet, are one of the principal causes of extirpation of native species. The *Washington Post National Weekly Edition* for October 17 to 23, 1994 ran a story about Yellowstone Lake in Yellowstone National Park, Wyoming. Yellowstone Lake supports a native cutthroat trout population that federal authorities have spent years protecting. As a result, Yellowstone Lake is one of the favorite sportfishing lakes in the Rocky Mountains. Now, however, lake trout have been discovered in Yellowstone Lake. Lake trout are native to the Great Lakes and not western rivers. Lake trout are larger than cutthroat trout, and one fish biologist referred to the species as "freshwater sharks." Fears are that lake trout will prey upon the smaller cutthroat trout.

Since the 1970s, the U.S. Fish and Wildlife Service has managed the Yellowstone cutthroat fishery actively. Every spring, the lake's cutthroats migrate up more than half of Yellowstone Lake's 124 feeder streams to spawn and thus provide grizzly bears with a source of protein after a winter of hibernation. Lake trout, on the other hand, spawn deep in the lake and thus do not become food for bears. Federal officials have offered a $10,000 reward for information leading to the arrest of those responsible for introducing the large lake trout. However, knowing who perpetrated the crime will not solve the problem.

Because some of its waters come from thermal vents, Yellowstone Lake is unusually productive for an alpine lake. Biologists have found several strains of cutthroat trout within the lake. Protecting the cutthroat trout also protects eagles, ospreys, and otters within the park. Montana has long been a magnet for sportfishing, and rivers like the Madison, Yellowstone, and Big Hole are known for their great fishing. However, the brown trout and rainbow trout for which the rivers are known are introduced species. Brown trout were introduced from Europe and rainbow trout were introduced from the Pacific coast. Native fish throughout North America are facing threats from logging, development, ranching, and introduced species. Until a species is declared threatened or endangered, with few exceptions, there are few laws that protect it.

Throughout the West, aquatic systems are collapsing. Grazing activities have had a devastating effect on many western streams and riparian areas. Among the native species in trouble are the bull trout, the westslope cutthroat trout, the fluvial grayling, and the white sturgeon, which was placed on the endangered species list. Many other species of freshwater fish are in trouble. In Colorado, the squawfish is about to be lost forever. Although many blame introduced fish species, habitat loss and dams contribute to the problem. Water, it seems, is a valuable commodity to humans.

AMPHIBIAN POPULATIONS DECLINE WORLDWIDE

Frogs, toads, and salamanders worldwide are in trouble, and populations are thought to be disappearing (Phillips 1994). Every continent has documented losses, except Antarctica, which has none. In Australia, 26 of 202 species of frogs have declining populations. Some frogs have not been seen in over a decade. In Costa Rica, the golden toad (*Bufo periglenes*) has not been observed in significant numbers since 1987. Scientists working in Brazil reported declining populations of amphibians in the early 1980s. Amphibians are thought to be sensitive environmental indicators.

Since the decline appears to be happening worldwide, factors such as climate change and increased atmospheric pollution have been blamed. Scientists are alarmed because even populations in protected areas have declined; hence the reasons sought are not always associated with habitat loss or other proximal causes.

At a meeting in Irvine, California, in 1990, scientists reported on declining ranges and populations (Blaustein and Wake 1990). Although amphibian populations fluctuate, sometimes wildly, it is the accumulation of evidence worldwide that indicates a global phenomenon. For instance, only 2% of the lakes in the Sierra Nevada Mountains of California that contained yellow-legged frogs (*Rana muscosa*) in the mid-1970s had them in 1989. Similarly, Blaustein documented the decline of three species in the American Northwest. Other species have disappeared from undisturbed lakes in the American Southwest.

A unique gastric brooding frog (*Rheobatrachus silus*) from Australia was discovered in 1973. The frog is interesting because its eggs develop and hatch in the mother's stomach. In 1981, it was thought to be extinct.

As Blaustein and Wake assert, amphibians are critical components of many ecosystems, often constituting the highest fraction of vertebrate biomass. Amphibians are carnivores that consume insects and are themselves preyed upon by birds, fish, mammals, and also some insects. They occupy the middle trophic level in the web of life. Acid rain, climate change, air pollution, UV radiation, and introduced predators have all been suggested as causes of amphibian declines.

RAPTORS

Swainson's hawks are birds of prey that spend spring and summer in western North American grasslands (Di Silvestro 1996). Before the North American fall, the hawks migrate to South American grasslands, where they spend another spring and summer. By migrating, the hawks live a life of springs and summers. Since the 1940s, the number of Swainson's hawks has declined by 90%, yet few dead birds had been found in the United States or Canada. Either migration was killing the birds or something was going on in their southern range. Most of the band recoveries from Swainson's hawks came from Argentina. To discover what was happening, two Swainson's hawks were outfitted with 30-g satellite transmitters in California. (As an aside, Swainson's hawks weigh from 700 to 1000 g. The federal weight limit for transmitters is that they must be less than 3% of the body weight of the bird.) The hawks were tracked to the Argentinian pampas.

Biologists visited Argentina, where the outfitted Swainson's hawks settled. Near La Pampa in north-central Argentina, introduced Australian eucalyptus trees grow in isolated stands called montés. Swainson's hawks and other birds of prey roost in the montés. One flock of hawks numbered 4000. An additional 700 birds were found dead. The deaths were suspected to be due to pesticide spraying. One of the dead Swainson's hawks was banded in Saskatchewan, Canada. In all, some 450,000 Swainson's hawks, nearly all the North American population, winter in Argentina.

La Pampa was mostly cattle country and alfalfa, but ten years ago agricultural crops were replaced with sunflowers. Recent irrigation developments have encouraged the development of cultivation. Large agricultural areas attract hordes of insects, especially grasshoppers. Swainson's hawks are attracted to the large swarms of grasshoppers. Unfortunately, the insecticide monocrotophos and other pesticides were sprayed to control grasshoppers in alfalfa and sunflower fields. A survey found 4100 dead Swainson's hawks. Total deaths were probably much higher.

Canadians also expressed an interest in the study. Two Swainson's hawks that had satellite transmitters attached in Alberta were at the same location as the dead hawks in Argentina. Of 12 leg bands that have been recovered from the dead Swainson's hawks, 9 originated in Alberta and Saskatchewan and 1 each in Colorado, Idaho, and California. Most of the dead hawks were in the white color phase, indicating that they were from Canada and the adjacent Great Plains of the United States, and 90% were adults.

The pesticides were sprayed from planes and air blasters towed by tractors. The hawks were following the tractors to catch disturbed grasshoppers. Not only did the hawks receive a direct spray, but they also received a secondary dosage from the ingested grasshoppers. The life cycle of the grasshoppers required multiple pesticide applications. Hence, hawks were probably subjected to multiple doses of insecticide. Unfortunately, many of the dead hawks were quickly scavenged by carrion-feeding birds such as caracaras and vultures. Efforts are now under way to find a replacement insecticide that kills grasshoppers but does not harm top predators such as Swainson's hawks. Other North American birds such as dickcissels also winter in South and Central America. These, too, are subjected to agricultural pesticides.

According to the *Albuquerque Journal* on April 23, 1996, the Forest Conservation Council, the National Audubon Society, the Carson Forest Watch, and the Forest Guardians warned the U.S. Department of Agriculture that timber sales in the Carson National Forest of New Mexico threaten wildlife habitat for the endangered Mexican spotted owl and the northern goshawk. According to the Department of Agriculture, 30% of the old-growth trees have been set aside as habitat, while only 18% is required. Both sides will convene in a courtroom to reach a decision.

Mexican spotted owls were added to the endangered species list in 1993. They are found in canyons and forests of Utah, Colorado, New Mexico, Arizona, and on the New Mexico–Texas–Mexico border. About 90% of the owls live within national forests.

NORTHERN WHITE RHINOCEROS

The northern white rhinoceros (*Ceratotherium simum cottoni*) is a rare race of white rhino. Its nearest relative is 1500 miles away in South Africa. Garamba National Park in the Democratic Republic of the Congo (DRC, formerly Zaire) was declared a national park in 1938 and is nearly 2000 square miles. Another 3000 square miles bordering three sides of the park is protected as a wildlife reserve with sustainable use of natural resources. In the mid-1970s, there were 400 to 800 rhinos in the park and 15,000 to 30,000 elephants (Karesh 1996).

By 1984, there were 15 rhinos and 4500 scared elephants. Poachers had decimated one of the last great herds in Africa. Shortly after that, conservation efforts led by Institut Zairois pour la Conservation de la Nature were able to increase elephant herds to 11,000 and the rhino population doubled to 30.

The DRC's economy was doing (and continues to do) poorly, and refugees from numerous warring nations east of the DRC have moved into areas surrounding Garamba. Park guards make five dollars a month and routinely risk their lives protecting the wildlife. International organizations are interested in increasing their activities in Garamba.

For the second time in 1996, poachers killed northern white rhino in Garamba National Park in the DRC. Garamba has the world's last 28 wild white rhinos. In mid-February 1996, poachers killed a male rhino and hacked away its horn and meat. It was the first confirmed rhino poaching in 12 years. About one month later, a pregnant female was found dead and hornless. Rhino horn can sell for $15,000 to $50,000 a kilogram in Yemen, where the horns are used to make dagger handles. In China, rhino horn is used to lower a child's temperature and hence is purchased for medicinal uses. There are some 7500 white rhinos left, most of which are in South Africa; however, these rhinos are in captivity.

Conservationists have urged the DRC government to stop the poachers. However, Garamba is in northeastern DRC, far from the capital. Unless con-

FIGURE 15.1 A mother rhino and her calf in Ngorongoro National Park, Tanzania. She is guarded 24 hours a day by an armed guard.

servationists are willing to provide funding to support an antipoaching patrol staff and increase security for the rhinos, nothing will stop the poachers save the supply of rhinos. Dehorning the rhinos might also be a solution. However, poachers who follow rhino tracks around for several days don't appreciate finding a dehorned rhino. So as not to waste time tracking them repeatedly, poachers kill the rhino in any case.

The penis of a rhino can sell for $600 and is used as an aphrodisiac. In India, one zoo sells rhino urine for 44 cents a liter. The urine is presumed to cure asthma and sore throats. At least the urine is a sustainable product and requires that the rhino be living! The World Wildlife Fund claims there are fewer than 11,000 rhino of all five species in the wild. Sadly, the two rhinos in Figure 15.1 may be the last of an ancient lineage.

MANATEES

The Florida manatee (*Trichechus manatus latirostris*), a subspecies of the West Indian manatee (*T. manatus*), is one of the largest inshore mammals of the North American continent. Adults typically weigh 1000 pounds and are about 10 feet long. They are completely aquatic and herbivorous. Manatees are found in shallow, slow-moving rivers, estuaries, canals, and coastal waterways. They are without natural predators and can live for 60 years. Some individual manatees migrate 1700 km annually (O'shea et al. 1995). Manatees are protected under the Florida Manatee Sanctuary Act of 1978 (which designated the entire state as a refuge and sanctuary for manatees), the Endangered Species Act of 1973, and the Marine Mammal Protection Act of 1972 and their amendments. The manatee was federally listed as endangered in 1967.

Federal management activities are directed toward improving the population status of the Florida manatee so that the subspecies can be removed from the endangered species list. To achieve this goal, healthy marine ecosystems where the manatee resides must be sustained. Twenty-five years ago, manatees were studied by a single scientist. Today, millions of dollars are spent annually on the research and management of the species. Much of what is known about the Florida manatee is presented in detail by O'shea et al. (1995). After perusing this publication, one might think that everything humans need to know to save the manatee is already known. In addition, in October 1989, Florida's governor and cabinet directed the Florida Department of Environmental Protection to work with 13 key manatee counties in Florida where over 80% of manatee deaths occurred to reduce injuries and deaths. Furthermore, anyone can call the Florida Marine Patrol manatee hotline (1-800-DIAL-FMP) anytime to report injured, dead, harassed, tagged, or orphaned manatees. How could anything go wrong? Apparently, everyone is behind saving the manatee and there is no wildlife issue here.

On March 28, 1996, the front page of the *Tampa Tribune* read, "Concern mounts as more manatees die." Dead manatees were being found in "alarming" numbers in nearshore waters off southwest Florida. The second paragraph of the article summed up the crisis: "If the current death rates continue, the population of manatees in this region could be wiped out by fall." The most recent counts pegged Florida's manatee population at 1822 and the population of the four-county area including Sarasota County at 726 manatees. The article went on to say that there had been 163 manatee deaths statewide in the first 87 days of the year. "At the current rate, nearly four deaths per day [sic], the population could theoretically be wiped out by the first week of September." Things often have a way of appearing to be worse than they actually are, especially when calculations are off by a factor of two.

As a result of the manatee deaths, Florida assigned 35 more professionals to investigate, doubling the number of researchers working on manatees. On April 4, 1996, the *Tampa Tribune* reported on previous manatee research. For all of 1995, 201 manatees died. This was less than the 206 manatees that died in 1990, when cold weather stressed the manatees. In 1994, 193 manatees died. Table 15.1 summarizes the causes of manatee deaths for 1994 and 1995.

TABLE 15.1 Causes of Manatee Deaths

Cause of death	1994	1995
Watercraft related	49	42
Floodgate/canal lock	18	8
Other human related	5	5
Perinatal (newborn)	46	56
Cold stress	4	0
Other natural	33	35
Undetermined	37	53
Verified, not recovered	3	2
Total	**195**	**201**

On April 15, 1996, the *Tampa Tribune* ran another article: "Influx of dead manatees crowds marine research lab." Manatees were dying in record numbers. The lab had run out of cold storage space for carcasses awaiting examination. Bodies were lined up side by side. The article hinted that scientists suspected a virus. Two hundred manatees had been found dead statewide, and this was one shy of all of 1995. The dead manatees were all from southwest Florida. The manatee deaths appeared "unrelated to a rise in the deaths of green turtles and cormorants on the Gulf of Mexico coast," one researcher reported. The endangered turtles are reported to have a cancer-like ailment that causes tumors. Cormorants are diving seabirds, and their ailment seemed to be the result of eating fish killed by red tide, a natural marine toxin that blooms in the Gulf of Mexico. As for the manatees, "The only permanent populations in the United States are in Florida, where a count this year [1996] put them at 2639." Between the beginning of 1996 and April 5, 1996, the population of manatees statewide somehow grew by nearly 800. Despite more than 200 deaths through 1996, population counts showed an all-time high. Moreover, aerial surveys have shown that since the 1985 census, when there were an estimated 1200 manatees, the population of manatees has grown steadily. The figure 2639 is the result of better counting procedures and therefore is the actual minimum number of manatees in Florida on April 5, 1996.

On May 2, Beeman (1996) reported that manatee deaths had tapered off. Red tide was blamed for the deaths. A finger of 68-degree water laden with so-called red tide, a highly toxic algae, was located 40 miles off the coast of Florida, due southwest of Tampa. A cold front in March may have been enough to stress manatees already weakened by the toxic algae. As the algae is washed near shore, it breaks up in the waves and emits a neurotoxin in the form of an aerosol. Because manatees come up for air, they inhale the neurotoxin. Their lungs also become irritated. Since January 1, 1996, a record 258 manatees died, 52 more than the previous record of 252 in 1990. About two-thirds of the deaths were blamed on red tide. However, there were also a record number of manatees living in Florida coastal waters.

What Is the Wildlife Issue Here?

From 1974 to 1992, Ackerman et al. (1995) analyzed mortality patterns and trends in manatees. As shown in Table 15.1, human-caused deaths of manatees outnumber all other causes of death after manatees are a few months old. Furthermore, destruction of prime manatee coastal and estuarine habitat contin-ues. Humans and their activities are the greatest threat to manatee populations. Humans and manatees interact along Florida's waterways, and rarely does the manatee come out on top. Wright et al. (1995) reported that 1376 sets of fatal or healed propeller wounds were found on 628 dead manatees from 1979 to 1991. Collisions with watercraft caused 406 deaths. Nearly two-thirds of the manatees were killed by humans! Watercraft impacts caused more deaths than propeller cuts. The total number of registered boats in Florida has steadily increased with the human population. Many manatees are identified by their healed scars from boats. Perhaps as many as 40% of the manatees have scars. Private boaters report seeing groups of manatees all scared by propellers.

Manatees are migratory mammals; hence, establishing a sanctuary where propeller-driven boats are prohibited will work only part of the time. As soon as the manatees leave the sanctuary, they become nothing more than speed bumps for boaters. Many boaters feel that their right to own a boat and drive it as fast as they desire is just that—a right. Apparently the grass roots cam-paign that stopped net fishing off the Florida coast cannot be repeated for the gentle manatee. Even a law requiring boaters to install something as simple as a flexible plastic propeller protector is too difficult to impose. We already know that habitat destruction cannot be outlawed.

Continued research on manatees will refine what we already know, and no surprises are anticipated. Destruction of wetland habitat will continue through-out Florida. Many of the fresh and marine grass beds where manatees graze have been reduced or entirely destroyed by water pollution, herbicides, and dredging projects. The human population of Florida will continue to increase by about 2.7% per year, and the number of boats on the water will increase at a rate of about 3.4% per year. The number of manatees dying from collisions with boats will probably increase at about 9.3% per year in the future. With less available habitat for manatees, the chance of an infectious disease spreading increases. Natural population constraints will take hold when the density of manatees in an area increases beyond its holding capacity. Long-distance migration will occur more frequently as the manatees become relatively more crowded.

As The Associated Press reported on May 21, 1996, "Manatee protection hurts tourism, some say." Seven manatee sanctuaries have been named by the federal government as off limits to people and boats from November 15 to March 31. This is peak tourist season in Florida, of course. Several other sites such as the Crystal River might also be added to the sanctuary list. Dive shop owners, developers, and others dependent on tourism are not likely to support the sanctuary idea. In Crystal River, some 30 to 40 manatees gather and attract 70 or more tourists at a time. One wildlife biologist accused the Save the Manatee Club of being extremist in backing the sanctuary idea.

Reintroducing Manatees

Wildlife introductions have occurred, and indeed continue to occur, worldwide. There is a worldwide homogenization of plants and animals taking place before our eyes. Prickly pear cactus, rabbits, foxes, and dingoes were all introduced in Australia. The results have been catastrophic. Twenty-one parrot species live and breed in and around Miami. The English house sparrow is found nearly everywhere in North America. Coyotes are now invading Florida. Native Hawaiian birds have been displaced by introduced species. Yet some introductions fail. In 1995, 19 manatees were returned to their native habitat. During 1994, ten were reintroduced to the wild. Injured manatees found in the wild are treated and released. Orphaned newborns are cared for and then released. Care for captive animals is expensive, time consuming, and at times heartbreaking. In the wild, a young animal learns much from its mother. Learning in a natural setting sometimes makes the difference between life and death. Because of the number of manatees in southwest Florida, the *Tampa Tribune* published an article on manatee introductions on April 6, 1996. Individual manatees were named by the researchers. Several of the released manatees were outfitted with radio transmitters.

A young female manatee (called MD) came to the Lowry Park Zoo in Tampa in 1992 after being hit by a boat. She was treated in captivity and then released in June 1993. In August, she gave birth. Later she was hit by a boat that crushed her satellite transmitter, disabling it. MD was lucky she survived. Baby J. Coral was brought in after being entangled in ropes and dragging around a concrete block. She was treated and released. Eight months later, she was found starved to death after being stuck in a culvert.

Naples was in captivity for seven years before being released in July 1994 with her captive-born two-year-old calf Andrea. Also released with her was four-year-old Timehri, an orphaned wild calf she adopted as a newborn. Naples is reportedly doing fine and gave birth to a new calf in August 1995. Timehri's transmitter quit in November 1994 but she is believed to still be alive. After setting off on her own, Andrea attempted to return to the place where humans had cared for her. When cold weather came, she showed no signs of leaving. Manatees are cold water intolerant, and she was moved by humans to a nearby power plant with artificially warmed waters. Researchers noted that she had survived the winter and was on her way to becoming the first introduced manatee to make the transition to a free-ranging lifestyle. Sadly, Andrea was found crushed to death in the summer of 1995 in the Caloosahatchee River's Ortona lock. Apparently she had not learned on her own about the dangers and hazards all free-roaming manatees face.

A young male manatee named Palmer was outfitted with a transmitter before his release. Unfortunately, Palmer was too used to humans and took advantage of boaters who wanted to pet him. He was caught in the act of begging junk food from the back of a boat. Palmer was addicted to attention from humans.

Round-the-clock rescue efforts at Lowry Park Zoo failed to save three of four cold-stressed manatees. A fourth was released after less than two weeks in captivity. Lucky Frank was injured and later released. New Bob was brought in for treatment in 1992 as an orphaned newborn. He was released at a special "halfway house" for manatees at the Merritt Island National Wildlife Refuge but was soon back in captivity. In 1995, four manatees between three and four years old "showed promise" but were unsuccessful in their return to the wild. Marjorie, her captive-born calf Valentine, and an orphan calf named Graham visited Lowry Park Zoo for three weeks on their way to freedom. They were flown to the Everglades and released in September 1995. Marjorie lost her transmitter but is believed to still be alive. Valentine and Graham are now also free-ranging manatees in the company of others of their kind near Ten Thousand Islands. Successes do happen.

Soon after his release, Moose, an orphaned manatee rehabilitated at Sea World in Miami, was killed by a boat. Three other captive-born manatees also failed to survive their releases. Harvey failed to find warm water during a cold spell and died. Foster was captured for failing to learn the same lesson and was sent back to captivity. Indie set off on a journey of 200 miles before he was hit and killed by a boat. Many dedicated people, specially designed facilities, rehabilitation centers, halfway refuges, expensive satellite transmitters, and flights to monitor the manatees every week are part of the program to save the manatee. Millions of dollars, ten of thousands of hours donated by concerned people, and the best research top scientists can produce are up against habitat destruction and mobile boaters. On the other hand, nuclear power plants have provided artificial warm-water sanctuaries that have acted to increase the population and create habitat. We can only wonder how the next generation will judge our efforts to save the gentle manatee from extinction.

During 1996, 415 manatees died—more than in any year since researchers started counting manatees in 1974. The previous record was 206 deaths in 1990. Over the last five years, about 175 manatees have died, on average, every year. Of the deaths in 1996, 151 were due to naturally occurring red tide outbreaks along the west coast of Florida, and 60 were attributed to collisions with boats—more than in any other year. The population of manatees was estimated to be about 2639 at the end of 1996.

RATTLESNAKE ROUNDUP

Fifteen species of rattlesnakes are found in the United States. The western diamondback rattler (*Crotalus atrox*) is perhaps the most recognized, although it is found only in New Mexico, Texas, and Oklahoma. Snakes are economically valuable species because of the services they perform for humans. Many snakes eat rodents and thus perform valuable ecosystem functions for farmers. The U.S. trade in rattlesnakes is unregulated, and TRAFFIC–USA estimates that up to 125,000 rattlesnakes have been traded annually since 1990. A three-foot diamondback is worth about $21. What is so curious is the so-called "rattlesnake roundup" that is an economically important event in some towns.

The Sweetwater (Texas) Rattlesnake Round-up is the largest and most publicized event of its kind (Weir 1992). Upwards of 35,000 spectators and 18,000 snakes could be seen over a weekend in 1989. The roundup is now in its 39th year. Gross income for 1989 was $55,000 and net income was about $30,000.

Rattlesnake roundups are held in New Mexico, Pennsylvania, and Texas, as well as other states. Prizes are offered to participants, and large crowds are attracted to the spectacle. The chance to behead a snake is not uncommon. When a snake is decapitated, the head and body continue to writhe because of the snake's slow metabolism. Some people find it interesting to bet on when the head and body will actually stop wiggling. Any number of snake curios and oddities can be had, such as snakehead earrings, snakehead belt buckles, and snakeskin boots. If none of these interest you, you might want to bungie jump with a snake or perhaps stand in a pit of snakes. All this fun can be had at a rattlesnake roundup for about five dollars.

Cash prizes are often awarded for the largest snake, the smallest snake, most pounds collected, and to the collector who brings in the most snakes.

Knowing this, some snake collectors precollect, or bring in the largest home-raised snake they can find. Rounding up rattlesnakes does require a bit of skill. Techniques for collecting are highly sophisticated. A common method in Texas is to carry around a tank of gasoline that can be pumped up with air and used as a sprayer. The tank is pressurized and the gasoline is sprayed into the snake's den, usually in some boulder field. Care must be taken not to spray in front of the den because the snake only backs into its den. If too much gasoline is sprayed, the snake simply dies in its den and cannot be recovered. If the spray is directed deep in the den, the snake will flee. With patience and experience, this technique works every time! When the blinded and confused snake emerges, a collecting pole is used to snare it around the neck. The snake is then placed into a sack. Even a pillowcase can hold a dozen snakes. Dazed, starved, dehydrated, and near death, the snakes are delivered to holding pens for public viewing.

Some rattlesnake hunts raise money for charity, others are motivated by the profit from admission fees, and some towns just need a reason to exist. No license is required to participate. No research has been done on the effect on snake populations. As for the ethical treatment of snakes that participate in the roundups, standards simply do not exist, and it is unlikely that any will anytime soon.

The impact on all snake species is, as one might expect, not healthy. Many types of snakes are taken during the hunts. For every 100 rattlesnakes taken, 40 harmless snakes are also killed. Regarding the rattlesnake hunt, Shelton (1981) states that the "purpose of the Sweetwater Jaycees Rattlesnake Round-up has never been to exterminate the specie [sic] but to aid nature in controlling overpopulation of rattlers in the area." In the same article, Shelton implied that deer hunting may actually have increased deer populations. Thus, Shelton's contradiction becomes clear. In 1982, 8166 snakes, weighing 17,986 kg, were taken. In 1991, 2031 snakes, weighing 4474 kg, were harvested. No one has studied small mammal populations in these areas to see if they are increasing. Rattlesnakes consume rats, mice, ground squirrels, and gophers that would otherwise damage crops and pastures. Perhaps an outbreak of plague must occur before people realize that snakes are beneficial.

ZEBRA MUSSELS

On October 8, 1995, the *Miami Herald* proclaimed in bold print: "Mussels strain water world." Zebra mussels are tiny striped mollusks that successfully invaded North America in the late 1980s, when they were detected in the Great

Lakes. The mussels are freshwater organisms but can tolerate salt water for short periods of time. They are believed to have been introduced from northern Europe after being transported on the hulls and bilges of ships. Within two years of being recorded in Lake Erie, some areas had 35,000 individual zebra mussels per square meter. They have shut down power plants, destroyed native clam beds, and disrupted natural aquatic functions. Like other mollusks, zebra mussels are filter feeders, taking in algae and bacteria and ejecting clarified water. Their feeding activities thus concentrate pollutants. However, they are not native to North America, and thus their long-term impact is not known. Two-thirds of all native freshwater mussels are endangered in the United States and Canada, perhaps because of exposure to contaminants.

What is known is that the zebra mussels are being transported across the United States. Boats and barges transport the mussels and their larvae up and down the Mississippi River. Pleasure boaters inadvertently have transported them to California. Meanwhile, the zebra mussels continue to adapt to their new environments. A mollusk's reproductive capacity is enormous; hence, adapting to a new environment is enhanced. Natural selection can quickly select against offspring that are poorly adapted. Thus, speciation is occurring in the zebra mussels. Boaters are being asked to check their boats for the mussels, but the zebra mussel's larvae are microscopic and hence nearly undetectable. It seems certain that it is just a matter of time before the entire United States is invaded by this adaptable menace. The biology and ecology of the zebra mussel will allow the organism to gain access to most, if not all, of North America's freshwater systems.

The Tennessee Valley Authority is emptying some of the lakes it manages to control the infestation of zebra mussels. The mussel population is so dense that water pipes become clogged by mussels. Screens installed to prevent fish from entering pipes have allowed mussels to completely close off water pipes. So far, the only way to eradicate an infestation is to drain the lake. However, the larvae follow the water out and settle elsewhere. From the zebra mussel in fresh water to introduced fish, birds, mammals, and plants, North America and most every other populated area is becoming homogenized.

GRIZZLY BEARS

Grizzly bears (*Ursos arctos horribilis*) existed across much of the western United States from as far south as the Mexican border up through Canada and into Alaska. The bear on the California state flag is a grizzly bear. Sadly, grizzly bear populations have crashed across their range. They have been extirpated from Oregon south and from Colorado south as well. In 1979, a grizzly bear report came from southern Colorado (Barrows and Holmes 1990). The U.S. Fish and Wildlife Service has created a grizzly bear recovery plan to aid one of America's most mythical creatures, a symbol of all that was wild and free and without any predators save primitive human hunters.

Grizzly bear populations in the lower 48 states now exist in five forested areas in the northwest. In Yellowstone National Park, an estimated 250 bears survive. Several hundred occupy Glacier National Park as well as an adjacent national forest in northern Montana and Canadian forests. Little more than a dozen live in the Cabinet Mountains on the Idaho–Montana border. A few dozen are believed to inhabit the Selkirk Mountains of northeastern Washington and Idaho. Perhaps grizzly bears use the Northern Cascade Mountains in north-central Washington. Other mountainous areas nearby, such as the Selway-Bitterroot Mountains in Idaho, are being considered as a reintroduction area.

Seeing a grizzly bear in Denali National Park in Alaska is an exciting experience. Most visitors see them from inside one of the many buses that take people back and forth across the park. However, this experience does not compare to seeing a grizzly bear on foot out in the middle of the park. Picture, for a moment, hiking across one of the braided gravel rivers found in Denali. Far above you and several miles away is the road where the tiny yellow buses move back and forth. You cross a stream on foot when you see a grizzly sow and three cubs making their way toward the stream. The cubs are trailing, turning over rocks and poking their noses into everything. Suddenly the wind shifts and the mother bear has picked up your scent more than a mile away. For a moment, fear washes over you like a warm wave. You freeze and your legs feel weak. The bear could make up the distance between you in less than two minutes. You turn and calmly walk directly away from the bear as she goes about her routine. Bear attacks on humans in Denali are extremely rare and almost always caused by humans stumbling onto bears. It's a good feeling to know that grizzlies are still around and we can experience them on their terms.

Once again, all the laws on the books won't save the grizzly bear unless something is done to save suitable habitat. Grizzly bears are top carnivores. Their requirements are enormous and, as in many, many other cases, their habitat is shrinking. Grizzly bears are known to have large home ranges and thus need large areas to exist. Unfortunately, roads in one form or another run through most areas. Large roadless areas are few and far between. The approach of modern forest management is to designate land as limited access or no access and then attempt to connect these large tracts of forest with smaller unoccupied corridors through which wildlife, including bears, can move. The

Kootenai National Forest, which includes the Cabinet Mountains, is a good example of what is wrong with the present management scheme. Since the grizzly was "promoted" to the federal threatened list in 1975, nearly 7000 miles of roads for loggers have been built in the forest and some 5 billion board feet have been removed. To be an effective bear sanctuary, the roads would have to be blocked off, logging would have to cease, and hunting would have to be discontinued.

Some people favor multiple use of public land. If the grizzly bears disappear from the lower 48 states, they can just as easily go to Alaska or Canada to see them. Others suggest that all logging on public land, including national forests, must stop for the foreseeable future. Millions of miles of roads in these forests must be removed and revegetated. Hunting would be restricted to certain areas known to have no bears. These people are not radical environmentalists—they are you and me.

SOUTHWESTERN WILLOW FLYCATCHERS

The *Arizona Republic* reported that Phoenix and five other area cities agreed to share the $440 million cost of raising Roosevelt Dam by 77 feet in the belief that the cities would receive more water. However, the habitat of 40 southwestern willow flycatchers (*Empidonax traillii extrimus*), an endangered species listed on February 20, 1995, would be significantly altered so as to be useless to the small birds. The birds migrate between Central America and the Southwest riparian habitats. About 160 willow flycatchers live along Arizona's rivers, and additional birds are found in New Mexico and southern California. No environmental assessment was undertaken by the Salt River Project (which manages the dam), and the city partners have already invested millions of dollars in the project. Willow flycatchers have "high site fidelity"; that is, they return to the same sites to nest and raise their young.

The Endangered Species Act not only may prevent the loss of the partners' initial investment but will probably force cancellation of the entire project. Creation of new habitat is very expensive and probably will fail in the long term.

Raising the level of the dam to impound more water is not a viable long-term solution. The dam "improvement" project would have allowed a million more people to move into the Phoenix area. This would have taken a decade. Then what would be done? Is more growth the answer? Is more water the answer? How much is enough? The long-term survival of the willow flycatcher seems inconsequential when addressing the heart of the issue. Are we really treating the disease or just its symptoms?

LOCUSTS

In April 1996, an enormous swarm of locusts moved across agricultural lands of the former Soviet Republic of Tajikistan and posed a serious threat to the economy of the nation. At least 250,000 acres of cotton and fruit crops were the insects' target. The locusts invaded from the southern plains of the Pamir Mountains to the east. The swarm was so huge that no attempts were made to stop it. Could it be that birds controlled these outbreaks in the past?

MEXICAN WOLVES

On May 2, 1996, The Associated Press reported that ranchers who see a reintroduced Mexican wolf attack and kill their livestock on private land can kill the wolf. The rule was proposed by the U.S. Fish and Wildlife Service because it wants to reintroduce wolves in New Mexico and Arizona. Al Schneberger, director of the New Mexico Cattle Growers Association, was quoted as saying that such proposals "aren't worth the paper they are printed on." The cattlemen distrust the federal government and believe federal authorities would renounce the agreement once wolves are reintroduced.

The Endangered Species Act has a special section that allows designation of the Mexican wolves as "non-essential and experimental," thus allowing more flexibility in their management. One problem cattlemen have is that if a Mexican wolf is attacking stock on public land, the wolf cannot be harmed. In the West, it is common for a rancher to own and live on one acre and pay less than two dollars per head per month grazing fees to use public land. Of course, grazing rights are a privilege and not a right. Thus, a simple solution would be to not grant any grazing rights to the public land occupied by the wolves.

The Arizona State Congress threatened to pass a law offering a bounty on any Mexican wolves shot in the state.

CALIFORNIA GNATCATCHER

The California gnatcatcher (*Polioptila california melanura*) lives in coastal sage scrub in southern California, south of Los Angeles. Developable land in this part of the United States can be extremely expensive. On March 30, 1993, the gnatcatcher was listed as threatened under the Endangered Species Act. However, many other species living in the coastal scrub are just as unique as the gnatcatcher, and all need protection in one of the world's most rapidly developing regions.

Mann and Plummer (1995) described an encounter between developers who purchased land, environmentalists who want the land left alone, and the U.S. Fish and Wildlife Service, which is responsible for protecting species. In the high-stakes world of modern development, tens of millions of dollars ride on housing layouts and access roads. Environmentalists argue that setting aside a connected sequence of reserves is essential to protecting species. An isolated sequence of reserves dissected by roads and billboards won't work, they say.

The gnatcatcher has already temporarily halted development on prime developable land. A southern California developer purchased a large tract of coastal scrub for $180 million in northern San Diego County. The developer planned to build 3000 upper-middle-class homes on the property. Oddly, the property was purchased by the development company *before* professional biologists surveyed it. Had the company paid biologists to survey first, the developers would have saved their money. Instead, the profit motive drove them to purchase the property first and ask questions later. This was a serious mistake as it turned out. Habitat conservation plans, amendments to the Endangered Species Act in 1982, and the formation of the Natural Communities Conservation Planning (NCCP) program in California in 1991 resulted. The NCCP was supposed to "inaugurate a new era of consensus" between all interested parties, developers and endangered species alike.

The very beauty of the California coastal sage scrub, which drew some people to live near it, is driving developers to destroy it. It is prime developable land. Meanwhile, more controversies over development across the United States are occurring. Are more shopping centers needed? How much habitat is going to be lost? There will be more and more issues similar to the California gnatcatcher, spotted owl, and manatee.

KILLER BEES

Four swarms of Africanized "killer bees" were reported in Nicaragua, the *Albuquerque Tribune* reported in April 1996. One swarm killed two elderly people. Eleven people were hospitalized after being stung during a Holy Week procession. In Costa Rica, one person died after being stung.

EXOTICS

During the 1960s, the New Mexico State Game Commission began a program to introduce exotic, big game wildlife into New Mexico to garner sportsmen's dollars (see Chapter 10). Barbary sheep from Africa, Persian ibex from Iran, Siberian ibex from Siberia, and oryx from Africa were introduced, and all succeeded. By the mid-1980s, the introduced species' populations exploded and overpopulation became a problem. The introductions may compete with native wildlife and may eat native plants, some of which are endangered. In some areas of the state, ibex can be hunted all year. Population control programs for the oryx were started.

PETS

More than ten million animals are euthanized (killed) every year in the United States. There simply are too many pets that are making even more pets. Pet overpopulation is an immense problem. Even pet owners turn in their pets to be killed. Our disposable society throws away even a family pet. People allow their pets to roam at large, without neutering them. Wild cats and dogs breed freely. Cats are wild birds' second worst enemy.

POACHING

Undercover agents working for U.S. Fish and Wildlife Service's Division of Law Enforcement see some gruesome activities. In a *Time* magazine article, Van Biema (1994) wrote about rampant poaching in U.S. parks. He suggested that 300 bears are poached every year for their body parts, including gallbladders. Not even bear cubs are immune from the illegal hunters. From spiders to elk and from butterflies to bighorn sheep, the slaughter in all 366 parks continues. Figure 15.2 shows the efforts of one poacher in a Rwandan national park.

Van Biema estimated that poaching of animals is a $200 million cash business. Species such as the hawksbill sea turtle, brown pelican, peregrine falcon, and Schaus' swallowtail are endangered but are shot or taken anyway. Of course, budget cuts make poaching a lucrative business. In 1994, there were 60 full-time rangers to patrol Yellowstone National Park, which is the size of Delaware and Rhode Island.

What drives poaching? The answer is money. Poaching is defined by federal law as hunting protected wildlife for a profit of $350 or more. A mounted bighorn sheep can be worth as much as $10,000. A grizzly bear can go for $25,000. A bear gallbladder can be worth $64,000 in Asia. Some poachers are notorious and are preoccupied with fame. One poacher was videotaped attacking a herd of elk in Yellowstone National Park. He was fined a mere $15,000

FIGURE 15.2   A poacher's bounty is confiscated in Rwanda. The skin of a cerval, an African cat, is held by the ranger in the center. On the left is a blanket of more than a dozen antelope skins.

and served only 30 days of an 18-month prison term. The courts, it seems, do not take wildlife crimes seriously.

Van Biema also reported on a snake-poaching ring. Snakes were caught for skins and for pet shops. Outbreaks of rats were reported in Big Bend National Park in Texas. Poachers had collected so many of the beautiful gray-banded king snakes that rat populations had exploded.

In Bandelier National Park in New Mexico, mule deer and elk poaching is a problem. Several years ago, a ranger disguised himself as a hunter. Four poachers entered the park from the western boundary in the Jemez Mountains. One of the hunters shot a large mule deer. The ranger heard the shots and located the hunters. The ranger turned on a hidden microphone and also took out his camera. Suspecting nothing, each hunter posed with the trophy deer. The ranger took their addresses so he could "send them pictures." Within a week, all four hunters were arrested for poaching. Sound, photos, and the ranger's testimony led to pleas of no contest by the four men. They were fined $100 each and their guns were confiscated for the remainder of the hunting season.

VOTING ISSUES AND HUMAN VALUES

Both Idaho and Michigan have referenda prohibiting the use of dogs when hunting bears. It is no small task to get a proposed initiative on a state ballot. In Idaho, at least 41,335 (10% of registered voters) signatures were required in 1996. The Idaho initiative reads:

> It is the intent of this act to prohibit the taking of black bears when female black bears are rearing their cubs. It is further the intent of this act to promote the concept of fair chase by eliminating the taking of black bears through the use of bait and dogs.

Voters will decide the issue. In 1995, 96% of all bear hunts in Michigan used dogs, and 30% of 5600 permittees were successful. We do not know whether male and female bears were taken. Oregon voters passed similar legislation in 1994. Only about half as many bears were killed in 1995 as in the previous year. Perhaps as a result, nuisance bear complaints were up 39% and 205 bears had to be destroyed. Are we certain that the reduced number of bears killed led to the increase in the number of nuisance bears? What did the bear population as a whole do? What did the human population do? Is it possible that there are fewer bears and many more humans, for instance?

Across the United States, voters are demanding that action be taken on a variety of issues ranging from gambling, term limits, child care, school choice, taxes, and even the legality of smoking marijuana. Wildlife issues are appearing on state ballots as well. In most cases, the long-term ramifications of successful initiatives will not be known.

Recall what the New Mexico state wildlife official said: "Scientists tell us what is going on, but the public tells us what to do." Typically, scientists gather facts, organize them, and make these facts available. Generally, scientists do not make policy decisions, although their recommendations are heard. In the coming months and years, voters will decide many wildlife issues.

According to high-level management in the National Rifle Association, the public is not educated enough on issues regarding hunting, wildlife management, and science to make informed wildlife management decisions. Animal rights groups suggest that the public is on their side when it comes to hunting and that it is the elected officials who will not carry out the will of the public.

Scientists can help educate the public and bureaucrats alike but cannot answer the moral questions that always arise. Now that humans have altered forever the delicate predator–prey relationships, humans must decide what to do. Of course, the best management strategy would be to manage ourselves into a position where no management decisions have to be made. This is the way things existed for hundreds of millions of years—before human impacts became the dominant force they are today.

Unfortunately, there are no easy answers. Albuquerque, New Mexico, seems an unlikely place to have too many beavers—but it does. Beavers live in bank dens along the Rio Grande. They routinely attack orchards and New Mexico's beloved cottonwood trees. Beavers have been blamed for the lack of regeneration of the cottonwoods, but, in fact, cottonwoods require spring flooding to germinate. Flood control programs have effectively prevented floods. Irrigation channels act as invasion corridors for beavers to penetrate to city limits, bringing beavers and pets into conflict. Beavers have become a wildlife issue in Albuquerque. Should hunters be allowed to take beavers?

Florida residents' fear of panthers is not shared by California voters who rejected a plan to allow hunting of the animals, even though a woman jogger was killed by a panther.

No doubt, more wildlife issues will find their way onto more state ballots. One result of a successful initiative will be that the number of nuisance animals will increase. There will then be a need to better define nuisance animals. When is a bear a nuisance bear or an alligator a nuisance alligator?

In Florida, a nuisance alligator is in the eyes of the beholder. If an alligator is sunning itself near a boat, it can be considered a nuisance. One call to the local state wildlife office is all that is required. A professional state-certified hunter will "harvest" the alligator at no charge to the state or the caller. In return, the hunter can sell the meat and hide and pocket the profit. The hunters are always successful.

In 1967, the American alligator was one of the original entries on the endangered species list. Today, there is a hunting season on alligators in Florida. Some 5500 nuisance alligators were killed in 1995, and the number of nuisance alligators continues to increase (see Figure 3.3). Indeed, the number of legally hunted alligators is less than the number culled as nuisances!

If a female bear cannot find enough food for herself, she might abandon her cub. The cub usually starves to death for lack of food. Is a bear cub a nuisance if it turns over a garbage can looking for food? Should it be destroyed? Should the female bear be destroyed if she raids a dumpster?

Normally, bears do not raid dumpsters unless they are driven to do so. Humans have, in many cases, destroyed the natural habitat of the bears. In the West, the forests have been logged and overgrazed. Humans have altered the fire regime and damaged the streams so they no longer support fish. Housing developments abut wilderness areas, and the prime lots are those on the perimeter of wildlife areas. In every way, humans have reduced the quantity and quality of wildlife habitat.

Much money is at stake in these initiatives. The National Rifle Association spent $100,000 trying to advance a panther hunt in California. The Humane Society spent $60,000 urging voters to prevent the hunting of panthers. These issues are seen in the larger framework by organizations on both sides. One side might see the issue as follows: "If panther hunting is prevented in California, all hunting will eventually be stopped in the state. Then 'they' will ask us to turn in our guns. The end is near." The other side might see things differently: "If we don't stop the panther hunt, 'they' will want to hunt everything. Soon the hunting season will be 12 months long and everything will be killed by any and all means." Fortunately, public beliefs are somewhere in the middle. A study by the U.S. Fish and Wildlife Service found that 75% of Americans approve of hunting, but a majority of those people also believe that a lot of hunters break laws. Hunters, its seems, are their own worst enemy.

WILDLIFE SERVICES

The federal agency responsible for damage and other problems caused by wildlife is called Wildlife Services (formerly known as Animal Damage Control). The agency exists because wildlife in the United States belongs to the people. Thus, when wildlife causes damage to private property, the people (i.e., the government) are responsible for managing the damage. According to the Wildlife Services Web page, its activities include dissemination of information to the public, research, and protection of endangered species. For instance, coyotes are reported to kill endangered San Joaquin kit foxes in California, and beavers have become abundant in Louisiana and through their activities threaten the habitat of endangered mussels. Wildlife Services claims it is not interested in eradicating predator populations; it simply manages the damage that individuals cause.

There can be no doubt that there are populations of various species that are extremely abundant and cause damage to private property. Aquaculturists, for instance, claim to suffer losses in their stocks due to bird predation. We also note in the Epilogue that snow geese in North America have become so abundant that they appear to threaten fragile arctic vegetation. It is a grim reality that protection and restoration of previously human-damaged habitats can lead to increases in some wildlife species and sometimes damage to natural resources and private property.

Unfortunately, Wildlife Services (http://www.aphis.usda.gov/ws) is not without its critics. The nongovernmental organization Predator Defense Institute of Eugene, Oregon (http://host.environlink.org.pdi.Default.htm) has criticized the agency ardently for killing predators (mostly coyotes) on public lands in the western states. Recently, another nongovernmental organization, Forest Guardians (http://www.fguardians.org), reported in its online newsletter *Frontline* (issue 23, June 24, 1998) that the U.S. House of Representatives had voted to cut funding for predator control by $10 million. According to the report, funding for the predator control program under the old name of Animal Damage Control had increased by 70% between 1983 and 1993, whereas the number of predators killed increased by only 30%. At the same time, the report also states that livestock losses due to predation "did not decrease."

In the western states, Wildlife Services uses several techniques to kill problem wildlife (mostly coyotes). In addition to the usual techniques of leg-hold trapping and shooting from aircraft, Wildlife Services also uses something known as the M-44. This is a spring-loaded device that is placed in the ground and baited with an attractive scent. When tugged on by a coyote (or other animal), it sprays sodium cyanide into the animal's mouth and nose. The animal then may die an excruciating death. A more hideous chemical once widely used was the poison known as 1080 (sodium fluroacetate).

Several questions come to mind. Why is the government killing coyotes on public land? Indeed, why is there still grazing on public lands? As articulated by Forest Guardians, livestock grazing on public lands amounts to a subsidy for private ranchers, especially those east of the Rockies. Is grazing on public lands a wildlife issue? You bet it is.

MANAGEMENT PRACTICES

Cattle grazing in unique riparian habitat along the Rio Grande in New Mexico, the so-called Rio Grande Bosque, is a big problem for outdoor recreationists. Cattle roam freely and are "destroying the wildlife habitat," one fisherman told the *Albuquerque Journal*. Some of the local residents claim they have grandfather laws that allow them to let their cattle roam at large. In New Mexico, fences are built to keep cattle out, not in. One duck hunter suggested that the Bosque looked like a corral. Locked gates were installed to keep people from driving in and dumping trash, cutting wood, and shooting guns and to keep cattle out as well.

The *Albuquerque Journal* reported that on May 20, 1996, Santa Fe–based Forest Guardians sued to force the U.S. Bureau of Land Management to conduct studies to determine how grazing on public lands was affecting several endangered birds and fish. Forest Guardians questions whether any level of grazing along streams and rivers is appropriate. Forest Guardians and other environmental groups have an ongoing injunction against logging in New Mexico and Arizona. It remains in effect until studies are completed to determine how the Forest Service's timber plans affect the Mexican spotted owl.

Both lawsuits demand studies on how listed species are affected by wideranging land management practices. The idea here is to look not only at the "content" of an area but also its "context." That is, grazing may not affect southwestern willow flycatchers in the immediate area, but a cattle-damaged mountain stream may silt up and foul water where there are willow flycatchers. In New Mexico, the Bureau of Land Management manages 500 miles of streamside land and 13 million acres.

COMMUNITY GROWTH

If growth is good for communities, more growth must be better. Most people are, in fact, in favor of growth. But is growth really good? Who benefits from growth? How many development projects are taking place within town limits across the United States? Such development projects are merely filling in areas and are not expansive. But who really benefits from such projects? Certainly the developer benefits. Local construction company workers benefit, temporarily at least. Do residents surrounding such projects benefit? The next time you are sitting in your car at a traffic light, ask yourself, "Am I benefiting from the growth of my community?" Chances are you are not. Your taxes have probably increased to pay for widening the streets to accommodate the increased traffic. Perhaps the police force had to be increased. New residents must pay taxes. How much comes back into the community? How have you personally benefited from this growth? Has this growth enhanced the quality of your personal life? How has growth affected wildlife?

EXERCISES

15.1 Make a list of the freshwater fish in several states in different parts of the United States. Where are many native freshwater fish in trouble?

15.2 Make a list of fish species introduced in the United States. Investigate some of these introductions in detail. What has been their role with respect to native species?

15.3 Freshwater fish have unique evolutionary histories. Compare freshwater fish in New Guinea to Australian freshwater fish.

15.4 Use the Web to search for information on migratory birds. What kind of information comes up?

15.5 Do a Web search using "salmon" as a keyword. Summarize some interesting findings.

15.6 Search the Web for information on sashimi and sushi bars. What meat do they serve?

15.7 Do a library and a Web search for "pupfish." How many species of pupfish are there? How did they get to be where they are today?

15.8 Investigate the introduction of Nile perch into Lake Victoria in Africa, where millions of people depend on the lake fishery for food.

15.9 Suggest how to test theoretical causes of declining amphibian populations using the scientific methodology discussed in Chapter 1. How can low-latitude effects be isolated?

15.10 Can the effect of UV radiation be tested directly on amphibians to determine their susceptibility? Suggest an experiment.

15.11 Formulate testable hypotheses regarding the issue of bears and humans. How would you gather data to test your hypotheses?

LITERATURE CITED

Ackerman, B.B., S.D. Wright, R.K. Bonde, D.K. Odell, and D.J. Banowetz. 1995. Trends and patterns in mortality of manatees in Florida, 1974–1992. *in* O'shea, T.J., B.B. Ackerman, and H.F. Percival (Eds.). *Population Biology of the Florida Manatee.* National Biological Service, Washington, D.C.

Barrows, P. and J. Holmes. 1990. Colorado's Wildlife Story. Colorado Division of Wildlife, Denver.

Beecher, W.J. 1955. Late-Pleistocene isolation in salt-marsh sparrows. *Ecology* **36**(1): 23–28.

Beeman, K. 1996. Manatees dying less frequently. *Tampa Tribune* May 2.

Blaustein, A.R. and D.B. Wake. 1990. Declining amphibian populations: a global perspective. *Trends in Ecology and Evolution* **5**(7):203–204.

Dewar, H. 1996. Not in your backyard. *Albuquerque Journal* April 21.

Di Silvestro, R. 1996. What's killing the Swainson's hawk? *International Wildlife* May/June:38–43.

Karesh, W.B. 1996. Rhino relations. *Wildlife Conservation* **99**(2):38–43.

Longstreet, R.J. 1955. Dusky seaside colony endangered. *The Florida Naturalist* **26**:78.

Mann, C.C. and M.L. Plummer. 1995. California vs. gnatcatcher. *Audubon Magazine* **97**(1):38–48, 100–104.

O'shea, T.J., B.B. Ackerman, and H.F. Percival (Eds.). 1995. *Population Biology of the Florida Manatee.* National Biological Service, Washington, D.C.

Phillips, K. 1994. *Tracking the Vanishing Frogs.* Penguin Books, New York.

Shelton, H. 1981. A history of the Sweetwater Jaycees Rattlesnake Round-up. pp. 95–234 *in* Kilmon, J. and H. Shelton (Eds.). *Rattlesnakes in America.* Shelton Press, Sweetwater, Texas.

Swardson, A. 1994. A loss that's deeper than the ocean. *Washington Post National Weekly Edition* October:124–130.

Van Biema, D. 1994. The killing fields. *Time* August 22:36–37.

Walters, M.J. 1992. *A Shadow and a Song: The Struggle to Save an Endangered Species.* Chelsea Green Publishing, Post Mills, Vermont.

Weir, J. 1992. The Sweetwater Rattlesnake Round-up: a case study in environmental ethics. *Conservation Biology* **6**(1):116–127.

Wright, D.D., B.B. Ackerman, R.K. Bonde, C.A. Beck, and D.J. Banowetz. 1995. Analysis of watercraft-related mortality of manatees in Florida, 1979–1991. *in* O'shea, T.J., B.B. Ackerman, and H.F. Percival (Eds.). *Population Biology of the Florida Manatee.* National Biological Service, Washington, D.C.

OCEANIC INTERNATIONAL ISSUES

16

n summarizing a recent conference, Baker (1995) concluded that "tensions
between commercial and recreational fishermen, between U.S. and foreign
fleets, and among fishing interests, environmentalists, and resource managers create a climate of finger-pointing rather than common purpose." Few trust
the government or regulators. "If there is uncertainty, we need to err on the side
of protection of the resource," said D. Hall, deputy administrator for the National Oceanic and Atmospheric Administration. It seems that politicians do not
want to err on this side. In truth, we are witnessing the end of a cycle of human
mismanagement of a world resource. With each harp seal death, biodiversity
decreases and more predatory fish that feed on young cod survive. The cod are
being hit with a "double whammy." They have been overfished by humans, and
now their underwater protectors are being decimated by humans. Without cod
to eat, predatory fish will switch to another resource, such as hake, and decimate their populations. A downward spiral will ensue, and the richness and
abundance of species that have co-existed for millions of years will have collapsed directly by the hand of humans.

HARP SEALS AND ATLANTIC COD

As noted earlier, the Canadian government blamed the harp seal (*Phoca
groenlandica*) for the decimation of one of the richest and most productive
fisheries on earth, although harp seals have been harvested commercially since
about 1700 (Busch 1985) (see Figures 16.1 and 16.2). In 1996, Fisheries Minister Brian Tobin announced that the total allowable harp seal harvest for the
year would be raised from 186,000 seals to 250,000 seals. Even more impor-

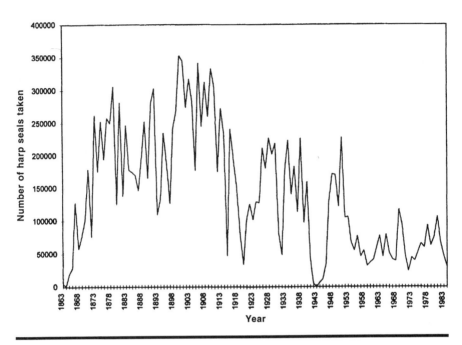

FIGURE 16.1　Annual Newfoundland harp seal harvest.

tantly, a 15 cents per pound bounty was placed on seals. The government also encouraged the use of large vessels to help hunters reach breeding colonies on offshore ice floes. In this way, large colonies of breeding females could be harvested, or massacred, depending on your viewpoint. Canada was not alone; Iceland, the Faroe Islands, Norway, and Russia also announced harvests of seals. An environmental group, Animal People, reported that Canada was close to signing an agreement with an Asian country that called for delivering 250,000 seal carcasses a year.

　　Environmentalists reacted strongly against the raised quota and the bounty. The International Wildlife Coalition and the International Fund for Animal Welfare, for instance, condemned the action. Clubbing the helpless seals seems like a senseless act of violence. Who among us has not seen the tiny, white ball of fur with coal-black nose and eyes staring so innocently? Is the hunt humane? Environmentalists say no. The hunt is inherently uncontrollable, because there are only a few inspectors and the range of the seals is enormous across the frozen North Atlantic. Monitoring the hunt is impossible.

　　The Canadian government and the environmentalists have faced off against each other in a public battle. Indeed, by focusing attention on the harp seal

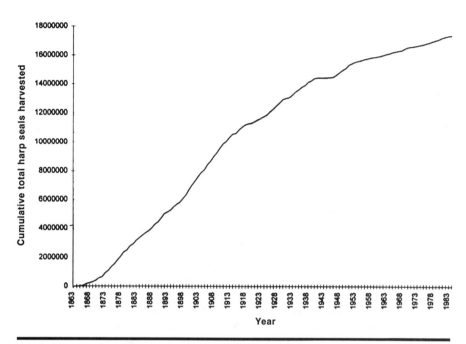

FIGURE 16.2   Cumulative number of Newfoundland harp seals harvested.

harvest, public attention is being deflected from the real wildlife issue. Is the real wildlife issue the clubbing of baby seals or the loss of the entire North Atlantic fishery? Is the real problem the Canadian government policymakers or the harp seals? How can we, as scientists, address this issue? Do politicians have the will to turn scientific information into policy?

The Canadian government closed the largest Atlantic cod (*Gadus morhua*) fishery on July 2, 1992 for a two-year period. The Newfoundland fishing industry was the backbone of the province's economy, and the closing was a disaster for fishing communities scattered along the harsh coast. For 350 years, the fishery meant life to these communities. Survey data from fishing ships showed a dramatic decline in fish size taken from 1989 to 1992 however.

Historically, northern cod catches peaked in about 1968 at about 800,000 metric tons. The historical average preceding that spectacular catch was 200,000 metric tons. Foreign trawlers—fish factory ships—were largely responsible for the additional catch. The catch of 1968 caused catch forecasts for other years to be raised. In 1977, the offshore fishery belonged to Canada when it declared a 200-mile fishery zone from its shores. During the 1980s, the fishing in Newfoundland strengthened. Cod population assessments by scientists led to lowered quotas. Many untestable statements were made, such as "the fish are hiding" and "unusually cold weather did them in." Another statement was testable: "the seals ate them." A 1992 survey indicated that the cod fishery was on the verge of commercial collapse and the fishery was closed. The usual quotations from sea captains appeared in the media: "I've been fishing cod out of these waters for 20 years and there are more cod now than there were when I got started." Despite such statements, the cod population has not rebounded.

Many theories have been offered to explain why the fishery collapsed. Some suggest overfishing by humans caused the population to collapse beyond its ability to sustain itself. The so-called *tragedy of the commons* is typically invoked to explain people's behavior toward a shared resource, where the attitude might be something like, "I better get mine before someone else does." Foreign floating fish factories were blamed for taking too many fish before the 200-mile limit was established. However, cod stocks located beyond the limit were then fished out.

Cod are omnivorous feeders and attain weights of up to 35 kg (Serchuck et al. 1994). These fish can live in excess of 20 years, although young fish make up most catches. Cod can breed as early as two years old and certainly by the age of four (Trippel 1995). The Georges Bank of Newfoundland and the Gulf of Maine cod fisheries have been heavily exploited until recently. The 1991 fishing season yielded 48,000 metric tons. In 1992, 39,500 metric tons of cod was harvested from these two fisheries. The total cod fishery take in 1993 was 31,400 metric tons. Both Canadian and U.S. harvests declined.

Survey data show a consistent trend toward catching, on average, younger and younger fish each year. Other results show that individual cod born in 1987 have been fished out almost entirely. Commercial landings in 1993 were dominated by the 1990 year class, which accounted for 55% of the catch by number and 41% by weight. Older fish were almost nonexistent in the catches and middle-age fish were marginally represented. Examination of on-board catches in 1994 showed that the year classes of 1991, 1992, and 1993 were the lowest on record. Recruitment into the population was extremely poor. Worse still, spawning stocks were expected to decline to unprecedented low levels. The Georges Bank cod stock "is at a very low biomass level, and is overexploited. Without substantial reductions in mortality, there is a possibility of stock collapse" (Serchuck et al. 1994).

Scientists were responsible for gathering data to make surveys and projections, but we now know that their research methods and assessments were flawed. For instance, separate cod populations were treated as a single unit. Natural mortality, recruitment, and other variables were treated as constants in models that then gave overly optimistic results. Scientists also collected data aboard ships. However, it was common practice to omit undersized fish whose numbers seemed to increase as harvested stocks decreased. Such blunders led to underestimates of mortality and overestimates of cod populations. The harvests continued until fishermen were shut down completely. In 1992, a way of life for hundreds of years was tossed into the dustbin of history. The Canadian government blames the harp seals for this demise.

"There's only one major player still fishing that stock. Its first name is harp; its second name is seal," said Canadian Fisheries Minister Brian Tobin. But are harp seals responsible for the collapse of the fishery? Can it be that unless harp and hooded seal populations are regulated by man, they could decimate the fishery?

The pinnipeds (fur seals, harbor seals, and walruses) are mammals that returned to the sea approximately 23 million years ago. Their former mammalian limbs became flippers. Enaliarctos (sea bear) is the earliest known seal and comes from the late Oligocene or early Miocene of the Pacific Ocean near what is now California (Janis 1993). Enaliarctos was otter sized and might have looked like an otter with flippers. Unlike modern seals, Enaliarctos lacked the specialized features of the inner ear that enable modern seals to deep dive. From the North Pacific, they dispersed and speciated into the pinnipeds we know today. Today, pinnipeds generally occupy polar and subpolar regions; however, exceptions to this rule are the Hawaiian monk seal (*Monachus schauinslandi*), the Caribbean monk seal (*M. tropicalis*) (now presumed extinct), the Mediterranean monk seal (*M. monachus*), and the Galapagos fur seal (*Arctocephalus galapagoensis*) (Lavigne and Kovacs 1988). The Caspian Sea, the largest inland body of water, is occupied by seals—evidence that this sea was once connected to the Black Sea and the Mediterranean Sea.

Adult harp seals are about 1.7 m (5.5 feet) long and weigh about 120 kg (285 pounds). Harp seal distribution and range limit are determined by pack ice throughout the North Atlantic (Lavigne and Kovacs 1988). Their distribution spreads across the North Atlantic from northern Newfoundland and Labrador, part way into Hudson Bay, along most of the coast of Greenland, and across the North Atlantic to Iceland, the Scandinavian countries, and the coast of Russia. In late autumn, the harp seals move in front of the advancing ice sheet. After a winter spent feeding in the frigid waters, adult females assemble by the thousands on their traditional whelping (pupping) grounds. Most give birth to a single pup.

The northwest harp seals give birth in late February and mid-March. There are three principal breeding grounds, and morphological studies show that there is little mixing of the populations. After pups are weaned, the adult females join the males for the annual mating season. Shortly thereafter, the harp seals molt. During the annual molt, the coat and surface layers of skin are replaced. The seals then migrate north to traditional feeding grounds. Since pups are born on the ice floes and molt on their whelping grounds, no one understands how the inexperienced pups migrate to the feeding grounds. During the summer months, seals disperse to feed.

Perhaps the most photographed animal in the world is the harp seal pup or whitecoat. Since a female will feed only her own young, an abandoned whitecoat faces certain starvation. A female harp seal defends the territory around her pup. The pup grows quickly from the female's milk, which is about nine times as rich as cow's milk. Within a week, the pup is a fat whitecoat. Within two weeks, the pup can reach 30 kg (66 pounds), at which time the female abandons it. The pup's call for food is unanswered, and eventually it becomes quiet. The pup then molts its long, fluffy white fur and replaces it with a so-called ragged jacket. When the pup finally enters the water, it typically feeds on crustaceans, not cod. The pup is known as a beater, perhaps because of its poor swimming skills. Harp seals reach sexual maturity at 4 years, and their normal life span is 35 to 40 years.

During the nursing period, the female does not feed. Thus, all the weight that the pup gains comes from the female's body. In addition, there is always a net loss of weight through body maintenance and thermoregulation. A female can lose up to 3 kg per day. Thus, the entire process of birth, whelping, and weaning must take place rapidly.

Cod and harp seals have shared the same seas for millions of years. If harp seals decimated the fisheries, there would be no harp seals. Indeed, because they have co-existed for so long, we might conclude that neither could exist without the other. If the seals had decimated the cod, then the seals would have switched to more abundant prey and allowed the cod to recover. Most likely, the seals feed on a variety of prey. The aquatic food web is complex, and we must appreciate that predatory fish also feed on other fish, including cod. Top predators such as the harp seal undoubtedly permit a higher biodiversity to exist by regulating the populations of predatory fish.

For hundreds of years, humans have hunted harp seals, and this is by far the greatest cause of death (Lavigne and Kovacs 1988). Certainly natural causes of death occur. Polar bears hunt seals, as do Greenland sharks and orcas. Archaic Indians living on the coast of Newfoundland hunted seals. The diaries of early visitors to the east coast of Canada, such as John Cabot (1497) and others, recorded that aboriginal people hunted seals for a variety of products. Subsistence living probably depended on abundant seal populations. It was not until Europeans arrived that serious slaughter began.

HISTORICAL SEAL HUNTING

The gray seal (*Halichoeruss grypus*) was also slaughtered by the tens of thousands from New England to Newfoundland. Hunting gray seals year round was a way to make a living. By 1860, the population had been nearly wiped out. By 1950, the gray seal was thought to be extinct in eastern Canada. As with the walrus, commercial uses led to its demise. The harbor seal (*Phoca vitulina*) was not commercially exploited but was nevertheless hunted by local people for food, fuel, and leather products. It, too, was easily massacred because of its habit of congregating. The Lake Ontario population of harbor seals was obliterated in 1824. That harbor seals survive at all today in sparse scattered populations is but luck.

The ice-breeding seals, the harp and hood (*Cystophora cristata*) seals, were also discovered farther north and hunted relentlessly. Because of their habit of remaining mostly in the water, these seals were the last to be exploited. However, because of their huge populations, they, too, have been commercially exploited since the 1600s. The local Innuit people used nets made of sealskin to catch seals, and this technique was adopted by Europeans. By the 1850s, a lucrative so-called "fishing" industry was built on harp seal exploitation.

As more was learned about harp seal behavior, seal hunts were expanded. By 1723, the harp seal hunt was making a contribution to the economy of Newfoundland, mainly from the export of seal oil. Between 1723 and 1795, hunts yielded between 7000 and 128,000 seals annually. When whelping ice occasionally carried pups close to shore, the young whitecoats were heavily exploited. In 1773, for instance, 128,000 seals were harvested, some 50,000 of which were whitecoats. At the beginning of the 1800s, seal hunting became a serious business. Sealers took to the sea in search of their quarry. The cod fishery was harvested in summer and fall, while the seals were harvested in winter and spring. Exploiting the bounty of the sea became a full-time occupation. Between 1803 and 1816, harvests averaged 117,000 seals per year. Between 1818 and the turn of the century, annual seal harvests totaled more than 200,000 seals for all but three years. In many years, the annual harvest exceeded half a million seals! By the middle of the 1800s, seal harvests were the second largest contributor to the Newfoundland economy. Cod fishing was first.

An indication of the immense number of seals comes from the years 1831, 1832, and 1844, when 680,000, 740,000, and 686,000 seals were harvested, respectively (Lavigne and Kovacs 1988). In 1857, more than half a million seals were taken. At the height of the Golden Age of Sealing, when 370 large sailing vessels and 13,600 sailors participated in the seasonal hunts, the number of seals hunted began to fall. From 1850 to 1862, the harvest went from 329,185 to 268,624. Despite technological improvements and capital investments, seal harvests had seen their limits. A century of relentless hunting had taken its toll, and the years of overkill finally did in the seemingly inexhaustible resource. From 1890 to 1895, seal harvests averaged 307,000. The total harvest for all seals during the 1800s on the North American seacoast was a staggering 33,000,000 seals. It did not cease, however.

During the early 1900s, annual hunts claimed 266,000 seals. In 1914, 234,000 seals were killed. Between 1915 and 1919, the average annual hunts yielded 143,000 seals. From 1920 to 1929, average annual hunts brought in 154,000 seals. From 1940 to 1945, the number of seals killed plummeted to a mere 20,000 annually due to World War II. During the remainder of the 1940s, however, catches averaged 200,000 seals annually. Between 1950 and 1959, 312,000 seals were harvested. In 1951, for instance, 430,000 seals were taken. During the 1960s, total annual catches of seals averaged 284,000. In 1971, the hunt brought in 230,966 harp seals.

Around the world, large numbers of seals are being killed to supply penises to the Far East for aphrodisiacs and to protect fisheries. A seal marketing strategy report paid for by the Canadian government confirmed that over half of the income from seals was derived from the sale of penises. The average price paid to sealers for a seal penis was about $23. In Namibia, Cape fur seals (*Arctocephalus pusillus pusillus*) are hunted for trophies and penises. In 1994, 120,000 seals died for inexplicable reasons and 55,000 were taken by hunters. Norway accounts for up to 20,000 seal deaths per year. In Russia, 30,000 harp seals have been taken every year for several years. The Mediterranean monk seal is one of the 12 most endangered animals in the world. The International Fund for Animal Welfare claims that 70% of Canadians are opposed to killing seals as there is virtually no market for their fur. Indeed, the marketing report also found no viable markets for seal meat or oil. Lack of a market was the reason used to explain why only 67,000 harp seals were taken in 1995 when the quota was 186,000, but that was before the bounty was established.

Scientific studies have shown that, on average, less than 1% of a seal's diet is cod. In fact, large predatory fish are probably the young cod's most formidable enemy. Removing harp seals that eat large fish will have an immediate and severe negative impact on Atlantic cod. Apparently, the logic of this statement has escaped the Canadian government. Harp seals are a keystone predator, acting to maintain a high biodiversity through their predatory activities. Because harp seals consume fish that prey on young cod, we suggest that there are more young cod when harp seals are present. This can be tested.

THE BOUNTY OF THE SEA

Harp seal hunting is directed mainly at whitecoats because the mothers are usually in the water near their young. Hooded seal mothers, in contrast, stay with their young and thus both are easily killed. In 1974, only 9995 hooded seals were taken, and the quota of 15,000 seals was not reached. Since 1976, annual quotas have not been exceeded. In 1987, for instance, a quota of 2340 was established, but 2008 were taken. As for the seal's voracious appetite for fish, scientists now suggest that Atlantic harp seal fish consumption has been overestimated by anywhere from five- to tenfold. The potential impact of harp seals on fish populations is far less than was previously claimed (Lavigne and Kovacs 1988).

A growing human population and advances in technology have led to increased pressure on fish stocks worldwide. In the Gulf of Mexico, habitat deterioration was blamed for decreasing harvests. Dams and logging in the Pacific Northwest have led to declines in salmon populations. Salmon populations in the Columbia River are 5% of their historic levels because of dams and habitat destruction from upstream logging and agricultural activities.

Many of the world's fisheries will continue to decline unless better management decisions are made and put into practice (Baker 1995). While scientists often understand the situation and know what needs to be done, policymakers are unwilling or unable to end overfishing until a crisis occurs. In the United States, 40% of fish stocks are overexploited (Baker 1995), and overfishing is destroying many fisheries worldwide. New England's Georges Bank fishery has been overexploited by people. The fishery was one of the richest in the world and was considered inexhaustible. J. Cato, director of the Florida Sea Grant, blamed the media, industry leaders, and policymakers for "cowardice" in the face of an obvious need to act (Baker 1995). In Alaska, on the other hand, when biologists say a fishery must be closed, there is no appeal. Hence, Alaska's salmon populations are at record levels. However, Alaskan rivers have not suffered the intrusion of dams and pollution from industry and agriculture. Habitat loss, however, should not be considered the scapegoat for overfishing.

WALRUS

Walruses (*Odobenus rosmarus*) were once abundant on the northeast coast of North America from Cape Cod through the St. Lawrence River and up the coast to the Magdalen Islands. Magnificent congregations bred and proliferated. Today, there are none. Their loss is a tragic story of merciless exploitation beginning in the 1600s. Lavigne and Kovacs (1988) summarize their complete and total demise. The last walrus in Massachusetts was killed in 1754. By 1680, the St. Lawrence population was wiped out. A lone walrus was sighted off the Magdalen Islands in 1800. The St. Lawrence walrus population was officially declared extirpated 187 years later by a Canadian committee on endangered wildlife. The walrus was not alone, however.

Looking back hundreds of years, it is clear that the bounty of the sea was a treasure of nature. Walruses by the ten of thousands sunned themselves from Cape Cod to the Magdalen Islands. Harp and hooded seals with populations in the millions occupied the sea waters and raised their young on ice floes. Tens of millions of cod, hake, shrimp, crab, and capelin were but a few of the undersea biotic riches. Native peoples with little or no technology plied the seas for a living and harvested seals as needed. Commercial interests simply did not exist. Something that took nature millions of years to create has been nearly wiped out in 300 years. Today, we are witnessing the death of a truly unique resource. Perhaps it will come back when the walrus returns to the North Atlantic.

EXERCISES

16.1 Do a search on the World Wide Web for "harp seal" and "seals." Pursue your leads into their biology and management. Summarize your results.

16.2 What other fisheries have been decimated since 1900? Find information on the Peruvian anchovy fishery.

16.3 Search in the library and on the World Wide Web for information on the world's fisheries. Write a brief summary of your findings.

16.4 Have any freshwater fisheries been decimated?

16.5 Investigate aquaculture more thoroughly using the library and the World Wide Web. What aquatic species lend themselves to aquaculture technology?

LITERATURE CITED

Baker, B. 1995. Is overfishing or habitat destruction the key culprit in fishery depletion? *BioScience* **45**(11):751.

Busch, B.C. 1985. *The War Against the Seals.* McGill-Queen's University Press, Montreal.

Janis, C. 1993. Victors by default. *in* Gould, S.J. (Ed.). *The Book of Life.* W.W. Norton, New York.

Lavigne, D.M. and K.M. Kovacs. 1988. *Harps & Hoods.* University of Waterloo Press, Ontario, Canada.

Phillips, K. 1994. *Tracking the Vanishing Frogs.* Penguin Books, New York.

Serchuck, F.M., M.D. Grosslein, R.G. Lough, D.G. Mountain, and L. O'Brien. 1994. Fishery and environmental factors affecting trends and fluctuations in the Georges Bank and Gulf of Maine cod stocks: an overview. *International Council for the Exploration of the Sea (ICES) Marine Science Symposium* **198**:77–109.

Trippel, E.A. 1995. Age at maturity as a stress indicator in fisheries. *BioScience* **45**(11):759–770.

AN IN-DEPTH STUDY OF IRIAN JAYA

17

Many wildlife issues are of international importance and affect all of us in one way or another. Many birds, for instance, summer in the United States but winter in South America, where they are subjected to agricultural pesticides that have long been banned in the United States. Combined with habitat destruction in the United States, these species face trouble at both ends of their range. Logging in Panama to supply the Japanese with select hardwoods for their bathrooms destroys habitat for U.S. songbirds as well. Gold mining by an Australian company on the Irian Jaya border is destroying immense tracts of pristine forest but is providing nearly 20,000 badly needed jobs. Seared kangaroo is a delicacy at a restaurant in Australia. Dogs are regular table fare in the Philippines, while millions of dogs and cats are euthanized and incinerated in the United States. How can we appreciate the wildlife issues of our time? Many issues are resolved by default before we have a chance to confront the problem. We never even learn about others until it is too late.

According to Sawhill (1995), Indonesia is a "true biodiversity hot spot." However, Indonesia has 210 endangered species, more than any other country in the world. Indonesia's rain forests and marine ecosystems are some of the most diverse areas on earth. Fully 10% of the world's rain forests remain in Indonesia on some of its 17,000 islands. Indonesia contains 10% of the world's plant species, 12% of its birds, 12% of its mammals, and 16% of the world's reptiles and amphibians.

PARADISE AND PARADISE LOST

As discussed earlier, a wildlife issue requires two people and a difference of opinion regarding a wildlife resource. In many cases, human invaders have caused extinctions (Steadman 1995). In other instances, humans have used wildlife resources for thousands of years without damaging them. In Irian Jaya, for instance, bird of paradise feathers are an important part of ceremonial life. Wild pigs are traded for brides. The men of some villages all sleep under one roof away from the women of the village, who sleep in the upper story above the pigs. The world still holds strange and wonderful surprises for those willing to appreciate them. However, there are some issues we never hear or read about because they are thought by those who bring us the news to be unimportant or uninteresting to us. Such is the case with Irian Jaya, a nation state of Indonesia.

During the Cretaceous, New Guinea and Australia were beneath sea level. New Guinea first emerged from the sea in the Eocene, about 40 million years ago, at about 35 degrees south latitude (Axelrod and Raven 1982). Since then, the island has been a changing landscape of emergent islands, and biotic events have been strongly influenced by changing sea levels, particularly during the Pleistocene. Throughout the geologic history of the island, the central mountain range has acted to separate biotic interactions. The flora of Australia and New Guinea is derived from Gondwana stocks. However, the Dipterocarpaceae so common in Indonesia are rare in New Guinea. Cores from the middle Miocene about 15 million years ago show pollen from vegetation that was common to Australia. The chain of islands extending southward from Indonesia contains a transition of predominately Asian flowering plants and placental mammals, while Australia contains marsupial mammals and different flowering forms. Plant geographers view New Guinea as principally Malaysian. Zoologists noted both marsupial and placental forms on New Guinea and also that the avifauna was more like Australian forms. During the glacial periods of the Pleistocene, extensive grasslands and savannas formed on New Guinea, which was joined with Australia. These regions contain dry-adapted, fire-resistant flora. New habitat was created when the central mountain range was uplifted during the Pliocene and early Pleistocene, and as a result, the extant floral and faunal forms further diversified. Wallacea, the region containing New Guinea, Timor, and Sulawesi, actually contains few mammals from either Australian or Asian stocks. Many plant families are shared, but most species are unique to New Guinea. Thus, New Guinea's tectonic history and interesting mixture of Australian and Asian forms make the island unusual.

Indonesia's population of about 200 million has had a direct negative environmental impact (Hardjono 1991). The most common cause of degradation is deforestation, followed naturally by soil erosion. Erosion has caused the sedimentation in some lakes to reduce their holding capacity and hence provide less extensive wetlands used for rice crops. The population growth of the so-called "inner islands" like Java is forcing agricultural lands to be converted to urban housing. Asian demand for wood products is now forcing open timber tracts on the so-called "outer islands" such as Irian Jaya and Kalimantan. After two decades of harvesting on Kalimantan, the forests have been completely logged over, and never again will the size and thickness of the trees approach the scale that once existed. New roads in Papua New Guinea have also destroyed pristine forests (Figure 17.1). The indigenous people such as the Dayaks of the northern islands have been relocated, so extensive is the damage (Potter 1991).

Irian Jaya is the western half of the world's second largest island, New Guinea; the eastern half is Papua New Guinea. Irian Jaya is part of Indonesia, the fifth most populous country after China, India, the United States, and Brazil. Indonesia is, in general, densely populated. Irian Jaya, however, is the least populated island, with only 1.35 million people. From 1848 to 1963, Irian Jaya was Dutch New Guinea and was a protectorate of the Netherlands. After existing under the United Nations flag for several years, West Irian, as the country was called, was integrated, against its own desires, into Indonesia.

The indigenous people of Irian Jaya have a long and unchanged history of subsistence fishing, agriculture, and living off the land. Many continue to

FIGURE 17.1 Tropical rain forest destruction for a new road in Papua New Guinea.

FIGURE 17.2   Human cultural diversity is celebrated at the Mt. Hagen Sing-Sing, a gathering of over 100 tribes, in Papua New Guinea.

maintain their culture, as do their neighbors in Papua New Guinea (Figure 17.2). Some live off Sago palm pulp, while others eat the larvae of the Capricorn beetle as a source of protein. Highland crops include sweet potatoes and taro. Pigs are raised for food and traded for brides. Well over 250 languages are thought to be spoken, so isolated are many of the local tribes. Even through the middle of this century, Stone Age people lived as headhunters and practiced cannibalism and ritual warfare. Superstition was a powerful force in their lives. Now, however, these beliefs and practices are becoming a thing of the past. Some lament their passing. Since 1969, many local natives have waged a war of freedom against their more heavily armed Indonesian oppressors (Osborne 1985, Whittaker 1990).

On February 17, 1996, an earthquake of magnitude 7.5 rocked Irian Jaya. Twenty-four people were killed and many more were presumed dead from numerous mud slides. Subsequently there were 19 aftershocks measuring 4 to 5 on the Richter scale. Only six days passed before another quake was felt. Remote villages in the interior of Irian Jaya have virtually no contact with the rest of the world; hence there is no way of knowing who needs help. Dysentery is always a problem after earthquakes and other natural disasters, and therefore, the number of dead undoubtedly rises. The tidal waves that struck the villages of Borarsi, Borobudur, Fanindi, and Wosi also killed many people and damaged houses. On February 27, 1996, news reached the outside world that a pig epidemic in the remote villages of Brome and Ndugwa, near Wamena, Irian Jaya, killed 177 people. Symptoms included breathing distress, vomiting blood, and then death. Medical teams were dispatched on receiving the news but were days too late. Because of a hostage case involving foreign graduate students, mail couriers feared leaving the villages. How much of this news reaches people in the United States?

The Indonesian Army Strategic Reserve Command sent Battalion 432 Airborne, an elite commando unit, to assist in the rescue of 3 British graduate students, 2 Dutch, a German, and 17 Indonesians. The graduate students had been researching local tribesmen, who still carry on subsistence living, in an attempt to learn more about an isolated area of high floral and faunal diversity, when they were taken hostage. In the past, these so-called Stone Age tribesmen used handmade long bows and arrows to defend their territory against Indonesia government helicopter gunships and paratroopers. The local people's knowledge of the region was no match for the more heavily armed government army. Within a few days, not a single paratrooper was alive. One result of the paratroopers' dropping in was that these primitive tribal people received a supply of the latest in automatic weapons, such as M-16s, and regular armed forces standard weapons and ammunition (Osborne 1985).

A team from the International Commission of the Red Cross was attempting to establish contact with a group of freedom fighters, or rebels as the Indonesians refer to them, who still withhold the whereabouts of researchers around Mapunduma Village, Tiom District. As of March 10, 1996, the whereabouts of the hostages was unknown.

No accurate vegetation map of Irian Jaya exists, even on a crude scale (Perocz 1989). Some hill tribes were not even discovered by the outside world until the 1930s (Archbold and Rand 1940). Indeed, before 1824, Europeans believed that birds of paradise were mystical creatures without legs that lived on dew and survived permanently on the wing until death. The discovery of birds of paradise by the European fashion industry led to the demise of over 50,000 birds for the feather trade shortly before 1900. For thousands of years, natives lived essentially in harmony with nature and the birds of paradise. The indigenous people use the birds for their headdresses as ceremonial ornaments. The more colorful sexually mature male birds are only taken after they have reproduced. Most local people over the age of ten know most of the birds in their area and are familiar with the natural history of the birds they value.

The fauna of the island of New Guinea lies at the intersection of the Oriental region and the Australian region, an area referred to as Wallacea, named after the great naturalist–explorer Alfred Russell Wallace. It contains a mixture of Australian and Oriental floral and faunal forms. According to Perocz (1989), Irian Jaya contains the richest concentration of plant life in the country of Indonesia, a "floristic diversity of tremendous proportions." A conservative list contains well over 20,000 species. An astounding 124 genera of flowering plants are endemic, compared with 59 for the island of Borneo, 17 for Sumatra, and only 10 for Java. Species endemism is extraordinarily high—90% of all species are endemic. Leach and Osborne (1985) report that this region contains the highest number of plants useful to humans. However, members of the family of trees Dipterocarpaceae are rare and hence the forests are generally not of interest to timber companies. Many indigenous uses of plants have yet to be described and will undoubtedly one day contribute to the health and welfare of the human race if we can only learn to appreciate the fact that these "primitive Stone Age people" have forgotten more about plants than most of us will ever learn.

Surprisingly, orchids make up nearly one-third of all flowering plants on Irian Jaya. Giant epiphytic anthouse plants can reach 2 m in length and house thousands of ants. Some pitcher plants are carnivorous, living on ants. From glowing fungi that emit an iridescent light from their gills, to the fruit of the *Amorphahallus campanulatus* that stinks like rotting flesh, Irian Jaya holds a fantastically diverse flora waiting to be explored.

As for land mammals, Irian Jaya has 2 monotremes, 57 marsupials, 63 chiroptera, a single carnivore, 2 artiodactyls, and 49 rodentia. These forms place the island of New Guinea at the intersection of two zoogeographic regions, with the first two orders being Australian and the remaining orders more closely related to Oriental forms. Several of the mammals are curious forms, such as the echidnas, the eight species of cuscus, pygmy possums, and four species of tree kangaroos. Large carnivorous predators are absent from New Guinea.

The avifauna of the island of New Guinea is legendary, with approximately 700 species known. Of these, 643 species occur in Irian Jaya, including 269 endemic species. The birds also show affinities to Australian and Oriental forms. "Birds have attracted the attention of visitors and voyagers to New Guinea since its discovery, and they are well known and utilized for many purposes by the indigenous peoples." Of course, the birds of paradise are unique among the world's birds, and many consider them the most beautiful birds on earth. For instance, the blue bird of paradise's mating call sounds like an electric guitar. The male hangs upside down from a branch, puffs up its plumage, and emits an almost electronic sound while it bobs up and down, displaying for the drab female. The nocturnal Papuan frogmouth sits with its mouth open, catching insects. Strange bower birds build large nests of sticks called bowers, and some even line the entrance with blue berries and flowers. The cassowary is a deadly flightless bird that has been known to disembowel a human with a single kick. Almost nothing is known about the birds of prey. We don't even know what the nests look like. They still remain unknown to us. No one has found a nest—ever!

The reptiles and amphibians are equally strange and beautiful. Although 250 reptiles are known, it is suspected that 100 snakes and 200 lizards may actually occur. There are legless lizards, arboreal dragon lizards, and the nonvenomous green tree python. Approximately 120 tree frogs are known to occur and more can be expected to be discovered. As for fish, there are between 6000 and 7000 species in nearby coastal seas. The surrounding seas have been the principal evolutionary center from which the entire Indo–Pacific region has been populated (Carcasson 1977). This total represents fully one-third of all species of fish. Allen and Cross (1982) note that of about 158 freshwater species, only 11 occur on either side of the central mountain range. Little is known about the insects, but estimates range from 80,000 to 120,000 species. The spectacular birdwing butterflies, the Ornithopterids, occur in Irian Jaya. There are some 800 spiders, including the giant bird-eating spider. Some stick insects are 25 cm in length. Of all groups, insects have been investigated the least, and biologists will undoubtedly be surprised by new discoveries. Colorful coral reef communities like none other on earth are found off the coasts of Irian Jaya. In most areas of Indonesia, mollusks have been depleted, but Irian Jaya has rich and diverse offshore resources. Among the offshore islands are also strange life forms, such as the world's largest terrestrial arthropod, the coconut crab (*Birgus latro*). Of the smaller organisms, scientists have not even ventured a guess as to how many exist. Indeed, the smaller the organism, the more likely it is that it has yet to be described.

Perocz (1989) estimated that over 80% or more of Irian Jaya remains in a pristine state. Traditional tribal practices maintained a low population density, and communal land practices prevented anyone from accumulating large land holdings. Now, however, development is proceeding at a rapid pace, and age-old practices are being condemned as standing in the face of progress. In the past, protected areas were never established because there was no need for them. Indeed, the concept of a protected area is a modern one because large-scale development creates a need to protect relatively small areas after the fact. These areas are usually, for one reason or another, not developable at the time. As the tribal value system is torn apart and the local people displaced, conservation areas become a popular subject.

Two species of crocodiles have been harvested in Irian Jaya for many years. The saltwater crocodile (*Crocodylus porosus*) is listed in the IUCN Red Data Book as an endangered species, while the New Guinea crocodile (*C.*

novaeguineae) is considered vulnerable. Most crocodile farms depend on wild crocodiles to stay in business. As of 1989, only one farm had been able to produce crocodiles in captivity. The crocodile industry exists mainly to supply a growing export market. It remains to be seen if crocodile farms can wean themselves from wild-caught crocodiles and wild egg exploitation.

The birds of paradise have also been harvested, perhaps for thousands of years. There are 26 species of birds of paradise, and the most beautiful are highly valued. Pigs are often traded for the finest feathers. All the birds of paradise are protected by Indonesian law. There are also 44 species of parrots, 5 of which are protected by law. Frequently, these laws are not enforced, and fine caged birds of almost any kind can be found in the local markets. Before the illegal pet trade was established, traditional uses of the birds, and especially the birds of paradise, were controlled by cultural barriers and social taboos. No birds of paradise are known to have been driven to extinction by the indigenous people. Perocz (1989) suggests that as the bird trade became more commercialized and markets outside Irian Jaya were established, certain species of birds of paradise were driven to extinction in parts of their range. The parrot trade has also thrived in spite of laws designed to prevent it. Once a commercial trade is established, market hunters must harvest birds continuously. As the birds become more rare and their value increases, market hunters are forced to drive them to extinction. The need for a quick profit in a competitive marketplace puts enormous pressure on resources, as we have previously seen. From orchids and butterflies to giant clams—all are under increasing pressure now, principally because of external market forces above and beyond traditional uses.

Marine vertebrates such as sea turtles and dugongs are also exploited for food. Irian Jaya has nesting grounds of leatherback (*Dermochelys coriacea*), green (*Chelonia mydas*), hawksbill (*Eretmochelys imbricata*), and olive ridley (*Lepidochelys olivacea*) sea turtles. All are now endangered. The dugong (*Dugong dugon*), the Irian Jaya form of the Florida manatee, is widely distributed and suffers from uncontrolled hunting pressure throughout its range. The dugong is considered a delicacy and its teeth are used for cigarette holders. Nietschmann and Nietschmann (1981) studied dugongs in the Torres Strait between New Guinea and Australia. Here the dugong is a part of the indigenous people's culture, tradition, economy, and religion. Thus, the dugong is conserved for perpetuity here. Elsewhere across Irian Jaya, netting and spearing these gentle creatures continue unchecked.

According to Perocsz (1989), many species are protected under Indonesian law. "Among these are whales, the three species of cassowary, all birds of paradise, all tree kangaroos and cuscuses, three species of marine turtles, and some fish....According to this law, the hunting, killing, capturing, trading (dead or alive), and even the mere possession of these species are forbidden. The law, furthermore, states that it is also forbidden to export these species from country to country or to possess or trade their skins, feathers, and other parts of the body, or the products made thereof, or to disturb, collect, trade, and possess their eggs and nests." Of course, having laws on the books and enforcing them are two very different things. In Irian Jaya, these Indonesian laws are totally ignored. Yet are we to blame the indigenous people for the commercial exportation of certain species? An International Institute for Environment and Development/Government of Indonesia (1985) report singled out the World Bank and its continued financial support for the transmigration program as the largest contributor toward the destruction of the forests on Kalimantan (Potter 1991). Will the same happen to Irian Jaya?

TRANSMIGRATION

With its low population, Irian Jaya is an attractive destination for the largest human transmigration program on earth. Hundreds of thousands of native Indonesians from the central islands surrounding Java have been relocated to the outer islands of the empire, such as Irian Jaya. As the world's demand for hardwoods increases, logging companies have established claims on 70% of Irian Jaya's virgin and virtually unknown forests (Perocz 1989). Social, political, and technological changes are rapidly wiping aside thousands of years of cultural diversity. Some suggest that "Indonesia's secret war" is nothing less than a genocidal war to remove the native Melanesian people and replace them with true Indonesians from the inner islands.

The massive transmigration program was announced to the world by Indonesia in 1963. Since then, hundreds of thousands of people have been relocated from the heavily populated inner islands of Java, Bali, Madura, and Lombok to the less heavily populated outer islands of Kalimantan, East Timor, Sumatra, and Irian Jaya. Often, missionary societies provide the initial foothold by penetrating the remote areas of Irian Jaya under the guise of religious reform. Of course, mission stations and airstrips must be established to secure outside supplies. For this privilege, the local people donate part of their land holdings. Thereafter, a small military outpost is established. With mission services flying into remote areas, military and aid-agency personnel are frequent passengers. According to Indonesia national law, all unimproved land is state property. Thus, the native forests that have supported indigenous people for thousands of years, by definition, are owned by the state. Transmigrants are usually Islamic, and cross-cultural conflicts are common. The Indonesian minister of transmigration admitted that an intended result of transmigration was that different ethnic groups would in the long run disappear (Colchester 1986).

Of course, land tenure in Irian Jaya is fundamentally different than in the West. Typically, a tribe holds communal property and people live on tribal lands. Living on the land confers certain rights and privileges but not outright ownership. If an individual "sells" the lands, he is merely conferring certain rights for a specific period of time. Hence, the land can be sold many times, much like renting property, for instance, with the provision that the land will revert back to the community after some time.

Ten percent of the transmigration funding was initially provided by the World Bank in the form of loans. After about 30,000 families were relocated, Indonesian and foreign companies invested money in the scheme in the hope of reaping future profits. According to Monbiot (1989), Indonesian army operations and transmigrants often make refugees of the indigenous people, forcing them to vacate their tribal lands and, in some cases, relocate to the neighboring independent country of Papua New Guinea. Unfortunately, without knowledge of the local growing conditions, the transmigrants' crops often fail, and without some local assistance, transmigrants go hungry. Both the new lords and the refugees became victims of grand plans and grim reality (Whittaker 1990).

NATURAL RESOURCE EXPLOITATION

Irian Jaya is rich in other natural resources, including timber, minerals (such as copper, zinc, and gold), and petroleum. The United States, Japan, and South Africa have been quick to take advantage of the prospects where almost no environmental laws are enforced. Some of the world's largest oil companies are already pumping oil. In the Cyclops Mountains between Irian Jaya and Papua New Guinea lays one of the largest mines in the world. In 1967, Freeport Indonesia Inc. received its contract to begin operations and produced its first copper in 1972. Oddly, in 1978, the Lorentz Strict Nature Reserve was created by Irian Jaya; it contained the entire mine, all the local housing, the airport, and the port facility far to the south on the coast. As of 1989, 45% of the mine was inside the park but underground. The reserve was designated a World Heritage Site in 1982 as one of the greatest natural areas on earth. Formerly, the area was remote and sparsely occupied. More than 17,000 people work at the mine in an area where good jobs are otherwise nonexistent. But very few jobs go to poorly trained and unskilled native people who are mostly laborers.

Perocsz (1989) reported that "slag and gravel disposal and particularly water contamination of the Otomona River will have an impact on the reserve through tidal transfer at the Arafura Sea. Untreated waters full of metal oxides will affect the well-being of the vast resettlement project downstream near Timka, the largest settlement program of its kind in the province." Just six years later, tons of toxic wastes may have been harming fish and other wildlife resources on portions of the river and contaminating tropical forest. One problem is that the wilderness is virtually off limits except for invited guests. The few visitors who have seen this region have remarked on the degradation that has occurred in the relatively small areas they saw, such as along the trails. The cost of implementing sound environmental awareness is incidental to a mining operation that unearths titanic amounts of ore each day. Figure 17.3 shows siltation in the Ok Tedi (Ok means river in Papua New Guinea) in Papua New Guinea on the border near Irian Jaya.

The South Africans and Indonesians pursued oil resources on Irian Jaya during the 1970s. Discovering oil-rich areas and then exploiting them is made

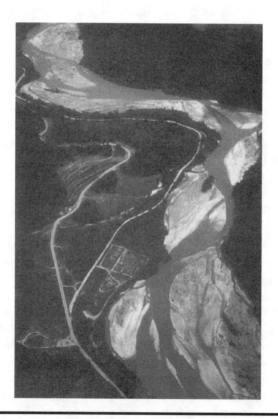

FIGURE 17.3 River siltation has destroyed the local fishery on the Ok Tedi in Papua New Guinea, near the Irian Jaya border.

more difficult first by the terrain and second by indigenous claims for compensation. Chevron, a U.S.-based company, had a pipeline damaged by the indigenous people over compensation for loss of communal lands and forests. Shell Oil Company ceased operations from 1983 to 1986 because of attacks by indigenous people.

Irian Jaya is nearly covered with what once was virgin forest, and now these forests are being heavily exploited. About 40% of the forests are in protected areas and World Heritage Sites; however, these designations apparently confer no special status. A variety of global multinationals are sharing the resources and neglecting native claims. Some 80% of the transmigration sites were carved from virgin tracts of forest (Secrett 1986). In effect, the reason we in the West are especially ignorant of the plight of the people of West Papua is that U.S. companies are exploiting these resources and enriching investors who want nothing more than to make a profit from their investments (Whittaker 1990). Traditional fishing stocks are being depleted by huge Japanese factory ships. Wildlife resources are being decimated by indiscriminate logging practices. Moreover, companies operating in Irian Jaya ignore the rights of the indigenous people and avoid compensating them for the loss of their forests, their wildlife, and the damage to their societies (Colchester 1986). These natural resources have been sustainably exploited by the indigenous people since humans occupied the island of New Guinea perhaps as long as 50,000 years ago. Now, the Western concept of "sustainable development" has replaced "indigenous sustainability." If the local people were given a choice, which would they choose?

REFUGEES FROM IRIAN JAYA ENTER PAPUA NEW GUINEA

People of the Awin Tribe of the Western Province, Papua New Guinea, are subsistence farmers and hunter-gatherers who live near the border with Irian Jaya. Awin tribal land is typical of tropical rain forest soil, and growing conditions are poor. All land belonging to the 20,000 tribal members is owned by the clans that make up the tribe. Individuals have claims to the land and can live on their own land, but they do not own, nor can they sell, the land. Traditionally, land is passed from father to son, but never to a daughter.

Although elaborate gardens are typical in the highlands of Irian Jaya and Papua New Guinea, most Awin tribal members gather fruits and nuts from the forest, hunt birds and pigs, and fish. Few crops are grown. The idea of conservation is new to most Awin people because they are not aware of what has happened beyond the Awin tribal land, much less the outside world. They are already conservationists—it is their way of life. The local people value their wildlife and natural resources, and their general knowledge of ecology is extensive. Furthermore, their knowledge of and enthusiasm for the wild birds is nothing less than fantastic and something tribal member are only too willing to share.

In 1995 and 1996, refugees from nearby Irian Jaya, fleeing an oppressive Indonesian rule, poured onto Awin tribal land. The government of Papua New Guinea and the United Nations High Commission on Refugees promised the Awin tribe "infrastructure improvements" if the more than 10,000 refugees were allowed to resettle on Awin land. The United Nations had no control over the refugees and failed to provide adequate resources for their relocation. Sadly, the refugees were not hunter-gatherers but gardeners who soon began practicing their slash-and-burn agricultural scheme.

The loss of forests and the wildlife and birds they contain threatens Awin tribal security. The infrastructure improvements promised, such as schools and health facilities, have not materialized. The Awin people are now dealing with several logging concerns interested in harvesting tropical hardwoods. Rather than see their trees felled and turned to ash by the refugees, the Awin people would rather receive cash for the trees and then build their own roads, schools, and health facilities. Ecotourism in this remote part of the world cannot supply enough money to the local economy to build and maintain the necessary infrastructure.

The addition of money as a trading medium will have a profound impact on most Awin people. People want money to buy what they cannot produce. Items such as Western clothing, more varied foods, fuel, outboard motors, and roofs for homes must be bought with money. The Awin once used home-crafted bows and arrows to hunt. Now they use guns and bullets, which must be purchased. Powered motorboats are replacing traditional canoes. As one tribal member put it: once money is understood, a want becomes a need.

For the Awin tribe and the 700 or so other tribes that live in Irian Jaya and Papua New Guinea, the world is changing rapidly. Having utilized their wildlife and natural resources sustainably for tens of thousands of years, we can only hope that the rush to a modern society does not force the abandonment of all their traditional values. Wildlife issues were probably unknown to the Awin tribe before the arrival of refugees from Irian Jaya. With the logging companies will come opportunities for jobs, and cash will become the medium of exchange. New values will replace age-old values, and Awin society will change forever. Perhaps the new generation will value its wildlife even more, for as the forests are cleared, the meaning of conservation will become clear and wildlife issues will become more frequent.

INDIGENOUS SUSTAINABILITY

Between Australia and the island of New Guinea in the Torres Strait, indigenous seafaring people have plied a living for centuries (Nietschmann and Nietschmann 1981). The Torres Strait islanders have hunted dugongs and green turtles that are otherwise endangered elsewhere. The survival of the indigenous society is intimately linked to the hunting of these animals and ultimately to the animals' food supply. The islands in the Torres Strait are thinly populated by about 4500 people and minimally developed. Seagrasses and algae are plentiful in the warm shallow strait, and dugongs and sea turtles concentrate in well-known areas. A large tidal range opens and closes some hunting areas, and the dugongs frequently follow the tides, feeding on the flooded grasses during high tides. In some areas, dugongs have stripped the vegetation clean.

Four species of sea turtles are found, but only the green turtle is hunted regularly. However, the Torres Strait may be the most important green sea turtle sanctuary in the world. The ecology of the green turtle in the strait is well appreciated by the Torres Strait islanders. The native people have a detailed, consistent knowledge of the ecology of the animals they hunt and the physical processes governing their abundance and distribution. There are more than 80 terms for tides and tidal and water conditions.

Hunters focus on middle-age female dugongs and turtles that are said to taste better than other animals. These are also hardest to catch. Through the use of certain closely guarded charms, hunters can increase their chances of success. After a successful hunt, the butchering of the prey is ritualized. Choice cuts go to the hunter and his extended family. As is usual in hunting societies, a successful hunter always gives away meat because he knows that one day he will accept meat from another hunter. The rest of the meat is shared equally among families. The remains are always fed to the sharks that gather.

The Torres Strait islanders have minimized their interactions with the rest of Australia and Papua New Guinea. Many of their ancient beliefs and practices continue to be maintained. As long as they are able to hunt, the Torres Strait islanders will continue to live in harmony with the natural world.

NOBEL PEACE PRIZE

The prestigious Nobel Peace Prize for 1996 was awarded to East Timor Roman Catholic Bishop Filipe Ximenes Belo and exiled independence activist Jose Ramos-Horta. East Timor lies off the north coast of Australia and is east of Java, Indonesia. Formerly independent, East Timor has struggled to regain its freedom from Indonesian domination since 1976. Human rights violations in East Timor have been a common occurrence. The world-class award left no doubt who was to blame for the widespread death, terror, and persecution after the Indonesian army and navy invaded the former Portuguese colony of some 650,000 people in December 1975. A year later, the Indonesian government declared that East Timor was Indonesian soil.

Fully 30% of the population of East Timor has since died from a variety of causes, including starvation and epidemics. The United Nations has never recognized Indonesia's control over East Timor, which has become a backward nation of people barely able to exist. East Timor is the only Christian area in Indonesia. The Indonesian transmigration program has aided an influx of Muslims from Indonesia, which is the world's largest Muslim country. Reports of events in Timor are nearly nonexistent in the United States.

Nobel awards chairman Francis Sejersted commented that the award was definitely also a criticism of the Indonesian government. The Nobel Prize honored the two men for their work toward a just and peaceful solution and in so doing focused world attention on Indonesia. While criticizing China and other nations, the United States has been noticeably silent with respect to calling attention to the plight of formerly free people in East Timor and also Irian Jaya. In the absence of human rights, wildlife issues cannot be dealt with satisfactorily. After all, what rights do wildlife have when people are starving? In 1998, President Suharto of Indonesia was replaced.

EXERCISES

17.1 Find some remote areas in the world and elaborate on their environmental problems and wildlife issues. Try searching on the World Wide Web for information on the Indonesian states of Timor, Sulawesi, Sumatra, Bali, and Java.

17.2 What are the wildlife and natural resource issues in the newly independent states of the former Soviet Union? Is there a common theme they share?

17.3 Locate a reliable source for more information on recent developments in Irian Jaya. Occasionally, newspapers have some articles of interest. Irian Jaya has been notably unsuccessful in its efforts to become an independent country like Papua New Guinea. Political events often overwhelm wildlife issues.

17.4   Try searching around on the World Wide Web sites at CNN and the BBC news services for interesting wildlife issues or news from far-away places.

LITERATURE CITED

Allen, G.A. and N.J. Cross. 1982. *Rainbowfishes of Australia and Papua New Guinea.* Angus and Robertson, Sydney, Australia.

Archbold, R. and A.L. Rand 1940. *New Guinea Expedition.* Robert M. McBride and Company, New York.

Axelrod, D.I. and P.H. Raven. 1982. Paleobiography and origin of New Guinea flora. *in* Gressit, J.L. (Ed.). *Biogeography and Ecology of New Guinea,* Volume 2. Dr. W. Junk, Publishers, The Hague.

Carcasson, R.H. 1977. *A Field-Guide to the Reef Fishes of Tropical Australia and the Indo–Pacific Region.* William Collins Sons, Glasgow, U.K.

Colchester, M. 1986. Banking on disaster: international support for transmigration. *The Ecologist* **16**(2/3):61–70.

Hardjono, J. 1991. The dimensions of Indonesia's environmental problems. *in* Hardjono, J. (Ed.). *Indonesia: Resources, Ecology, and Environment.* Oxford University Press, Oxford, U.K.

International Institute for Environment and Development/Government of Indonesia. 1985. A Review of Politics Affecting the Sustainable Development of Forest Lands in Indonesia. Jakarta, Java.

Leach, G.J. and P.L. Osborne. 1985. *Freshwater Plants of Papua New Guinea.* UPNG Press, Port Moresby, Papua New Guinea.

Monbiot, G. 1989. *Poisoned Arrows: An Investigative Journey Through Indonesia.* Michael Joseph Publishers, London.

Nietschmann, B. and J. Nietschmann. 1981. Good dugong, bad dugong: bad turtle, good turtle. *Natural History* **90**(5):54–63.

Osborne, R. 1985. *Indonesia's Secret War: The Guerrilla Struggle in Irian Jaya.* Allen & Unwin, Winchester, Massachusetts.

Perocz, R.G. 1989. *Conservation and Development in Irian Jaya.* E.J. Brill, Leiden, Germany.

Potter, L. 1991. Environmental and social aspects of timber exploitation in Kalimantan 1967–1989. *in* Hardjono, J. (Ed.). *Indonesia: Resources, Ecology, and Environment.* Oxford University Press, Oxford, U.K.

Sawhill, J.C. 1995. The Nature Conservancy and biodiversity: 1993 in review. *Nature Conservancy* **44**(1):5–9.

Secrett, C. 1986. The environmental impact of transmigration. *The Ecologist* **16**(2/3): 77–88.

Steadman, D.W. 1995. Prehistoric extinctions of Pacific island birds: biodiversity meets zooarchaeology. *Science* **267**:1123–1131.

Whittaker, A. 1990. *West Papua: Plunder in Paradise.* Anti-Slavery Society, U.K.

AN IN-DEPTH CASE STUDY OF GORILLAS

18

A visit to a habituated group of gorillas is certainly one of the most exciting adventures any wildlife enthusiast can experience. One of the authors (JGS) visited the mountain gorillas in Rwanda in November 1985, shortly before the murder of Dian Fossey. At the time, a trip to see the gorillas cost $100 and most openings were filled six months in advance. The registration center was located in the small village of Kinigi on the volcanic slopes of Mt. Sabinyo. Four groups of gorillas had been habituated to the presence of humans. Each day, 24 people visited these gorillas for one hour only, six people and their guide per group of gorillas.

The trek to the edge of the forest was two hours at an altitude of 8000 feet. Before entering the Parc des Volcans, the guide and his assistant recited the rules governing all tourist visits to the gorillas: no pointing at the gorillas, no smoking near the gorillas, no staring at the gorillas, no eating near the gorillas, no flash photographs, no running away if the gorillas approach, no breathing on a gorilla if it approaches, and so forth. Everyone was anxious to get started. Soon the darkness of the forest enveloped the tourists. The trail was muddy and slippery and the vegetation foreign, with stinging nettles and strange mushrooms. Some people walked for several hours before encountering a group of gorillas. Others encountered them in ten minutes. Most people remember the moments just before contact because their excitement level was especially high as the guide came back and ushered them forward. The guide also made low rumbling sounds to comfort the gorillas and make them aware that their daily routine was about to be interrupted.

381

By the time the first contact is made, the tourist's heart is pounding with excitement and anticipation. An hour with the gorillas is certainly one of the fastest hours anyone can experience. At all times the visitors are kneeling and watching, listening, and taking photographs. A few juvenile gorillas are playing, a silver-back male is eating in front of you, several females are feeding together, and a young might be nursing. It is surely as if you have invaded the privacy of an intimate family in their own home; while they go about their daily routine, you are crawling around taking pictures of them. When the hour is almost up, most people are ready to return and feel grief or guilt that they have interrupted the lives of these peaceful creatures. Many people liken the visit to a surrealistic voyage to a distant time and place far away from the middle of poverty-stricken Africa, to a place primeval where animals have not yet learned to fear humans, the way it once must have been. Most who visit cannot experience the gorillas only once. Some cannot get enough.

CONSERVATION, SCIENCE, AND WILDLIFE ISSUES

Today, scientists recognize three subspecies of gorillas in Africa. The western lowland gorilla (*Gorilla gorilla gorilla*), the eastern lowland gorilla (*G. gorilla graueri*), and the rarest of the three, the mountain gorilla (*G. gorilla beringei*), all live in forested regions of central Africa. As shown in Figure 18.1, the western lowland gorilla is located in Cameroon, Nigeria, Congo, Gabon, and the Central African Republic. The eastern lowland gorilla is located in eastern Democratic Republic of the Congo (DRC) and southwestern Uganda and is at least 2000 km from the western lowland populations. The mountain gorilla lives on isolated and remote mountains in eastern DRC, Rwanda, and Uganda. Yet 150 years ago, the existence of gorillas was only a vague rumor and a frightening one at that. Their dense, remote habitat and their rumored strength and fearlessness made human contact without a gun impossible. Today, tourists enjoy visiting gentle gorillas. Some enjoy gorilla heads, hands, and feet as trophies, and others find gorilla flesh good eating. Certainly these different interest groups cannot all be satisfied for long. What are the current wildlife issues concerning wild gorillas, and how can science be used to meet the conservation needs of these unique creatures? Again we must take a historical perspective.

Since the discovery in 1847 of what is now known as the western lowland gorilla, gorilla taxonomy has been controversial. The mountain gorilla was not even discovered until 1903, when Captain Oskar von Beringe shot two "large black monkeys." Coolidge (1929) and Schouteden in 1947 both accepted a single species. In 1961, Vogel proposed that the western lowland gorilla and

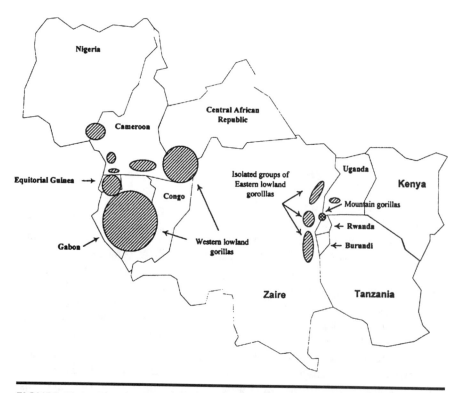

FIGURE 18.1 The distribution of gorillas in Africa. Habitat destruction in western Africa and political instability in eastern Africa threaten the existence of gorillas.

eastern lowland gorilla were separate species. He further divided the eastern lowland and mountain gorillas into distinct subspecies. Schaller (1959) accepted Coolidge's classification when he ventured into central Africa to perform ecological studies of gorillas. Fortunately for all later generations, Akeley (1923) was able to convince the ruling Belgians to set aside a sanctuary for the gorillas in 1922. In 1929, the boundary of the Parc des Volcans was enlarged to include much of the known mountain gorilla habitat. Groves (1970) analyzed 747 gorilla skulls and concluded that there were three gorilla subspecies.

Kahuzi-Biega National Park in eastern DRC presumably protects the eastern lowland gorillas, and several parks in Cameroon, Central African Republic, and Gabon protect western lowland gorillas. In principle, all gorillas are protected under international treaties. In addition, since the present populations of gorillas are recognized and differentiated only to the level of subspecies, we could, without loss of biodiversity, lose one or two subspecies completely without adverse impacts. For instance, if the remaining mountain gorillas vanished, under present taxonomic status there would be no loss of biodiversity since biodiversity is typically measured at the species level. Harcourt (1996) estimates there are 125,000 gorillas in Africa, double the previous estimate. Some of the areas may contain more than a thousand individuals. With the present parks, it would appear to the casual observer that the gorillas are in fine shape. Nothing could be farther from the truth.

MOUNTAIN GORILLAS

When civil war exploded in Rwanda in early 1991, conservation organizations feared the worst for the mountain gorillas of the Parc des Volcans. Harboring the last 300 mountain gorillas on earth, Parc des Volcans straddles three of the poorest countries on earth—Rwanda, Uganda, and DRC. Refugees began leaving Rwanda by the tens of thousands. Fighting took place within the Parc des Volcans. Land mines were planted within the park by both factions and no maps were kept. People leaving Rwanda moved into the forested part of the Parc des Volcans in eastern DRC. In Uganda, the rebel invaders camped in the Parc. Conservationists quietly pleaded with both sides to leave the gorillas alone. Akegera National Park in Rwanda was completely sacked and overrun. Most of the wild animals were eaten for food by Rwandan rebels. Those that were not consumed during the war were consumed afterward. The park was destroyed. Mercifully, the Parc des Volcans was spared, although at least one gorilla was killed when it surprised rebel forces.

In the spring of 1995, four gorillas were killed in the Bwindi Forest in Uganda, site of the Bwindi National Park and home to some 300 eastern lowland gorillas. These gorillas were speared by poachers to reach a nursing infant, such as the one shown in Figure 18.2. Suspected poachers were detained, but no arrests were made. Four months later, three mountain gorillas were killed in the DRC in Virunga National Park (the DRC part of the Parc des Volcans) just hours after tourists had visited them. In 1995, seven mountain gorillas were known to have been killed by poachers. The actual number of slain gorillas is unknown.

On February 1, 1996, news reports out of the DRC indicated that 750,000 Rwandan refugees were creating an "environmental hell" in Africa's oldest World Heritage National Park, the Parc des Volcans. Rwandan Hutu tribe refugees were destroying the Virunga National Park and carrying away an estimated 1000 tons of trees per day to be used for fuel and charcoal. Twenty thousand acres of prime bamboo forest, a staple gorilla food, had been completely destroyed to make mats and furniture.

FIGURE 18.2   Mountain gorilla mother and infant find sanctuary in the Parc des Volcans, Rwanda. The infant is sucking its thumb.

The environmental coordinator for the United Nations High Commission for Refugees was reported to have said, "Some things disappear and never come back—only in 200, 300 years." Hutu militants are now suspected to be training in the Parc des Volcans to begin a cross-border rebellion inside Rwanda. New attempts to smuggle infant gorillas outside Rwanda to buyers in Asia and the new Soviet republics might be made under the guise of being surprised by gorillas in the forest and accidentally killing them. Six Italian tourists were killed in the DRC in August 1995 by an unidentified gunman. Near the refugee camps, previously planted mines explode daily. Antelope, buffalo, and other bush meat is sold regularly for food.

Meanwhile, tourist visits to the gorillas increased. Driven by greed, visits to habituated gorillas in the DRC doubled by allowing two groups of tourists to visit daily. A single human with a sniffle can wipe out a gorilla group because gorillas have no built-in immunity to human colds. A tourist with the measles could wipe out the entire population of gorillas. This is why a distance of 5 m is kept at all times between humans and gorillas. In the past, this minimum distance was honored. Now, for a small tip, tourists can have a closer encounter with a rare and endangered species.

Part of the Virunga National Park in the DRC is a grassland, and the Rutshuru River was once known for its huge population of hippos. In March 1996, reports from the former park headquarters described a slaughter of horrific proportion. The banks of the river are "littered with hippo bones," evidence that the traditional snares and hand-pounded metal spears have been replaced by modern automatic weapons. Is it any wonder park rangers look the other way?

On the Ugandan side of the Parc des Volcans, Klaus-Jürgen Sucker was responsible for preventing incursions into the forest and maintaining gorilla patrols. Klaus made certain people and agriculture were kept out of the forest. He lived in a makeshift tented camp in the foothills of Mt. Gahinga before he relocated to the small town of Kisoro after two years. When the Rwandan civil war began, Klaus stepped up patrols in the forest. Rwandan rebels intent on the overthrow of the government established camps around Mt. Gahinga. Klaus met with them and asked them to safeguard the gorillas and not to shoot if by chance they surprised a gorilla in the forest. On June 20, 1994, Klaus was found dead in his home. There was a noose around his neck and his feet were on the floor. No arrests were ever made.

Klaus's death is shrouded in mystery. A few days before his death, he was ordered by the Ugandan government to vacate his position as director of the Mgahinga Mountain Gorilla Project. He was told to report to work in Kidepo Valley National Park in northeastern Uganda, about as far from Kisoro as a person can get and still be in Uganda. Powerful forces and big money were at work undermining years of Klaus's work. Certainly Uganda National Parks, a government agency, was aware of his achievements.

Klaus was no stranger to life on the edge. Naturally, he was not popular among the poachers and wood cutters in the region. His patrols put many of his enemies out of work. During the first year of his project, his tented camp was attacked by a small unit of unknown people armed with Russian Kalashnikov machine guns. Two were aimed directly at Klaus in an attempt to scare him off. "You go ahead and shoot," he said. Later, a small park headquarters was built. Mgahinga was declared a national park. Klaus stepped up patrols on dedication day; he knew the park's enemies only too well.

After the Rio de Janeiro Earth Summit, several multiple-use projects were planned for Mgahinga National Park by several U.S. organizations. Klaus opposed these projects because they allowed harvesting of honey, herbs, and bamboo, an important gorilla food. No controls or limits on so-called subsistence gathering were ever discussed. After all, he reasoned, Mgahinga was only 35 km², of which 10 km² was severely degraded and just recovering and therefore highly susceptible to disturbance. Together with Uganda National Parks and international scientists, Klaus argued effectively against subsistence use. Except for a small number of highly controlled tourists, no one was to enter the forest except to patrol. When the survival of the gorillas was at stake, Klaus believed, people would just have to take second place.

A month before his departure, Klaus discovered his mail was routinely being opened. The chief park warden of Mgahinga had a key to the post office in Kabale where the mail was received. Klaus never received a confidential letter mailed only to himself and one other person. Yet at least one employee of a widely known donor organization had a copy—marked confidential. His departure from Mgahinga was a guarded secret. However, it seems others knew about it. In the end, it cost Klaus his life. His death was not the first and will not be the last. Gorilla protection is a serious business. The gorillas in Uganda, Rwanda, and the DRC are listed as endangered. Harcourt (1996) discusses this classification.

EASTERN LOWLAND GORILLAS

The Hollywood movie *Gorillas in the Mist* was filmed in Rwanda and the DRC in the 1980s. The film portrayed the life and death of Dian Fossey (see Figure 18.3), who is widely credited with saving the mountain gorillas of the Parc des Volcans in Rwanda (Fossey 1983). One of the stars of the movie was a gorilla named Maheshe (Figure 18.4), a silver-back male eastern lowland gorilla living with his group in the DRC's Kahuzi-Biega National Park. Maheshe was a

FIGURE 18.3   The grave of Dian Fossey in Rwanda among the gorillas fallen to poachers.

FIGURE 18.4 Maheshe, an eastern lowland gorilla in the DRC, appeared in the movie *Gorillas in the Mist.* In 1993, he was killed so his head, hands, and feet could be sold as souvenirs.

gentle gorilla who took command of his group in 1975 when he was 15 years old. His group was eventually habituated to the visits of humans before he became a movie star.

Maheshe means *the chief* in the local tribal language. In 1987, when Maheshe was in his prime, his group consisted of 28 gorillas. In 1990, the group had 24 individuals; in 1993, 16 members were protected by Maheshe. In November 1993, Maheshe disappeared suddenly. He had not shown any signs of deteriorating health, and poachers were suspected. In August 1995, his corpse was located and unearthed after a tip-off from a local person. Maheshe's head and hands were missing, indicating that he had been killed for the so-called "trophies." Today, his hands are probably used as an ashtray or perhaps a soap dish. His moth-eaten head may grace a wall or perhaps his skull is used to hold a candle. From wastebaskets made from elephant feet to brooms made from monkey tails to ashtrays made from gorilla hands—native African wildlife is used in many interesting ways.

Eastern lowland gorillas (*Gorilla gorilla graueri*) are found in isolated forests in eastern DRC and Uganda along the border with Burundi and in Rwanda. No wild gorilla populations exist in Kenya or Tanzania, although similar habitats are found. The Bwindi–Impenetrable Forest National Park in Uganda has about 300 eastern lowland gorillas. Kahuzi-Biega National Park on the shores of Lake Kivu contains about 1000 eastern lowland gorillas. Mt. Tshiaberimu on the northwest shore of Lake Edward in the DRC supports an estimated 16 eastern lowland gorillas (Figure 18.5). Other populations of unknown numbers of gorillas exist on isolated mountains in the DRC.

The greatest threat to the eastern lowland gorilla populations is habitat destruction. These isolated mountaintops are surrounded by agricultural fields, and the gorillas have been pushed ever higher up the slopes of the mountains. Logging for fuelwood is extensive in and around these mountaintops. Bush meat hunting is a way of life for many rural people in eastern DRC. Even

FIGURE 18.5 All across Africa, parks are surrounded by agricultural fields. On one side is the Bwindi–Impenetrable Forest National Park in Uganda, home to some 300 eastern lowland gorillas.

where protected areas exist, constant vigilance is necessary to keep loggers and poachers out. Often the brother of an anti-poaching patrol officer is one of the most efficient poachers.

Wages for guards can be five dollars a month, and bush meat hunting is thus essential to feed their families. Who can blame them? Life in modern DRC is difficult for rural people. Often the reason parks and reserves are created is because the soils are too poor to support crops. Nevertheless, trees can still be harvested for building material and fuel.

Rwandan refugees fleeing the civil war in Rwanda in 1994 relocated to many of the forested areas containing eastern lowland gorillas. Organizations such as the Institut Zairois pour la Conservation de la Nature have tried to minimize the damage. This is a difficult task. In July 1994, over 700,000 refugees fled Rwanda and entered North Lake Kivu District, DRC (Lanjouw et al. 1995). In spite of a United Nations decree against locating refugee camps near national parks and protected areas, the United Nations High Commission on Refugees established refugee camps on the edge of Africa's oldest national park, Parc National des Virunga, a World Heritage Site.

Mt. Tshiaberimu is also located within Parc National des Virunga. Once famous for its large concentrations of hippos, Cape buffalo, and elephants, the park has since been decimated. Poachers armed with machine guns destroyed the hippos in a few months. Poaching is now rampant, and park guards are reluctant to intervene. Why should they? It is already impossible to keep people out of the park. The slopes of Mt. Tshiaberimu are now occupied by small numbers of refugees just trying to stay alive. Various conservation organizations are trying to reduce the impacts of the refugees.

In 1995, poachers killed four lowland gorillas and captured an infant in the Bwindi–Impenetrable Forest National Park in Uganda. The infant was never recovered. More lowland gorillas were killed in Kahuzi-Biega National Park in the DRC. Although gorilla tourism brings so-called hard currency into cash-strapped DRC and Uganda, the murder of gorillas continues. The international border between the DRC, Rwanda, and Uganda is a high-risk zone. Armed refugees, former Rwanda regular army soldiers, poachers, fuelwood gatherers, and hungry people all look to the forests for relief. It is all that is left.

Anti-poaching patrol guards were issued bullet-proof jackets and other essential field gear. The gorillas in some areas were receiving 24-hour protection. The situation continues to be precarious at best. Tourism is nonexistent and the flow of foreign hard currency has dried up. People are hungry and hence willing to do whatever it takes to obtain food. Gorilla populations across Africa are under the most intense pressure ever. Unlike the bush meat trade that threatens the western lowland gorillas, the eastern lowland gorilla's greatest present threat continues to be habitat destruction brought about by man's inhumanity toward man. The gorillas are simply caught in the middle.

WESTERN LOWLAND GORILLAS

Until recently, the gorillas of western Africa had been relatively safe. Now there is a connection between a beautiful, new wood table made in Cameroon and a western lowland gorilla. In the western stronghold of the gorillas, bush meat is a delicacy and gorillas and chimps are eaten as food. Gorillas were protected more by remote location than by any laws humans might choose to make. Now, however, their forests have been invaded and destroyed by French and German logging companies. With the new logging roads cut into pristine forests came the bush meat hunters. However, bush meat, when plentiful, is heavy, and transportation is always hard to come by in the forest. Enter once again the loggers and their heavy trucks. European loggers have not only opened huge virgin tracts of timber but have solved the transportation problem that formerly made bush meat hunting a losing proposition.

Drivers of the logging equipment purchase fresh bush meat of gorillas, chimps, and forest antelopes directly from the hunters on the logging roads. In makeshift shanty towns and villages, the loggers act as middlemen and market the fresh meat to waiting buyers. International treaties cannot be enforced in remote logging tracts, and the profit motive is a powerful incentive. Now there is a burgeoning market for fresh gorilla and chimp bush meat. Demand exceeds supply and prices are high. The problem here is that we do not know how many western lowland gorillas exist. Some reports suggest three gorillas and three chimps are now being killed each day to meet the demand for bush meat. This harvesting cannot be sustained.

Pictures of young orphaned gorillas and chimps (Figure 18.6) are finding their way onto the World Wide Web. These young primates will almost certainly die of starvation without their mothers' milk for nourishment. Some conservation organizations use these pictures as scare tactics to raise funds to support some sort of conservation effort. Nothing seems to be working. Most wildlife conservation organizations are mute. If objections are raised over the issue of bush meat consumption, the conservation organization's operations might be curtailed in that country. If no objections surface, all is assumed to be fine. These organizations have learned not to object too strongly lest they lose "part of their portfolio" of wildlife projects—the very projects they use in mailing campaigns to raise money. Typical requests for funds are quite emotional: "Orphaned as a by-product of the logging concessions, their parents shot

FIGURE 18.6   George and Martha, chimpanzee infants stolen from the wild in the DRC and then confiscated. The rest of the family of 40 chimps was killed trying to protect them.

dead for bush meat, these tiny babies will die a slow, excruciating death by starvation unless we raise $50,000 to create a primate orphan sanctuary..."
However, they are correct and factual.

The assault on Africa's elephants, rhinos, lions, hunting dogs (Figure 18.7), gorillas, chimps, and other animals continues. Powered by greed, demand for food is outstripping supply. The plight of the gorillas is a less familiar story. How can science be combined with conservation efforts to offer a solution to the gorilla's plight? That is the question facing a conservation biologist. No less than the survival of all the gorillas is at stake. If killing gorillas and chimps is acceptable, how big a step is it to killing humans? Thus, our own fate may be inextricably linked to the survival of the gorillas. What can be done? What are some possible solutions?

RECENT TAXONOMIC STATUS

New genetic evidence suggests that the western lowland gorilla (*Gorilla gorilla gorilla*) and the eastern lowland gorilla (*G. gorilla graueri*) appear to be more distinct from one another than are members of two established chimp species.

FIGURE 18.7   Wild dogs in Serengeti National Park, where they no longer roam the savannas.

The mitochondrial DNA differences in the two gorilla subspecies led to the conclusion that gorillas separated from chimps between eight and ten million years ago. Humans and chimps separated six million years ago. Morell (1995) reported that the eastern and western gorillas separated about three million years ago and hence should be considered separate species. Others suggest that humans, chimps, and gorillas diverged at the same time. In any case, genetic information is accumulating and perhaps the questions will be resolved before any of the gorilla populations are destroyed.

Esteban Sarmiento of the American Museum of Natural History has looked at morphological differences between the two mountain gorilla populations in the Parc des Volcans in Rwanda and the Bwindi Forest in Uganda (Sarmiento, personal communication). He reported that Bwindi gorillas live at lower elevations and warmer temperatures, are much more arboreal, have longer day ranges and larger home ranges, and eat much more fruit and pith and less bamboo and leaves than Parc des Volcans gorillas. Bwindi gorillas have smaller bodies; longer limbs, hands, and feet; shorter trunks and thumbs; big toes and tooth row lengths; and narrower trunks and orbital breadths than the other gorillas. Sarmiento suggests that the Bwindi gorillas should be considered a separate taxon from the Parc des Volcans gorillas. But are the Bwindi gorillas related more closely to eastern lowland gorillas or mountain gorillas?

Ecological, morphological, and behavioral differences between the Parc des Volcans gorillas and the Bwindi gorillas show that the gorillas have responded to different selective forces. Perhaps the Parc des Volcans gorillas are indeed only mountain gorillas and their population of 300 is all that remains on earth. Perhaps they are a unique species that requires special attention. Mammalian taxonomists will almost certainly recognize the western lowland gorilla and the eastern lowland gorilla as separate species. The mountain gorillas of Rwanda will most likely be recognized as a subspecies of eastern lowland gorillas.

Thus, at least two and perhaps three species of gorillas will emerge, and each will demand its own conservation strategy. The loss of a gorilla species will not be permitted by anyone, and conservation organizations will unite and stop worrying about being ejected from a country for bringing the international community's attention to wildlife atrocities. A new conservation ethos will emerge from the rubble of previous failed efforts. Conservation biologists will be supported by logging concessions to create parks and reserves within logging tracts to harbor gorillas and other fauna in western Africa. In the tri-border area of Rwanda, Uganda, and the DRC, gorilla tourism offers a hope for conservationists. Gorilla tourism was formerly a lucrative business, bringing in much needed foreign currency. Indeed, visits to the Rwandan mountain gorillas provided the second largest source of so-called hard currency in Rwanda (Vedder and Weber 1990). Cross-border agreements will limit tourism and increase the price of visits. Revenues will be shared among the three nations according to rules set in a treaty governing gorilla visits. Moreover, a broader tourism base will off-load gorilla visitation with birding and hiking trips. Concessions run by local cooperatives will allow more local people to participate in the revenue stream. Local hotels and eateries will be established. A holistic, comprehensive conservation plan will be created. Or is this wishful thinking?

Of course, community-based conservation education programs could create the groundswell of local support that all successful conservation programs require. Scientific research opportunities abound in the remote African forests. For instance, almost nothing is known about the mammals, insects, and plants of the gorilla strongholds. Thus, science can play a vital role in constructing and maintaining strong research programs that ensure a steady stream of researchers. Scientific understanding of local ecological processes is essential to the conservation of the local resources, gorillas included. There is simply no substitute for a strong scientific base. Scientists have demonstrated again and again in widely varying parts of the world that overexploitation of resources has potentially catastrophic consequences. Once the resource is depleted or destroyed, recovery is usually long and slow under the best of circumstances. Too many tourists getting too close to the gorillas will almost certainly lead to their destruction. Sustainable tourism is possible. The price of a visit to the gorillas should be driven by market forces. Just as OPEC controls the price of world oil, so too can these countries control the price of a visit.

PRIMATE CONSERVATION CRISIS

A major conservation crisis has developed over the bush meat trade in Central Africa. Since 1990, Karl Amman has been researching the commercialization of the bush meat trade, visiting various Central African countries on a regular basis. He is convinced that what is happening on the bush meat front is symptomatic of events and trends in Central Africa in general. The unsustainable utilization of wildlife and other resources (such as forests) will lead to more shortages and famine, which in turn will lead to more migration, more social unrest, more war, more starving children on television screens in the West, and eventually to millions of dollars in aid to try and do something—anything—to alleviate the problems. Karl, a businessman by training, believes a more business-oriented approach to conservation is needed.

Karl had the opportunity to discuss bush-meat-related topics with many conservation executives. Coming from a business background, what surprised him more than anything else was the lack of ways to measure results on the conservation front; no attempt was made to establish criteria against which performance could be assessed.

In his hotelier days, Karl was responsible for properties in several African countries. All general managers worked to specific targets and financial budgets. Independent quality assessors would visit unannounced, with long questionnaires to be completed. Guests would be encouraged to send in their comments. If the management did not live up to expectations, its Africa tour was often short-lived. In countries where even good managers could not produce acceptable results, management agreements were terminated. This is the way business works worldwide.

Many conservation organizations with operations in the countries concerned have budgets similar to those of large hotels, but there seem to be no real targets against which to evaluate the performance of the managers in capital cities or field workers out in the provinces.

Before it degenerated into its present state, Congo Republique was one of the more organized countries in Central Africa. Several large conservation organizations had offices, even head offices, in the capital, Brazzaville. Karl started visiting the Congo regularly in the early 1990s, mainly to document the operations of the three great ape sanctuaries there. (Two cater to chimpanzees and one to gorillas. All of them care for dozens of "bush meat orphans"). Some of the facts and observation Karl compiled on these trips are as follows:

- Bush meat from a wide variety of species was available for sale in all the major markets, irrespective of whether the hunting season was open or closed.
- While the meat of protected species was disguised in some markets, it was openly on display in others.
- For a while, elephant steaks, frozen and vacuum-packed, were on sale in the capital's most upscale supermarket chain. (When Karl questioned the French manager, he was told the meat had been imported from Chad. The manager thought that solved the problem. Obviously, he had never heard of CITES.)
- The prime minister went on television, during the closed hunting season, to encourage all school children to spend their holidays hunting and fishing.
- When some concerned individuals in the West responded to the initial publicity by writing letters to the Congo embassy in Washington, they received the following reply: "There is no poaching problem in the Congo."

At the Conkouati Wildlife Reserve, Karl filmed a truck being loaded with bush meat, right next to a conservation agency vehicle. When Karl interviewed one of the traders and asked why the cost of the meat doubled by the time it reached the coastal town of Pointe Noire, he was told that the government rangers manning the roadblocks required "payment." When Karl asked how much, he was told the more protected the species, the higher the price.

On his first and only evening in Ouesso, the gateway to the renowned Nouabale Ndoki National Park, Karl filmed a truck carrying tons of bush meat, including the carcass of a silverback gorilla. A Western researcher was dutifully recording yet another dead gorilla in his "bush meat book." His job was to assess the sustainability of the trade—not report it to the authorities or take significant action.

The next day, the police chief kicked Karl out of town, asking him to charter a boat to take him to neighboring Cameroon. Karl was escorted by an armed guard and assumed this was for his own protection. In the first village out of town, the guard stopped to load a large bag of ivory, which was to be "escorted" to Cameroon.

Two years later, an ABC crew filmed an elephant graveyard halfway between the Nouabale Ndoki National Park and the Odzala National Park. They counted 280 carcasses.

The Reserve de la Chasse de la Lefini is the largest protected reserve in the Congo. It is also the site where the first group of orphaned gorillas has been rehabilitated. Karl visited twice and walked for hours in the savannah and forest without seeing any trace of wildlife. The local trackers informed him that there were two hippos left. The last chimpanzees and gorillas had been shot in the 1960s. There is no human pressure or habitat loss in this region. There is also no encroachment. Market hunting for the capital, Brazzaville, some two hours away, resulted in the wildlife being wiped out. Regular flights from Ouesso carry bags of fresh meat as the main cargo, and it is easy to anticipate what this supply and demand will do to wildlife in the longer term, even in the more remote parks and reserves.

Karl wondered if there was any law enforcement with regard to poaching and wildlife. He asked Dr. Oko, the personal assistant to the minister, to see the records of poacher arrests. There were none.

Karl writes:

> Where is the hope for conservation when poachers and illegal loggers are not arrested or lose their licenses? What is the point when the Minister of the Environment eats bushmeat at every official function, and ministry officials rent out guns to poachers to supply the restaurants they own in logging concessions? (This happened in Cameroon, but I am sure the story is not so very different in Congo.) What do you tell a villager who happily suggests that, before you ask him to stop cutting trees or shooting gorillas, you first go to the capital and stop the wealthy individuals who continue to loot the national resources and economy in a big way? Are all of us who are concerned about the future of wildlife and habitats in Central Africa simply wasting our time and a lot of somebody else's money?

A prominent conservation organization to which Karl offered an exposé on bush meat for its in-house publication wrote back and said:

> The chief drawback, of course, was our firm conviction that publishing your article with your compelling photographs would have wide repercussions that certainly would adversely impact our scientists in Africa. An essential and exhaustive part of their job is to maintain good relations with the governments and indigenous people so that the Society's conservation projects will be permitted to continue.

In other words, the conservation organization is between the proverbial rock and a hard place. If it raises the issue of bush meat, it will be "escorted" out of the country. If the organization says nothing, the trade continues. This is a conservation failure.

The International Monetary Fund and other donor organizations regularly pull out of countries, especially if there is no political will. And they do not make excuses. Currencies collapse and politicians shout, but unless structural changes occur, the International Monetary Fund simply leaves. When has a conservation organization pulled out of a country in protest? Does the quiet, diplomatic approach achieve more than shouting and screaming? The governments concerned like to hide behind the fact that these major organizations have hung out their signs in their capital cities. Where is the problem? As long as these prominent groups are involved, everything must be fine.

The rate of loss of habitat, natural resources, and species in tropical Africa is now higher than ever before. The quiet, diplomatic approach has totally failed, and a lot of time and money have been lost.

As for the bush meat trade, it has now been commercialized to the point where it has become an integral part of the economy. The problem has now gone far beyond the capabilities of the conservation organizations. Even the giant logging companies are powerless to deal with this crisis. One executive of a major French firm told CNN that they were now afraid of the poachers, who had automatic weapons. Some German loggers, concerned over the bad publicity resulting from their facilitation of the bush meat trade, recently asked the transporters of their timber to tell their drivers to stop carrying bush meat. The drivers went on strike, and the loggers and transporters had to give in.

The Congo Republique has now disintegrated, which is not surprising considering the lack of law enforcement on the conservation front. In Gabon, a prominent German logging firm has just started operations in the biologically important Lope Forest Reserve. Cameroon appears to be going the way of the Congo Republique as the conservation situation continues to rapidly deteriorate. In the DRC, loggers are frantically looking for US$50 million to link the Central Congo River Basin to the logging-road infrastructure of the Congo, the Central African Republic, and Cameroon.

A prominent French logger in Cameroon went on camera and stated that what was happening now was "total destruction" and there was no point in counting on the government, the loggers, or the conservation community to effect any kind of meaningful change. He felt only a major international outcry would make a difference. As long as the conservation community needs to publicize its very limited success stories to survive and as long as the "quiet, diplomatic" approach persists, there will be no such outcry.

U.S. politicians, as well as most others, govern by opinion polls. When the public speaks, the decision makers listen. The ivory crisis, whale hunting, and seal clubbing became major issues through public concern. What will it take to turn the large-scale, unsustainable slaughter of chimpanzees, gorillas, and other forest wildlife into a similarly emotional and effective campaign? If we can do nothing for our closest animal relatives, what hope is there for other lesser known creatures?

Karl believes that the Western donor community is still taken seriously in tropical Africa. Large sticks and carrots are our best hope. As such, our best bet is to link donor funding to environmental performance in the same way that human rights issues are linked to donor assistance.

In 1998, the Indonesian economy had to be bailed out with tens of billions of dollars in donor assistance, and every human rights organization spoke out and asked for severe pressure to be put on the authorities to effect a change. There was no evidence of environmental groups working to link these huge loans to better environmental performance—and this was while huge forest fires were still burning. No conservation organization took advantage of this opportunity to "persuade" President Suharto to cancel the Rice Bowl Project, in which 10,000 km^2 of prime orangutan habitat was being cleared for a rice-growing scheme (using US$150 million from the National Reforestation Fund).

In January 1998, Karl visited the Yaounde bush meat market and bought two gorilla arms. He then acquired an equivalent amount of beef. Next, he bought the frozen head of a chimpanzee and matched it with a much bigger pig's head. He took all this back to his hotel and put price tags on his purchases to illustrate that beef and pork were less than half the price of gorilla and chimpanzee.

Clearly, there is a question of supply and demand. The supply of great ape meat, and meat from other wild species, satisfies the taste buds of a growing urban middle class willing to pay a premium for the product. The problem is that this practice is no longer sustainable and has not been for some time. Increasing demand and decreasing supply will inevitably result in price increases. With a limited resource, this will continue until the supply is exhausted.

Supply, demand, and pricing are the domains of economists, businesspeople, and now, apparently, conservationists. Should the three domains be separate or fully integrated?

CURRENT ISSUES IN GORILLA CONSERVATION

An unprecedented number of gorilla killings was witnessed in 1995 in Bwindi–Impenetrable Forest National Park in Uganda and in the Virunga/Parc des Volcans DRC–Rwanda–Uganda World Heritage Parks. Accusations and recriminations worldwide failed to frame the relevant questions necessary to address this problem and define solutions that enabled better protection for the gorillas. We asked two leading authorities to address several critical questions: H. Dieter Steklis of Rutgers University and director of the Dian Fossey Gorilla Fund and Esteban Sarmiento of the American Museum of Natural History. Both have extensive experience with the gorilla projects.

What is the motive behind the gorilla killings?

Dr. Steklis—The presumed motive in all cases (Uganda and the DRC) is capturing immatures for sale on the black market. But it may be more complex in some cases; that is, politics may also be involved. Political infighting and power struggles have been suggested as motives in Uganda, and in the DRC at least one death may have been caused by angry refugees.

Dr. Sarmiento—The most often quoted motive, that the killings were perpetrated for the capture and sale of gorilla infants, has never been substantiated. There is no evidence in any one of these incidents that infants were actually taken and sold. At Bwindi, group number counts taken before and after the killings account exactly for the four recovered corpses. Furthermore, given the trade restrictions, it is no longer possible to freely move gorillas out of any one country without alerting government authorities and a multitude of watchdog organizations. Notably, none of the supposedly abducted infants has surfaced, further arguing against the likelihood that they were sold or traded. Unfortunately, the autopsy report does not mention the condition of the slain female's uterus to confirm whether she had recently given birth and substantiate the possibility that the infant taken may have escaped census.

Given their abundance, political motives are most likely to be behind the gorilla killings. Disgruntled employees, job availability, uneven distribution of donor funds, and restrictions on local use of forest resources are all viable motives at the local level. Possible motives at the national and international level are as an attention-seeking device to resolve other political issues or to increase funding and public awareness and as a means of sabotaging tourism revenues. The perceived economic worth of Virunga and Bwindi gorillas, created by heavy funding of these areas and the associated media attention, has made them vulnerable to political maneuvering and terrorism and put these populations as well as the tourists who visit them in jeopardy. The one to two million Hutu refugees in the DRC between Goma and Rutshuru and the political problems in Rwanda, therefore, may presently pose the biggest threat to the safety of these gorillas. In part, this threat can be traced to the large sums of external aid Rwanda received during the 1970s and 1980s and the population explosion and social inequality that such funding helped to precipitate. Unfortunately, there aren't any readily clear solutions to the multitude of political problems in the area. Developments over the last 20 years, however, strongly suggest that throwing money at or around the forests in which gorillas live without carefully planned long-term goals is more likely to exacerbate the problem of protecting gorillas than solve it.

Why weren't the killings prevented? Is it a problem with the present policy or with its implementation? What changes in policy or implementation could be undertaken to better protect the animals?

Dr. Steklis—In the Uganda cases, internal politics cannot possibly prevent or protect gorillas from getting caught in the middle of some personal dispute among the very officials charged with their protection. Otherwise, better patrols in the area should be sufficient to prevent killings. This strategy has worked in Rwanda.

In the DRC, the refugee situation has heightened the structural chaos that normally prevails there. Moreover, the presence of the former Rwandan soldiers in the park has made it virtually impossible for Institut Zairois pour la Conservation de la Nature (IZCN) patrols to work safely (that's why the IGCP [International Gorilla Conservation Project] outfitted them with bulletproof vests!) or efficiently. Until the refugee problem is solved and they are gone from the park area, little can be done effectively without starting a small war.

Dr. Sarmiento—For all these killings, it is easy to blame the park rangers for either falling short of their responsibility to patrol the forests or promptly reporting to their superiors the suspicious behaviors of trespassers. The warden in charge is also easily criticized for not having the necessary presence, rapport, and/or control over his subordinates to ensure that they do their jobs. Without a strong voice in the media or professional "spin-meisters" working on their behalf, neither wardens nor the rangers when blamed have much to say in their defense. If their voices could be heard, they would complain as to the lack of equipment, working communication systems, and transportation necessary to do their job effectively. Undoubtedly, they would also complain about their salary. If they received a wage sufficient to feed their families and send their children to school, they wouldn't need to sacrifice their working schedule by taking on other jobs.

Such is the economy of Third World countries that they are unable to provide funds for adequately protecting these World Heritage Sites. Supposedly, this is why these sites have been designated World Heritage Sites in the first place and why so much donor money each year is funneled into them. Both the taxes collected by Western donor nations and donations to NGOs [nongovernment organizations] by an environmentally conscious populace are supposed to ensure that these sites are protected from such incidences. Bwindi–Impenetrable Forest National Park and the Virungas volcanoes both have their fair share of NGOs touting conservation (IGCP, CARE, and World Wildlife Fund at Bwindi; IGCP, CARE, Dian Fossey Gorilla Fund, and Morris Animal Fund at Virunga).

Given that the resources for conserving these areas exist, why weren't the killings prevented? The answer is simple: the NGOs aren't doing what they proposed to do with the donations they are busy collecting. Why? Primarily, because there is no motivation for them to do so. There is no accountability. They would receive future funding whether or not the jobs they proposed to do are done very well or are not done at all. It is logical, therefore, that instead of investing the majority of their resources to see their proposals through, NGOs invest the majority of their resources to come up with new proposals and collect more money. Ironically, the killings have been used as a rallying cry by NGOs to ask for more money, instead of an indication to donors that these organizations are not providing the necessary protection to these areas. Thus, the problem is not one of implementing the existing conservation programs but of formulating effective policy that will not only help protect the gorillas and their habitat, but will also assure us NGOs will do what they propose to with the funds they receive. At the moment, there simply is no carefully thought out, long-term tested policy for implementing, or the motivation to formulate one.

To protect rhinoceroses, the Kenyan parks department assigns armed guards to each animal 24 hours a day. Couldn't gorillas also benefit from this type of protection?

Dr. Sarmiento—The World Bank endowment to Bwindi can ensure an armed guard for each and every gorilla in the forest 24 hours a day. With limited funds, however, the expense to patrol individual gorillas may not be worth it. In the case of Bwindi and the Virungas, no one could convincingly argue that these past killings, if they don't continue, will have any effect whatsoever on either population. The Virunga population was heavily hunted at the turn of the century, with records for over 50 specimens collected in the 15 years following Captain Beringei's discovery. The actual number hunted was probably much higher. Population numbers, however, were never shown to be adversely affected. It is estimated that approximately the same number of gorillas exist in the Virungas now as existed when Beringei originally discovered them. More importantly, there is no certainty that keeping poachers away from gorillas or locals from using the forest in the long run will ensure their survival. It is conceivable that in some cases limited use may be maintaining the conditions that everyone wants to conserve. Deterioration of their present habitat with population growth, increased intensity of agriculture, and development of surrounding areas poses a much larger long-term threat to the gorillas.

The long-term threats to gorilla survival that these killings alerted us to point to a much more serious shortcoming of conservation policy. Such threats also plead for a more precise definition of conservation, aside from a word that evokes sympathy and justifies the expenditure of vast sums of money. Unfortunately, there is no known magic policy that can ensure the protection of these animals and their habitats. Both Bwindi and the Virungas are relatively small tracts of land, surrounded by relatively dense populations containing relatively large mammals (i.e., elephants, Cape buffalo, and gorillas). Being small tracts of land, they have a proportionally larger perimeter relative to their surface area and hence are much more susceptible to the influx of exotic plants, animals, and/or pests and degradation from adjoining populated areas. Furthermore, because these areas are relatively small, the fauna may be expected to have a more profound impact. Conservation methods used in American national parks that are relatively large tracts of land often buffered by large uninhabited expanses and generally contain smaller sized fauna are unlikely to prove effective without modifications. In this regard, whatever management policies are implemented, they have to be tested for long- and short-term effects.

In the long run, what role will ecotourism have in conserving gorilla populations? What must be done to ensure that tourism and the development that goes along with it will have a positive effect?

Dr. Sarmiento—Although eagerly embraced by donor countries looking for justification to pour in Western funds, Weber and Vedder's proposal [see Vedder and Weber 1990] that revolves around making gorillas an economic asset to the local population has never been adequately tested. It is uncertain whether following a policy of this type would protect the gorillas in the long run. If the recent situation in the Virungas and Bwindi is any indication, this policy is likely to cause more problems for protecting these animals than solving them. The same applies to the oxymoron currently popular with donors: "development through conservation." While debate on the merits of these policies can be argued eternally, careful testing is the only way to gauge their effectiveness. In this regard, the conservation methods presently used at Bwindi and the Virungas are experimental and have to be carefully and painstakingly researched. It is mind-boggling how many NGOs and donor countries flaunt policies of not supporting research with the tacit suggestion that research is a frivolous expenditure when setting out to do conservation.

Prior to undertaking research on the success of implemented conservation programs, research as to systematics and ecological relationships of the flora and fauna that presently exist in these parks has to be undertaken. In this regard, the Virungas area is far ahead of Bwindi, but there is still much work that needs to be done in both. Once systematic and ecological relationships are clear, careful monitoring of the park's flora and fauna can gauge the effects of implemented programs within the forest (i.e., tourism, limited use of forest resources) or in the areas surrounding the forest (crop changes in adjacent fields, more intensive agriculture, new roads, dams, waterways, etc.).

Who is to take responsibility and make the necessary changes to ensure that these killings do not continue?

Dr. Steklis—The international community needs to take a far stronger, more concerted stand on the DRC's tolerance of the Rwandan army on its soil. The present regime in the DRC is basking in aid dollars that have once again poured in because of the refugees, so the government has no immediate motivation to end the crisis, especially as dozens of agencies are making high profits themselves seeking solutions to the environmental degradation caused by the refugees. All are "band-aid" solutions, like cutting wood for the refugees elsewhere (virgin forest) rather than getting rid of the refugees. If all aid to the DRC were stopped today, the refugees would be gone tomorrow! Unfortunately, it is all big business for governments and NGOs alike, because both profit from crises!

Dr. Sarmiento—Ultimately, the donors have to, since they control the assets the conservation NGOs seek to attract and thus are the only ones that hold the rein of accountability. This points a finger at each and every one of us that donates to gorilla conservation groups and also to USAID [U.S. Aid for International Development], which is one of the largest donors in the Bwindi and Virunga areas.

All of us clearly agree conservation of gorillas and their habitat is a good cause, but that is not enough to justify our donation to any group that uses all the "politically correct" but hollow catchphrases. NGOs worthy of funding should have a proven track record of implementing conservation programs successfully. They should have clearly set long- and short-term goals and a feasible carefully outlined conservation program for each and every area they propose to conserve. More importantly, they should not confuse proven versus experimental programs and implement aggressive research policies in cases where program implementation is experimental. As individual donors, to hold NGOs accountable we have to take the time to familiarize ourselves with their accomplishments, ideologies, and conservation policies. As noted, blindly throwing money at both Virunga and Bwindi areas may be causing more harm than good.

To implement a system of accountability, major donors or donor countries must have a working knowledge of what is needed to establish and carry through effective conservation programs in areas they plan to fund. This should include estimates of the cost of carrying out such programs. This would effectively eliminate NGOs from working at cross purposes in any one area. Because they are donating this money in the name of taxpayers who supposedly want to support conservation, donor countries should have a clear and precise policy as to the types of conservation programs they are willing to fund for each area and how effective each program must be. For example, will it fund only tested, proven methods of conservation or will it fund experimental ones? Because USAID is part of the U.S. federal government, if you are an American voter, you have some say through your district representative as to how this organization donates money. If you want to help, write to your congressional representative and tell him or her you want the government to implement programs of accountability in international conservation policy. These programs should impartially gauge the effectiveness of NGOs in implementing proposed programs, cease funding NGOs that aren't successful in carrying out their proposals, and hold them accountable for funds if the proposed work was not satisfactory. Add that you do not want untested conservation programs to be implemented unless accompanied by rigorous scientific testing.

Fortunately, if these gorilla killings do not continue, what damage has been done is negligible. Like every ominous grey cloud that mars a perfect blue sky, the killings have a silver lining. In this case, they alerted us to the fact that our efforts to conserve Virunga and Bwindi gorillas are far from satisfactory. Contrary to what the popular press would have one believe, the major reason for this has not been lack of concern or funds, but the lack of holding those who receive these funds accountable.

EXERCISES

18.1 Run a search for "gorilla" on the World Wide Web using your favorite search engine. There are official and unofficial Web pages. Often, other Web sites such as news networks have information on gorillas. Besides habitat destruction, what are the latest threats to gorilla populations?

18.2 What other information on primates is available on the World Wide Web? Is there any information on New World primates such as tamarins and marmosets?

18.3   Look for more information on primates. How many species are recognized today? What genera have converged with other genera on different continents?

18.4   Compare and contrast the plight of gorillas and other primates on a country-by-country basis. Use the Web to check on countries such as the DRC, Nigeria, and Equatorial Guinea. Is there information on chimpanzees?

LITERATURE CITED

Akeley, C. 1923. *In Brightest Africa.* Garden City Publishers, Garden City, New York.

Coolidge, H.J. 1929. A revision of the genus *Gorilla. Memoirs of the Harvard Museum of Comparative Zoology* **50**:293–381.

Fossey, D. 1983. *Gorillas in the Mist.* Houghton Mifflin, Boston.

Groves, C. 1970. Population systematics of gorilla. *Journal of Zoology* **161**:287–300.

Harcourt, A.H. 1996. Is the gorilla a threatened species? How should we judge? *Biological Conservation* **75**:165–176.

Lanjouw, A., G. Cummings, and J. Miller. 1995. Gorilla conservation problems and activities in North Kivu, eastern Zaire. *African Primates* **1**(2):44–46.

Morell, V.W. 1995. Will primate genetics split one gorilla into two? *Science* **265**:1661.

Schaller, G.B. 1959. *The Mountain Gorilla.* University of Chicago Press, Chicago.

Vedder, A. and W. Weber. 1990. The Mountain Gorilla Project (Volcanoes National Park). *in* Kiss, A. (Ed.). *Living with Wildlife: Wildlife Resource Management with Local Participation in Africa.* World Bank, Washington, D.C.

CAN HUMANS MANAGE WILDLIFE?

19

With wildlife in the United States making a comeback, all people who enjoy the outdoors have a lot to be thankful for. What a thrill it is to see wild turkeys skulking through the eastern hardwood forests of Pennsylvania, to hear bull elk bugle their mating calls in the high meadows of New Mexico, to see wolves and grizzly bears in Denali National Park in Alaska, or to see pine martens hunting squirrels in Yellowstone National Park in Wyoming. Some people may never get to see bighorn sheep but nevertheless share an enthusiasm for wildlife. The success of the National Wildlife Federation's Backyard Wildlife Program attests to the fact that Americans are outdoor people who enjoy wildlife.

At the close of 1996, a U.S. Fish and Wildlife Service press release reminded all Americans that wildlife was coming back:

1996: Duck Populations Soared, Buffalo Roamed Free, Condors Flew in the Southwest

The continued recovery of the Nation's duck populations after decades of decline is just one wildlife success story in a year that offered many bright spots for species from buffalo to butterflies.

"The American people are making an impressive effort to restore wildlife across the Nation. At year's end, we like to take a moment to reflect on some of the good news that people may have overlooked during the busy year," said Acting U.S. Fish and Wildlife Service Director John Rogers.

Among this year's good news stories:

An estimated 90 million ducks flew south from their northern nesting grounds, the highest figure since the Service began estimating the "fall flight" in the 1950s. Several years of plentiful rain and snowfall in primary nesting areas of the north central United States and south central Canada, along with restoration and conservation of millions of acres of wetland habitat, has boosted the duck population by 34 million since 1990.

In early December, six California condors were released into the wild in northern Arizona after an absence of 72 years. The six condors, which were bred in captivity in California and Idaho, were held in acclimation pens at the release site for several weeks before they were set free in mid-December.

The huge birds nearly became extinct during the 1980s and have been restored through captive-breeding in zoos and releases to their former range in California and, now, Arizona.

For the first time since the mid-19th Century, buffalo are again home on the range in Iowa amid the tall prairie grasses of the Walnut Creek National Wildlife Refuge near Des Moines. A total of 14 buffalo were relocated to Iowa from herds at Wichita Mountains National Wildlife Refuge in Oklahoma and Ft. Niobrara National Wildlife Refuge in Nebraska. The Service hopes the Walnut Creek herd will reproduce and eventually number 100–150 buffalo.

Reintroduction of the gray wolf in Yellowstone National Park in Wyoming and in central Idaho has been so successful that no new releases will be made in either area in 1997. In addition to the wolves released in both areas in 1995, the Service released 20 wolves in central Idaho and 17 in Yellowstone National Park in 1996. There are now 52 wolves in Yellowstone and 40 in central Idaho. Wolf recovery team leaders say that further wolf releases will be considered on a year-by-year basis.

Eighty-two young bald eagles were fledged from 58 active nest sites at the Upper Mississippi River National Wildlife and Fish Refuge in Minnesota. Nine of the nests produced triplets. By comparison, in 1986, there were nine active nests, each producing only one young.

In August, biologists reported the first recorded breeding of northern fur seals on the Farallon Islands off California since 1817. Biologists observed a bull, several females, and a pup on West End, a wilderness area of the Farallon National Wildlife Refuge, 30 miles west of San Francisco's Golden Gate Bridge. San Miguel Island in the Channel Islands is the only other northern fur seal breeding colony in California.

Mississippi Sandhill Crane National Wildlife Refuge, where most of the remaining wild population of this endangered crane resides, reported a record 13 nesting pairs in 1996, the highest number of nesting pairs recorded in 30 years of monitoring. Today there are 95 Mississippi sandhill cranes in this country, 23 of them hatched in the wild, compared to only 30 in existence in 1975.

About 170 whooping cranes are expected to arrive this year at Aransas National Wildlife Refuge in Texas, up from 158 last year. Only 16 whooping cranes were left in the wild in the 1940s.

Despite losses last winter caused by red tide along Florida's southwest coast, manatees are doing well at the Crystal River National Wildlife Refuge in Florida and populations may top last year's record high of 304 animals. By the end of November, 283 manatees had already congregated in the warm waters of the Gulf of Mexico with more expected by the end of December. The Service helped return a wandering manatee named Sweet Pea to the wild in Florida after her sojourn and rescue near Houston.

More than 250 endangered Schaus swallowtail butterflies were released into their historic habitat near Miami. The butterfly, which occurs only in Florida, was on the verge of extinction in 1991; populations have been on the rise since mosquito spraying was halted on northern Key Largo during the Schaus breeding season.

Endangered black-footed ferrets have been discovered in the Shirley Basin of Montana, where the species was reintroduced between 1991 and 1994. Surveys at Charles M. Russell refuge near Lewistown, Montana, confirmed the presence of approximately 20 black-footed ferret kits in 7 to 9 new litters. At least two of the litters were born to last year's wild-born females.

The U.S. Fish and Wildlife Service is the principal Federal agency responsible for conserving, protecting, and enhancing fish and wildlife and their habitats for the continuing benefit of the American people. The Service manages 511 national wildlife refuges covering 92 million acres, as well as 72 national fish hatcheries.

What's the bad news? On April 1, 1997, the U.S. Fish and Wildlife Service issued the following report:

Report Warns That Snow Goose Population Explosion Threatens Arctic Ecosystems

In the mid-1980s, wildlife biologists and conservationists struggled to reverse a sharp decline in duck populations by restoring wetlands in key nesting areas. The effort was successful. Boosted by 3 years of plentiful rainfall and millions of acres of restored wetlands, this fall's duck migration was estimated to be the largest on record.

A decade later, biologists are facing a completely different challenge. Instead of too few ducks, the problem today is too many snow geese—so many, in fact, that they are causing ecological havoc on their arctic breeding grounds.

A recently published report by the Arctic Goose Habitat Working Group, comprised of U.S. and Canadian biologists, found that even liberalized hunting seasons for snow geese have failed to stop the population explosion and, by the most conservative estimates, the number of birds is rising at 5 percent a year.

The long-term impact of the population explosion is still uncertain, the report said, but the possibility exists that the overabundance could cause a decline in other species that nest in the same arctic region. These include semipalmated sandpipers, red-necked phalaropes, yellow rails, American wigeons, northern shovelers, and a variety of passerines.

"The geese are literally consuming their own habitat," said Paul Schmidt, chief of the U.S. Fish and Wildlife Service's Migratory Bird Management Office and co-chair of the Arctic Goose Joint Venture of the North American Waterfowl Management Plan. "They break open the turf and uproot plants, especially grasses and sedges, leading to erosion and increased soil salinity. In turn, fewer plants grow and you have a vicious cycle with habitat conditions growing worse each year. The end result is a degradation of the fragile arctic ecosystem. It is an ecosystem in peril."

The Working Group's report cited changes in agricultural practices that have increased food supplies and reduced the winter mortality rate among snow geese. In addition, the growing availability of Federal and state refuges has expanded the suitable habitat for the birds and dispersed geese over wide areas, increasing survival rates.

"Action needs to be taken soon," Schmidt said. "The damage to the ecosystem is not only severe but it also has the potential to be long-lasting," he said. Experiments show it takes at least 15 years for grasses to begin to come back on damaged, hypersaline soil.

While hunting is certainly part of a solution, the report said that more recreational hunting as governed by current regulations and treaty obligations is unlikely to solve the problem by itself.

Possible solutions cited in the report include loosening regulations on baiting, electronic calls, and concealment during spring "snow goose only" seasons; expanding late season hunting before March 10; and negotiating a revision to the Migratory Bird Convention with Canada to allow appropriate hunting of migratory birds between March 10 and September 1.

"These are uncommon solutions, but these are uncommon times and we can't sit by and ignore this problem," Schmidt said. "We expect to discuss the problem during the coming year and develop an effective strategy in 1998."

NUISANCE WILDLIFE

As we have seen, however, some wildlife species are becoming a nuisance in some areas. On the front page of the *Washington Post,* Lipton (1996) reported on efforts by suburban Washington, D.C. governments to control their burgeoning white-tailed deer populations, a common problem in the Northeast that is spreading to the Midwest. About 1000 deer were living on 3600 acres of public land in Montgomery County, Virginia. Officials suggested that the population of deer was ten times greater than the land could support. Other nearby counties were facing a similar problem. Deer were allegedly leaving parks and eating anything they could digest, including azalea bushes, dogwood trees, and holly. Residents were putting fences around their shrubbery to protect their investments. In other areas, sales of deer fences, repellents, and ultrasonic whistles have increased rapidly.

Many believe the absence of hunting and natural predators has led to an increase in deer populations. Others suggest that habitat conversion from mixed hardwood forests to golf courses, shopping malls, apartment buildings, and commercial developments has caused deer to crowd into the only land left to them. As we have seen, this problem is shared by other states as well. The problem usually goes unnoticed until people's pocketbooks or health are impacted. A growing number of cases of Lyme disease, caused by the spread of ticks carried by deer, was reported in Montgomery County, for instance. There were 465 cases of the disease in Maryland in 1995 and 55 in Virginia. In Montgomery County, the number of car accidents involving deer rose from 782 in 1992 to 1244 in 1995. When automobile insurance rates increased because of deer-related traffic accidents, people naturally blamed the deer for the rate increases. And when deer hoofprints on local golf course putting greens made three-putting routine, golfers and other residents began to appreciate that there were too many deer.

Is it possible to have "too many deer" or "too many raccoons"? Can we evaluate statements that suggest "there are more deer in the United States now than when the Pilgrims landed"? Is it acceptable to have an overabundance of wildlife such as deer and raccoons if these species could be kept off roads and out of residential and agricultural areas? In this chapter, we argue that there is indeed an overpopulation of certain wildlife species and that it is not acceptable to allow the situation to persist under any circumstances. Once the problem has been recognized, a cost-effective solution must be found. What makes the overabundance of wildlife so interesting and challenging is that scientists are just now discovering and deciphering the causes and resulting effects of the overabundance problem. The general public does not read scientific journals, and thus the results of long-term experiments are not yet common knowledge. Indeed, many scientists and wildlife managers are just learning of new results and must also be educated. As is often true of new knowledge, the scientific method played a key role in the design of experiments to test hypotheses, as we will see.

TOO MUCH OF A GOOD THING

When deer became a problem in Fairfax County, Virginia, residents debated many possible solutions. Hunting, reintroduction of predators, hiring sharp-shooters, and using deer contraceptives were options considered to control the deer population. Cost effectiveness is not always the determining factor, however. Lipton (1996) claimed that Fairfax County officials "laughed" at suggestions to reintroduce wolves to control the deer population. In heavily human-populated Fairfax County, hunting in suburban areas may also not be an option. Officials estimated the cost of oral or injected contraceptives to be between $500 and $1000 per deer.

Hanback and Blumig (1993) reported that more than 100 studies directed at wildlife birth control methods were being funded by a variety of agencies, including the Humane Society of the United States, the National Park Service, and the U.S. Department of Agriculture. The article states that wildlife contraception is necessary because "humans continue to encroach upon critical wildlife habitat at an alarming rate." In 1990, the authors reported that 35 of 50 states considered ungulates such as deer, moose, and elk to be a significant problem. Although wildlife contraception continues to be an active area of research, the technique would be difficult to administer in free-ranging wildlife.

A hunt was executed in Fairfax County. As elsewhere, animal rights groups reacted predictably and protested against planned hunts in nearby Montgomery County. In the end, Montgomery County officials selected a cost-effective managed hunt. Winners of a lottery were assigned to certain "firing zones" and allowed to shoot three antlerless deer. The public parks were closed during the hunt to protect residents. Deer, however, are not the only wildlife problem humans have.

Raccoons and Tortoise Eggs

What's wrong with this scene? Picture a raccoon on a deck chair on a back porch somewhere in suburban America (Gilman 1990). Another raccoon stands under the chair. "The raccoons...climb onto the deck to frolic, sit in the comfortable chairs and generally make themselves at home." The owner "always leaves a snack on the deck for them." Sometimes the raccoons "become downright rambunctious" and try to get the homeowner's attention "by banging on the door with a rock. Occasionally at night, they throw everything off the deck."

The answer is given later in the story, as follows. "A distressing instance of their curiosity occurred last spring when the resident box turtle dug a hole, laid her eggs and covered them." The homeowner "eagerly anticipated the eggs' hatching, only to find the raccoons feasting on them the next day!"

Raccoons are not wildlife when they bang on the door expecting to be fed. Raccoons are expert nest predators, especially of birds. Where raccoon densities are high, few birds or turtles can successfully nest because raccoons prey on the eggs and young of birds and other species. Is it any wonder songbirds are declining when people are feeding raccoons? Raccoons are examples of meso-mammals. The above raccoons would more appropriately be called nuisance animals that should have been destroyed.

Blue Jays and Warblers

A University of Texas graduate student studying songbirds has concluded that bird feeders are inadvertently aiding blue jays (*Cyanocitta cristata*) and thus harming more diminutive birds such as the golden-cheeked warbler (*Dedroica occidentalis*), whose populations are declining. Providing extra food for the more aggressive jays enables them to expand their populations and the habitat they occupy. The warblers are, in turn, displaced from the habitat. This means that bird feeders may be doing more harm than good as many warbler populations are declining.

Warblers avoid nesting in more developed areas, but the reason for this avoidance was not previously known. Research has determined that development brings in bird feeders, pet food, garbage, and other sources of food that blue jays prefer and other birds avoid. Consequently, development wipes out habitat for warblers and increases habitat and food resources for blue jays and other "junk" species. Moreover, development nearby warbler habitat is sufficient to cause warblers to decline as the jay's invasion is aided by humans.

Special feeders are available that exclude jays and other "junk" birds. Also, installing a wire mesh large enough to permit warblers and exclude jays is a viable alternative to not providing any food. Pet food and garbage would also require more care. Once again, the misguided hand of humans is responsible for the decline of another species and the increase of a species that we already have enough of. Sadly, most people believe that they are helping save birds by having feeders.

Canada Geese

Canada geese were always a favorite among hunters. By 1962, Canada goose populations were recognized to be small and declining. In the mid-1960s, wildlife reserves and breeding programs were established for Canada geese because they were rare and populations were declining. The population in Michigan rose from 9600 in 1969 to 200,000 in 1994. The Michigan program was successful. Similar success stories were reported elsewhere. Farmers have now been complaining of crop damage, and lakes without geese are becoming scarce. Golfers and park visitors complained of smelly, slippery goose droppings littering areas surrounding fairways and lakes. Population studies have shown that the Canada geese population has doubled about every five years thanks to the protection program and the nearly complete elimination of their natural predators. In 1995, Minnesota became the first state to obtain a permit from the U.S. Fish and Wildlife Service to kill geese for food. Michigan and New York followed Minnesota's example.

In New Jersey, Canada geese are also considered a nuisance at times. Management plans require that communities adopt a "no feeding of waterfowl ordinance," encourage goose hunting, modify habitat so as to discourage geese, and utilize harassment techniques. Industrial complexes with attractive lakes have proven a haven for Canada geese, and some populations have stopped migrating farther south. Instead, geese overwinter in more heavily populated areas of New Jersey. A special hunt timed to occur before migratory geese arrived harvested an estimated 5877 Canada geese in 1994. In 1995, an additional 895 geese were killed. Sheepdogs run to exhaustion chasing geese from ponds in front of the most prestigious corporate headquarters.

Predictably, the geese harvests have sent animal rights groups into federal court seeking injunctions to halt the practice. Studies in Illinois and Minnesota have shown that the geese are safe to consume in spite of the fact that many of the geese feed from suburban lawns. State wildlife biologists now round up adult Canada geese for the cooking pot and release the goslings. Each year these gosling will return to the same lakes where they learned to fly in order to breed. Fortunately, the species has been successfully rescued. The remaining step is to reintroduce its natural predators to keep the populations in check more naturally.

Beavers

In New Jersey, 145 complaints of beaver damage were reported to the Wildlife Control Unit during 1994. New Mexico is an unlikely place for nuisance beavers, yet many residents in Albuquerque suggest that there are too many Rio Grande beavers. Orchard growers report that beavers have damaged apple trees. Beavers routinely travel up irrigation ditches, which provide invasion corridors for dispersing beavers in the spring. Residents blame beavers for the loss of cottonwood trees along the Rio Grande. On the Rio Grande through Albuquerque, there are bank beaver dens approximately every half kilometer.

Coyotes

Coyote complaints totaled 47 in New Jersey in 1994; however, the state is preparing for more complaints, especially from sheep ranchers. The American Sheep Industry held workshops to demonstrate the latest technology in coyote management, such as shooting and trapping. There is little doubt that concentrations of resources attract animals which prey upon them. Without natural predators, the coyote population will increase. Unfortunately, coyotes generally do not hunt deer.

Exotics

Most people agree that exotic species should be prevented from becoming established. The brown tree snake on Guam and the Cuban tree frog in Florida are known to be especially insidious invaders that cause great harm to native species. Exotic species including gamebirds have been deliberately introduced for the pleasure of hunters. One ubiquitous example is the ring-necked pheasant. In a 1996 brochure, the Pennsylvania Game Commission indicated that it raises and releases more than 200,000 pheasants yearly. Some hunters, in fact, hunt only pheasants and no other animals or birds. Huntable populations of pheasants exist in almost every state, including some of the Hawaiian Islands. Most people probably think that the ring-necked pheasant is a native species (it is a native of Asia). If people were educated to the fact that the pheasant is an exotic species, would they agree to the eradication of the pheasant?

By 1987, hunters from across the United States were applying for limited permits in New Mexico to hunt exotic species such as Barbary sheep, ibex, and oryx which were introduced using Pittman–Robertson Act funds. The problem with these exotics is primarily that they eat native and sometimes rare plants and exclude native herbivores that co-evolved with the natural vegetation of the region.

The Ecology of Too Many Deer

Simply stated, too many deer reduce the biodiversity of native forests. The suspected decline in songbird populations might well be attributed to there being too many deer. We do not mean to suggest that deer are bad for forests. However, ecologists have documented that too many deer have deleterious effects on forests. We agree with Jones et al. (1993): "The task as we see it, however, is not to find a way to reduce deer populations...It is rather to convince those for whom there cannot be 'too many deer' of the serious and permanent consequences of not reducing deer numbers—thereby removing impediments to what we believe are necessary actions." Deer were consuming forest resources faster than the forests could regenerate.

The states manage their deer populations actively. Management problems are compounded because hunters like to hunt deer, and when they go deer hunting, they enjoy shooting deer. Not all hunters are successful, however; therefore, from the perspective of an unfulfilled hunter, there are not enough deer. Wildlife managers have focused on managing deer populations, so much so that species negatively impacted by large deer populations have been neglected or forgotten. When one species sequesters most of the resources, other species decline. For example, the declines of songbirds and some forest gamebirds have been blamed on habitat loss. Perhaps this is the case, but habitat loss can be caused by many agents, including humans. Large deer populations cause the loss of forest understory, for instance; thus, we should expect that birds that use understory, for whatever reason, are declining. Indirectly, deer may be causing the disappearance of some birds. Recall that the fourth member of the Evil Quartet is chains of extinction, set off when one species adversely affects another species.

Crop damage by deer is already high. However, forest damage by deer is incalculable and largely unnoticed. Two clear-cuts were created in a forest in Pennsylvania. In one region, deer density was about 10 per square mile, whereas in another, there were upwards of 80 deer per square mile. Forests recovered far faster in the area with fewer deer. Deer eat seedlings before they can become established, thus preventing forest regeneration. In areas where deer densities were high, almost no understory existed, with the exception of ferns and grasses that deer avoided. For example, black cherry (*Prunus serotina*) was a valuable hardwood that was abundant in the hills of Pennsylvania. Because deer ate black cherry seedlings, black cherry trees were not regenerating. Once ferns and grasses covered the forest floor, no hardwoods, including black cherry, could grow. In other words, ferns and grasses were poorer competitors than hardwoods for space and light, but because deer ate hardwood seedlings, ferns and grasses could become established. Once established, they spread and prevented hardwoods from sprouting (Jones et al. 1993).

Deer are generalist herbivores that consume many plants. However, deer prefer such hardwood trees as ailanthus, aspen, beech, black cherry, prickly ash, and striped maple. Thus, we might expect to see many young seedlings in a forest of such trees—presuming deer densities are low. If deer densities are high, no preferred seedlings will be found. How, then, can the forest regenerate? In a forest loaded with deer, the most abundant plant species will be plants deer find unpalatable. Jones et al. (1993) gave a specific example. In an old-growth forest in Pennsylvania, 27 woody species were found in 1920 and deer density was less than 20 per square mile. In 1990, deer density had doubled and there were only 11 woody species. Did deer directly kill 16 species? The answer is no; instead, they preferentially consumed all seedlings of 16 species over a period of 70 years. The old trees died, and new ones of the same species did not replace them. Moreover, species composition was dramatically, and perhaps permanently, altered. This is the ultimate human irony. With the best of intentions, humans protected the deer and brought their populations back. Indeed, we have done too good a job, and deer are now destroying our forests and, importantly, the other species that depend on the forests.

Deer have also altered the character of the shrubs and wildflowers in the understory. If the understory of an eastern forest is empty and you can see clearly through the trees, you probably have deer to thank. As for threatened and endangered plants, their future is grim if deer favor them. With the loss of the natural character of our forests and the elimination of the understory, which accompany too many deer, forest songbirds lose nesting sites and protective cover. Songbird populations across the East are suspected of declining. Small forest mammals also suffer losses in abundance, and some species may disappear completely.

In Pennsylvania, deCalestra (1994) found that the mean richness of intermediate-canopy nesting birds declined 27% from the lowest to the highest deer density. Four intermediate-canopy nesting species (eastern wood pewee, indigo bunting, least flycatcher, yellow-billed cuckoo) were not detected when deer densities exceeded 7.9 deer per square kilometer. The American robin and eastern phoebe were not detected at deer densities greater than 14.9 deer per square kilometer, whereas they had been detected elsewhere where deer densities were less. Cerulean warblers, upper-canopy nesting birds, were not detected at deer densities greater than 14.9 per square kilometer. Richness of ground-nesting birds and other birds in the study area was unaffected. deCalestra also found that abundance of intermediate-canopy nesting birds declined 37% from the lowest to the highest deer densities, whereas ground-nesting and canopy-nesting bird abundance was unchanged.

deCalestra (1994) concluded that white-tailed deer densities greater than 7.9 per square kilometer reduced intermediate-canopy nesting species richness and abundance by reducing the height of woody vegetation in the intermediate canopy. Three intermediate-canopy nesting species (Carolina wren, warbling vireo, yellow-breasted chat) and two ground-nesting species (golden-winged warbler, worm-eating warbler) were not present at any of the study sites but had been previously reported as present. Smith et al. (1993) noted declines in abundance of several intermediate-canopy nesting species in the northeastern United States, including the eastern wood pewee, least flycatcher, and yellow-breasted chat, species that either disappeared with increasing white-tailed deer density or were already absent from deCalestra's study area.

deCalestra (1994) also concluded that by altering critical nesting habitat for intermediate-canopy nesting species in fragmented forests, where they were already exposed to increased predation and nest parasitism, high deer density further endangered bird species.

Certainly, ecologists were not suggesting that all deer be eliminated from forests. Such a strategy would invariably push the complicated ecological pendulum in a different direction altogether. The ecological carrying capacity of forests varies considerably. Some wildlife management specialists suggested that seven deer per square kilometer was a rough estimate for deer carrying capacity. Others suggested that artificially increasing foraging for deer lowered the effective density of deer. In this scenario, high-quality forage for deer was additionally provided so that deer did not overconsume forest resources (Jones et al. 1993). For example, adding nitrogen to the soil increased the growth rate of seedlings and, presuming they grew faster than the deer ate them, the forest could be regenerated. On a regional scale, such a strategy would be costly and, at best, temporary.

By one means or another, deer populations must be contained. Increased and sustained hunting pressure is the most cost-effective method known today. No one is suggesting that everyone must enjoy and participate in hunting. Hunting is an acceptable management strategy that is cost effective for controlling an exploding deer population that is threatening our forests. Hunters are willing and able to pay for the privilege, and thus the cost to the public that does not hunt is near zero. Deer contraceptives are an acceptable alternative; however, hunters might not be so willing to shoot deer with syringes and not take home the results of their efforts. Thus, the contraceptive alternative is likely to cost us all, and these costs are likely to be extracted from critically needed conservation funds. The public should have no objection, however, to allowing those who favor the use of contraceptives to pick up the tab.

As animal rights groups protest, deer in the eastern United States, elk in the western United States, red deer in Scotland, and moose in Sweden are damaging forest habitats for diminutive but no less important creatures. The ecological costs are unacceptable because they are long term and destructive. Somehow the public must be educated to this fact. Budiansky (1994) described an experimental exclosure facility built by Smithsonian researcher W. McShea to formally test the ecological changes deer enable.

McShea constructed six ten-acre deer exclosures in two different forests. He periodically compared mammals living inside the exclosures with those living just outside. He reported that in years when oaks produced few acorns and food resources were scarce, competition between deer and small mammals forced small mammal populations to fall. Small mammal populations fell to as low as 10% of what McShea found inside the fences where deer were excluded. Furthermore, migratory songbirds began showing up inside the deer exclosures where, he speculated, insect populations were higher perhaps because the understory was present.

Other research results obtained in Wisconsin showed that deer were changing the tree species composition. Once-dominant hemlocks and white cedar had failed to regenerate because deer ate the seedlings that emerged. The changes did not occur rapidly and therefore were not obvious; however, over tens of years massive, widespread changes occurred. Where deer populations were controlled, hemlocks and white cedar continued to thrive. The ecological consequences of deer overpopulation were a serious threat to biodiversity.

Reiger (1994) reiterated that hunters pay for the privilege of hunting deer and hence controlling deer populations. Hunters, in part, pay the federal excise tax that supports the Pittman–Robertson Act, which grants funds back to the states to support wildlife programs. Hunters, Reiger claimed, were the most cost-effective means of managing the deer population, especially in the absence of predators. He also pointed out that hunters in Germany used rifles and were highly trained before they were allowed to hunt. As a result, hunters were able to shoot deer in city parks in populated areas. The meat from culled deer was then sold to local markets at low prices. In contrast, Reiger lamented that the discharging of firearms in suburbs in the United States was completely illegal, and citizens sometimes preferred to spend scarce conservation dollars on expensive deer relocation or contraceptive programs. Millions of deer were killed each year by hunters. Are conservation efforts directed toward proliferating the species or sparing an individual from pain?

THE COMPLEXITY OF NATURE AND THE AMERICAN WOODCOCK

Found in the eastern half of the United States, the American woodcock (*Scolopax minor*) is a chunky, short-necked, short-legged, long-billed gamebird with large black eyes set high on its large striped head. Although they appear to resemble snipes, sandpipers, dowitchers, phalaropes, and other shorebirds, woodcocks live and nest in moist forests, woodlands, and thickets. They feed mostly on earthworms. Woodcocks are nocturnal, secretive, and rarely seen. However, once observed, their spectacular mating display is unforgettable. The spring mating display of the male woodcock takes place at night. The bird circles high in the sky while constantly making a twittering sound, then plummets to the ground while wind rushes through the wings, making a whistling sound. At the last instant before impact, the male pulls up and lands on its feet in the tall grass, where it sits hiding, often emitting a *peent* sound, while waiting once again to launch upward to impress the females. Sadly, the woodcock's population has been declining since the 1960s. No one knows why. The U.S. Fish and Wildlife Service has been unable to reverse the slide even though it has been seriously studying the woodcock since 1990, when the American Woodcock Management Plan was launched.

The reasons for the bird's decline are not understood. Experimental ecology with treatments and exclosures may reveal the answer. As our discussion proceeds, think about what kind of exclosures could be constructed and in what habitat these exclosures might be placed to isolate and test various mechanisms suggested as causal agents for the bird's decline. Remember that the scientific method can be used to unravel this mystery, and your ideas and thoughts can contribute to saving the woodcock.

Robinson (1995) suggested that "a dry summer, which caused the ground to harden and prevented young birds from being able to scratch up worms, was indicated as the primary reason why the population [of woodcock] had tumbled" in 1994. "Contrary weather patterns have made birds especially hard to find" (Bourjaily 1994). U.S. Fish and Wildlife Service scientists blamed the population decline on the loss and degradation of habitat. Bourjaily (1994) states that each year "precious cover is lost forever, cleared and paved over or plowed under. Less readily apparent but of greater consequence is the widespread degradation of habitat that occurs as timber matures. Woodcock are birds of early succession. Where abandoned farms and orchards in the North once provided ideal habitat, many of these areas have now grown up into mature, open woods. And these areas are far from idyllic woodcock cover." These statements beg two questions: Why do mature forests of the Northeast lack the understory associated with early successional forests? How did the birds exist before humans cleared the forests and enabled succession?

Woodcocks use their long bills to probe for earthworms in the shaded, moist soil. Dry soils become hard, and earthworms leave. Woodcocks prefer early successional forests that have plenty of complete understory and a moist forest floor so they can probe for earthworms. Bourjaily (1994) blames the loss of habitat for the woodcock's decline, yet farms are returning to forests in the Northeast. Others also blame "the uncontrolled loss of prime habitat" (Robinson 1994). However, one of the conservation success stories of the last half of the century is the return of the forests of the Northeast. Indeed, there is now more forest in the Northeast than existed just 25 years ago. Habitat loss could still be occurring in the forest, however, if the understory does not exist. Thus, the loss of forests is not the reason for the decline of woodcock populations that has occurred for nearly a quarter of a century, but the loss of prime habitat *within* the forests might be.

Mature, open woods are typical of the present Northeast forests. Such forests are open because they are missing an understory. Without an understory, leaf litter dries out and soil hardens. But why is there no understory?

Bourjaily (1994) also suggested that acid rain might be a problem. "Given the continuing acid rain problem, biologists have little idea how worms are holding up." How can we test this idea? Bourjaily also suggested that "one of the most important land management techniques [for increasing woodcock populations] is selective clearcutting," as suggested by a U.S. Fish and Wildlife Service biologist studying woodcocks. One has to wonder how this bird survived to this day without help from people! A ban on clear-cutting was offered on the November 1996 ballot in Maine. The legislation threatened "to eliminate such controlled clear-cutting on public lands and managers would then be helpless to create the new growth woodcock need," says Bourjaily (1994) quoting the biologist. Clear-cuts permit seeds in the soil to sprout and receive sunlight and hence allow the rapid creation of an understory necessary for woodcocks. Maine voters shelved the attempt to stop clear-cutting. The central question still remains, however: Why is there no understory in the forests of the Northeast?

Robinson (1994) presented the line of reasoning followed by the biologist: "During the peak woodcock years, early in this century, the Northeastern states were loaded with lands that had been cleared for farming but were no longer cultivated. The rich soil, high in nutrients that support a thriving food chain, became overgrown in young timber and thick shrubbery. These patches of dense shrubs, which keep soil moist and shade out low-growing plants, provide ideal conditions for woodcock, which must consume about a pound of earthworms a day to maintain their adult weight of 6 to 8 ounces.

"But since the 1960s the best of that choice land has been lost to development. Abandoned farmland is often the first chosen for residential, industrial, or commercial development or for another attempt at farming. And where it has not been developed, the best of the old abandoned farmland has now grown into mature forest, which no longer provides the understory of protective shrubbery and bare, moist soil that woodcock require." Once again, we are back to blaming mature forests that apparently do not support an understory. Yet woodcocks thrived before widespread human interference. These two statements are conflicting.

We hypothesize that deer are at work continuously consuming acorns and seedlings before an understory can be created. We also suggest as evidence the

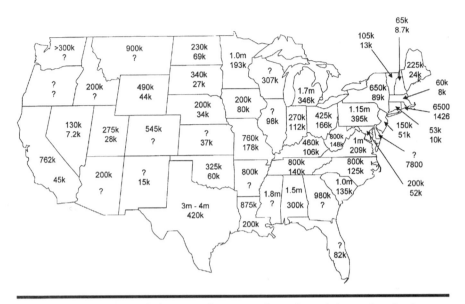

FIGURE 19.1   Estimated deer populations and number of deer taken by hunters in 1994–1995 season (Almy 1995).

idea that most forests in the Northeast contain few young trees because deer have prevented them from becoming established. Deer are ubiquitous in the Northeast (Figure 19.1) and effectively overpopulated the region. As long as there are too many deer, woodcock and other species that depend on an under-story will continue to decline. How can our hypothesis be tested using the scientific method? How can humans manage wildlife when we can't even manage deer, arguably the most studied semi-wild animal ever known?

UNDERSTANDING THE PAST AND PREDICTING THE FUTURE: THE RETURN OF PREDATORS

One possible management scheme is to manage people first and then wildlife. Certainly human numbers are increasing, and, as we learned earlier, wildlife issues are really human issues. The living organisms we see today are the result of hundreds of millions of years of evolution. Human management schemes have managed to bring back some species of wildlife from the brink of extinction. However, fine-tuning nature to permit the highest possible biodiversity seems a daunting task for even the most creative wildlife managers. Suppose that instead of managing game species, we were to change our management strategy to managing just the top natural predators. First, predators may act to limit the growth of their prey populations. Because predator numbers are fewer than prey numbers, we would have to manage fewer animals overall. Humans could manage top predators, and these predators could manage all species below them in the food chain. Thus, one management strategy would be to first manage people so that natural predators could be reestablished and managed. The predators would then act as *de facto* managers and manage prey species, including deer.

THE HISTORY OF WILDLIFE

The sequence of events in human/wildlife interactions seems clear. Early on, wildlife was considered an inexhaustible resource. No limits were established on harvesting wildlife. The Northwest Atlantic walrus, sea mink, prairie chicken, passenger pigeon, Carolina parakeet, and ivory-billed woodpecker were all driven to extinction. Other populations such as deer, moose, buffalo, elk, and especially the large predators were reduced to extremely low numbers. As food production through agriculture increased and society industrialized, less time was spent gathering food. More food became available at a lower price. People began to notice that wildlife populations were low and some species were on the verge of extinction. Conservation ethics evolved as a consequence of improvements in people's lives. Reintroduction programs began.

The sequence of reintroductions also seems predictable. First, hunting populations of herbivores are established. Hunters purchase their equipment and thus provide funding through the Pittman–Robertson Act, pay fees, and enjoy their sport. Indeed, hunters act as the front line of herbivore population management. Herbivores are fairly simple to establish because of their basic biology and their location in the trophic structure. There is also no long learning period, especially when carnivores are absent. Birds of prey such as bald eagles, peregrine falcons, and, to a lesser extent, ospreys are charismatic birds that present no threat to humans and attract attention because of the Endangered Species Act. Reintroductions of small carnivores require more patience and dedication. The reintroduction of the large predators such as panthers, wolves, and grizzlies, in addition to being more difficult biologically, is exponentially more difficult politically. After all, consider what happened to a tourist in Montana's Glacier National Park.

On June 5, 1996, Ken Larsen was attacked and mauled by a grizzly bear in Glacier National Park, Montana. Although he survived the attack, he has said he will never again hike where grizzly bears live. As Larsen walked quietly in the early morning along a trail, he saw the grizzly some 20 feet away. The bear was charging. Larsen turned to run but got only ten feet. The attack lasted no more than 30 seconds and left him with a broken leg with a chunk of flesh missing and numerous puncture wounds.

Larsen was hiking alone quietly—two things a person should never do in grizzly bear country. The grizzly was accompanied by a younger bear that immediately ran upon discovering Larsen. The larger bear charged Larsen. After instinctively mauling Larsen, the bear sniffed him and then ran away, never turning back. Generally, bears avoid contact with humans, and in this case the evidence suggests that Larsen surprised two bears traveling through the forest. Once the larger bear realized that Larsen was a human, the bear fled the scene. As a result of investigative work, park rangers decided to spare both bears' lives. In the past, most bears have not fared as well. Humans have managed grizzly populations—right to the brink of extinction in the lower 48 states, where fewer than 1000 grizzly bears remain. They are considered threatened. Their range formerly included most mountain ranges in the Rocky Mountain west.

We have seen that deer populations have become a serious problem in much of the United States. Populations of so-called meso-mammals such as coyotes, armadillos, raccoons, and opossums are also increasing. Coyotes, once unknown in Florida, have invaded. Other states report rabies outbreaks in raccoon populations because their numbers are too high. In Alaska, wolves are still hunted by state wildlife control officers who believe that wolves are responsible for reducing caribou herds—a situation hunters find deplorable. The natural balance within wildlife populations that existed 200 years ago, when the human population was much smaller, has now been completely upset. Deer populations in the East and Midwest have exploded without natural predators. Human management schemes are politically, not biologically, motivated and are largely ineffective. If humans cannot manage semi-tame wildlife like deer, how can they expect to manage other species? Can humans really keep nature in an intricate ecological balance?

The only cost-effective approach to biological problems is through biological, not political or mechanical, means. The answer becomes apparent: bring back the big predators. The authors are confident that this will inevitably occur because large predators are the most cost-effective, biologically sensible way of controlling wildlife populations. Furthermore, only a few states need to allow large predators to become established deliberately or naturally. Where adequate resources exist, predators will find them. These top predators could then manage prey populations naturally.

Burke (1990) commented on the reintroduction of natural predators such as wolves and panthers in New Jersey: "Although wolves or mountain lions could possibly re-establish themselves over a period of several years in limited areas, the cost of purchasing the predators from a state willing to live-trap them and the transportation to the release points would make this alternative extremely expensive (Weise et al. 1975). Also, the reintroduction of these large predators in a state as densely populated as New Jersey could result in many undesirable repercussions. They would be reintroduced into an unfamiliar area. Their behavior could be quite abnormal and they may react aggressively in encounters with domestic animals or humans. Because of the expense and the unknown reaction of the predators, the reintroduction of natural predators as an alternative to the proposed action is not recommended."

Burke (1990) continued: "Bobcat, coyote, and black bear already inhabit the state of New Jersey with no measurable impact on the deer population. Forty-eight thousand deer are not going to be taken by a few predators. These animals scavenge deer carcasses more than they prey on live deer. When a live deer is taken by a predator it is usually a one- to two-month-old fawn or an injured deer that falls prey."

Will humans have to fear for their lives if top predators are reintroduced? Experience in California, New Mexico, Arizona, Nevada, Montana, Wyoming, Colorado, and Canada clearly shows that seeing a large predator in the wild is an extremely rare and unlikely event. Even for outdoor enthusiasts, seeing a panther is usually a once-in-a-lifetime never-to-be-forgotten event. When large predators and humans mix, the predators almost always flee first. The fact of the matter is that large predators avoid humans. Both authors spent more than 25 years each in Colorado and New Mexico. Although both of us were avid outdoorsmen, neither saw a panther in the wild, although both observed fresh tracks. While on foot, one author saw wolves and grizzlies in Alaska. Fascination, not fear, describes the emotional response. States such as California, Idaho, Minnesota, Montana, Wyoming, and North Carolina are leading the progressive front in reestablishing and protecting predator populations. We saw earlier that ranchers' claims that Yellowstone wolves would menace their livestock simply did not materialize. One can hardly consider California sparsely populated. Panthers, in fact, live within walking distance of Los Angeles.

Idaho's wolf reintroduction program is doing better than expected. Gray wolves were reintroduced into the wilds of Idaho and are forming packs and staying alive in greater numbers than anticipated. The U.S. Fish and Wildlife Service announced in late 1996 that no new wolves will be necessary to supplement populations. In 1995, 15 Canadian gray wolves were introduced. Another

20 were added in 1996. When the state of Idaho chose not to be involved in the wolf recovery effort, the Nez Perce tribe agreed to coordinate the effort. Native Americans, it seems, take pride in native wildlife. In 1996, three breeding pairs produced six to eight pups in Idaho. There were eight reports of livestock predation in 1996; wolves were involved in three cases. This seems a small price to pay to have the wolf back. An executive from the Idaho Cattle Association was quoted as saying, "I expect to see wolf cubs all over the place this next season." Certainly, this can be scientifically tested. Wolves are capable of only raising one litter per year. Most cattle grazing in Idaho takes place on public land. To protect a majority of Idaho cattle from wolves, perhaps cattle grazing on public lands should be discontinued altogether.

Gray wolves are making a comeback in Minnesota, Wisconsin, and Michigan. In Minnesota, wolf populations have increased from 700 to about 2000 (Mladenoff et al. 1997), and the wolf has been downgraded to threatened. With federal protection and education programs, wolf populations have increased in Wisconsin and Michigan.

In 1980, the red wolf was declared extinct in the wild in the United States. Three captive breeding sites were established at Cape Romain National Wildlife Refuge in South Carolina, Horn Island in the Gulf Islands National Seashore, and the St. Vincent refuge on an island of Florida's panhandle. Some 300 exist in zoos. Red wolves have been released in Alligator River National Wildlife Refuge and Great Smokey Mountains National Park in North Carolina. The population of red wolves in the wild is about 60. Red wolves weigh 50 to 60 pounds and are smaller than gray wolves. They prey on small mammals such as rodents. Slowly, ever so slowly, wolves are coming back. What is needed are landowners who want them back and are willing to have them released on their lands.

The "lobo" or Mexican wolf, once the rarest subspecies of wolf, was extirpated from the American southwest by the mid-1900s. The Mexican wolf was on the first endangered species list in 1976. The smallest of the wolves, adults weigh 50 to 90 pounds and average 4.5 to 5.5 feet in length and reach 26 to 32 inches at the shoulder. There were 149 lobos in captivity in 1996.

In 1996, the U.S. Fish and Wildlife Service moved ten captive Mexican gray wolves from eight states and the Republic of Mexico to a new holding and breeding facility at the Sevilleta National Wildlife Refuge near Socorro, New Mexico. Officials said the new facility was necessary to avoid crowding at several other captive breeding sites and to afford greater flexibility in managing the captive Mexican wolf population in general. The facility at Sevilleta consists of five pens ranging in size from one-quarter to three-quarters of an acre, arranged around a central sixth pen of about one and one-half acres. The enclosures are heavy-gauge chain link and are supplemented with solar-powered electric fencing. The facility provides the wolves with a secure and isolated environment. Wolves transferred to New Mexico were from captive breeding facilities in Missouri, New Mexico, Ohio, Michigan, Illinois, Minnesota, California, Texas, and the Republic of Mexico. There are 24 captive breeding sites in the United States and 5 in Mexico managing the Mexican gray wolves. Releases are being planned for Arizona and New Mexico. The Arizona legislature promised to enact laws reestablishing wolf bounties to protest the U.S. Fish and Wildlife Service's plans to reintroduce the wolves.

The service plan, if approved, calls for Mexican wolves to be released first in eastern Arizona in the Apache National Forest; they would then be permitted to disperse into the Gila National Forest in New Mexico. This region would become the Blue Range Recovery Area. The service also floated plans to release Mexican wolves into the White Sands Missile Range onto what would become the White Sands Wolf Recovery Area. No land use restrictions or prohibitions on using private and tribal lands were placed on the wolves.

Education and understanding are vital to present wildlife reintroduction programs. That is unlikely to ever change. Despite the incredibly low odds of even catching a glimpse of a large predator, the animals are viewed by many as a direct threat to humans. We have presented examples of a jogger attacked and killed by a California panther and a hiker in Glacier National Park attacked and mauled by a bear. Both of these events are extremely rare, but they did occur. Many states with highly successful, admirable, and progressive reintroduction programs still will not consider reintroducing top predators. One state actively hunts them down. However, this phobia, too, shall inevitably pass, and just as the buffalo will some day roam the Midwest and the walrus will return to the northwest Atlantic, so too will panthers and wolves roam the forests of Maine, spill into New Hampshire, and one day show up in Pennsylvania. Why must this be so? The answer may be that where there is prey, predators will eventually find it. Our long-term management strategy should be to put humans in the position of allowing wildlife to manage wildlife populations. In other words, the goal of our management strategy will be to have humans manage humans first and predators second, so that humans do not have to manage all wildlife. If humans can manage comparatively fewer predators, predators can manage the rest of the wildlife, and humans can spend more time educating other humans about how wildlife should be managed.

TRENDS IN THE POLITICS OF WILDLIFE AND HABITAT

By any measure, 1996 was a banner year for conservationists. Some 70% of statewide conservation ballot measures passed and more than $4 billion in bonds was approved. There were also some noticeable defeats. By analyzing those issues concerned with wildlife and habitat, we can attempt to predict the future of the conservation movement in the United States.

Although all states allow legislature-referred ballot measures, only 24 states have provisions for statewide initiatives and referenda placed on the ballot by citizen petitions. An *initiative* is a proposed law or resolution placed on a ballot as a result of a petition drive among registered voters. Voters go to the polls to accept or reject the initiative. A *referendum* is a decision by the legislature that is put to the voters to accept or reject. Citizens can petition for a referendum or state legislators can introduce one.

Requirements for placing initiatives and referenda on the ballot vary widely from state to state. On November 5, 1996, there were 243 statewide ballot measures, more than in any others year. Less than 40% of these ballot measures were accepted, making the rate of 70% of conservation measures passed even more incredible. Although losses in Maine, Florida, Montana, Idaho, and Oregon may have made the headlines, the trend is clear. Voters are demanding cleaner air, cleaner water, and more parks and recreational land, and most often they are willing to pay for them. First, we must understand why certain issues were defeated. We can then analyze why other issues passed. Then we can attempt to extrapolate to the next round of measures.

Montana voters defeated Proposition 122, which required new or expanding hard rock mines to remove toxins, carcinogens, and metals from wastewater before it enters groundwater, streams, and rivers. Some 1200 miles of streams in the state have been severely damaged by mining. Canadian mining companies financed a statewide opposition campaign, and the measure was defeated by a vote of 52% to 48%. The fishing might be great in Montana, but you better release any two-headed catch. Oddly, Proposition 125 was accepted by voters by a margin of 54% to 46%. This proposition outlaws corporate contributions to ballot measure campaigns. It seems clear that Proposition 125 will severely limit the amount of money that can be spent on defeating the reincarnation of Proposition 122 on the next statewide ballot in four years and that Montana residents will reclaim their freshwater resources and their reputation for great trout fishing.

In Florida, voters recently approved Proposition 1, which requires a two-thirds majority to pass all new tax issues. This measure has profound implications. Voters defeated Proposition 4, which attempted to tax Big Sugar one cent per pound of sugar produced to pay for the Everglades cleanup, and that was done with a simple majority. The next round in four years will require a two-thirds majority to pass—but that will not happen. Proposition 5, which requires water polluters of the Everglades to pay for the pollution abatement, and Proposition 6, which creates a trust fund for the Everglades cleanup, were both passed into law easily.

A ban on clear-cutting forests in Maine was defeated. A bill to prohibit livestock from watercourses was defeated in Oregon. Many issues that directly address wildlife concerns were decided by voters in 1996.

In Alaska, voters passed Proposition 3 to ban same-day airborne hunting of wolves, foxes, lynx, and wolverines. In March, Proposition 197, which over-turned protection for the panther, was soundly defeated in California. Colorado voters passed Proposition 14, to prohibit the use of steel-jaw leg-hold traps and other body-gripping devices, by a margin of 52% to 48%. Massachusetts voters cleaned house on their Proposition 1 by a vote of 64% to 36%. Proposition 1 in Massachusetts outlawed the use of dogs to hunt bears and bobcats, elimi-nated the statutory requirement that hunters and trappers must dominate the state Fisheries and Wildlife Board, and also banned the use of body-gripping devices, including leg-hold traps. At the same time, Idaho voters decided in favor of Proposition 2 to continue to hunt bears with bait and dogs in the spring. Michigan's Proposition D to outlaw bear hunting with dogs and the use of bait was defeated. In Oregon, Proposition 34, which overturned a previous ban on bear baiting and the use of dogs to hunt bears and mountain lions, was defeated by a margin of 44% to 56%. By a vote of 63% to 37%, Washington voters passed into law Proposition 655, prohibiting bear baiting and hunting of bears, panthers, bobcats, and lynx with dogs.

Clearly, the ethics of hunting bears and panthers was on the voters' minds. Some voters, such as those in Washington and Oregon, decided that it is un-sportsmanlike to hunt bear while the animal's head is stuck in a box of jelly donuts. California voters do not want their panthers hunted at all. Voters in Michigan and Idaho decided that hunting bears with dogs or baiting them is acceptable.

Past elections also help set the stage for future ballot initiatives. In 1990, for example, Californians passed Proposition 117 to ban trophy hunting of panthers and set aside $30 million a year for each of the next 30 years for habitat to benefit threatened and endangered species, deer, and panthers. In 1992, Arizona voters defeated Proposition 200 to ban the use of body-gripping traps, including steel-jaw leg-hold traps, on public lands. Two years later, Arizona voters over-whelmingly approved Proposition 201 to ban all body-gripping traps, including steel-jaw leg-hold traps, on public lands. Public lands account for 83% of the land area in the state of Arizona.

In 1992, Colorado voters passed, by a vote of 70% to 30%, Amendment 10 to ban the hunting of black bears during the spring and outlawed any hunting of black bears with bait or dogs. In 1994, Oregon voters approved Measure 18 to ban the hunting of black bears with bait and outlawed the hunting of black bears and panthers with dogs.

Based on the 1996 election results, results of previous elections, and new initiatives, certain trends become clear. Most importantly, more Americans are being asked to decide wildlife issues at the voting booth. This is indeed the Golden Age of Direct Democracy, as Roy Morgan, president of Americans for the Environment, put it. The central theme of wildlife issues concerns wildlife and is not an assault on American values or traditions. The assault is on what is viewed as animal cruelty and inhumane treatment of wildlife. Have American values changed? You bet they have. The use of body-gripping devices and leg-hold traps for hunting is being voted out of existence. In short, trapping for fur and fun is finished in states where voters get a chance to have a say. Although certain states still permit the use of dogs to hunt certain wildlife and baiting to hunt bears, these ways of hunting are being found to be unacceptable to the majority of the voting public. In state after state, this type of sport hunting is being outlawed. Clearly, the states that permit the use of dogs to hunt certain animals and allow bear baiting will be targeted by organizations dedicated to putting an end to cruel and unusual animal torture, as they see it. Even 44% of hunters in Colorado voted to support the ban on bear baiting and the use of dogs to hunt bears. More importantly, a majority of voters in California put an end to hunting panthers and, moreover, demanded that large amounts of money be spent each year to acquire more habitat for these top carnivores, thus allowing their populations to further increase. People in California want more panthers, and they are willing to pay for it.

Do these votes represent trends against so-called "American values" or unalienable rights? Is this all-out war on individual freedom and the American way of life? Importantly, only 13% of Americans hunt. The majority of American voters have never voted to make hunting wildlife illegal, although they clearly have the power to do so. Hunting remains a privilege that requires training and a license. Hunting most species is not an issue with voters. However, what is an issue is how and under what circumstances hunting is considered, by the majority of people, humane. In this very real sense, when the majority of the voting public makes something illegal, then we can consider whatever was outlawed to represent past American values that are no longer part of the American way of life. In other words, American values change to represent the new majority. This is not to say that these values have changed forever, for they can be voted on again and again. Finally, not all states allow voters to sign petitions to create statewide initiatives. Voters in these states must depend on their state legislators to bring forth current issues—a sometimes difficult task, indeed. In these states, old traditions sometimes seem laughable and live on for better or worse, anachronisms in an otherwise modern state.

The following states do not allow statewide voter-petitioned initiatives or referenda:

Alabama	Louisiana	Rhode Island
Connecticut	Maryland	South Carolina
Delaware	Minnesota	Tennessee
Georgia	New Hampshire	Texas
Hawaii	New Jersey	Vermont
Indiana	New Mexico	Virginia
Iowa	New York	West Virginia
Kansas	North Carolina	Wisconsin
Kentucky	Pennsylvania	

Direct democracy refers to the ability of the public to collect enough signatures to bring an initiative or referendum before a state's voters. Note in the above list that most western states allow citizen petitions, whereas many eastern states still do not permit such political activism. In the early 1900s, all states in the West except New Mexico allowed both initiatives and referenda. New Mexico permits referenda today. A bill to allow statewide initiatives was introduced in the state legislature in Pennsylvania in 1991. Two years later, it was dead. When people found state legislatures dominated by special interests, they used grass roots campaigns to collect signatures and force a vote.

EXERCISES

19.1 The U.S. Fish and Wildlife Service successfully introduced 19 panthers into Florida in 1995. Track down the cost of capturing the cats and transporting them to the release site. Create a strategy to reintroduce panthers in another state. Plan on spending the bulk of your proposed budget on education.

19.2 State-sanctioned wolf hunts in Alaska continue. Build a case against this practice by considering the sensitive ecology of the tundra plants. The assumption is that wolves keep caribou populations artificially low. If wolf populations are reduced, caribou populations will grow. Increased caribou populations will increase grazing pressure on the tundra, causing plant species composition to change. Plants favored by caribou will disappear and be replaced by plants that caribou will not eat.

19.3   Jaguars (*Pantera onca*) are known to have occurred in Florida during the Pleistocene and were found in the southwest United States as late as the early 1900s. In 1996, two jaguars were seen and verified in Arizona. One was shot and killed by a hunter who was not charged with any violation because jaguars were known to be extinct in the United States. Comment on this situation. Should jaguars be reestablished in the Southwest?

19.4   What is the status of the Mexican wolf reintroduction program in Arizona and New Mexico? Call or write the U.S. Fish and Wildlife Department in Albuquerque, New Mexico, for information. Would you hike in a national forest where wolves were known to exist? Also search the Web for information on the Mexican wolf.

19.5   Armadillos crossed the Mississippi River when bridges spanning the river were built. Now they have become established in Florida. How can armadillo populations be managed? Some biologists suggest that armadillos have caused skunk populations to decrease. As armadillos spread into Florida, skunk populations were also decreasing. Is this coincidence or competition?

19.6   Raccoon populations have exploded in many states and are upsetting the balance of local faunas. Design a management strategy to control their populations. Search the *Journal of Wildlife Management* for relevant papers on raccoons. Estimate the cost of such a program. Will coyotes take raccoons?

19.7   Design a set of exclosures that differentially test deer–raccoon–songbird interactions. Propose hypotheses and show how the exclosures you created will test them. For example, a hypothesis might be that species richness is highest when no deer or raccoons are present in a forest. How long does the experiment need to run before you can determine the results?

19.8   Deer overpopulation is a serious threat to the biodiversity of our forests. However, forest change takes place over decades. Design an educational program that educates lay people on the potential threat that deer overpopulation poses to our forests. Create a Web page with your message.

19.9   Design a set of experimental procedures to test if acid rain affects woodcock populations. What is the null hypothesis?

19.10   Design a set of exclosures that tests whether forests of the Northeast support an understory with and without squirrels and deer present.

How are some mammals differentially excluded? How will bird counts be done in the exclosures?

19.11 What other birds in the Northeast depend on a rich understory to survive? How are their populations doing? If these populations are holding steady or growing, perhaps deer are not the problem for woodcocks.

19.12 A healthy wolf has never attacked a human in North America. How many people are injured by cattle each year?

19.13 How many registered voter signatures are required to send a petition to the ballot in your state? What percent of the registered voters in the state is this? Is the number apportioned by county? Does your state allow paid petitioners to collect signatures? If so, how much does it cost per signature, on average, to collect the requisite number?

19.14 To learn more about ballot measures across the United States, check out the Web page of Americans for the Environment (AFE). See http://www.ewg.org/pub/afe/hmepage.htm for more information on ballot measures with environmental components. Also included on the AFE Web page is information about programs and publications. AFE's e-mail address is afedc@igc.apc.org.

LITERATURE CITED

Almy, G. 1995. 1995 book of bucks. *Sports Afield* **214**(2):88–101.

Bourjaily, P. 1994. An ebbing of autumn's upland tide: woodcock bird population falling. *Outdoor Life* **194**(2):42–43.

Budiansky, S. 1994. Deer, deer, everywhere. *U.S. News and World Report* **117**(20):85.

Burke, D. 1990. An Assessment of Deer Hunting in New Jersey. New Jersey Division of Fish, Game and Wildlife, Trenton.

deCalestra, D.S. 1994. Effect of white-tailed deer on songbirds within managed forests in Pennsylvania. *Journal of Wildlife Management* **58**(4):711–718.

Gilman, E. 1990. A close-up view of wildlife. *New Jersey Outdoors* **17**(1):30–32.

Hanback, M. and C. Blumig. 1993. Now it's deer on the pill. *Outdoor Life* July:68–69, 90–93.

Jones, S.B., D. deCalestra, and S.E. Chunko. 1993. Whitetails are changing our woodlands. *American Forests* November/December:20–25, 53–54.

Lipton, E. 1996. Hunters, wolves to rescue? Suburbs battle growing deer herds. *The Washington Post* p. A-1.

Mladenoff, D.J., R.G. Haight, T.A. Sickley, and A.P. Wydeven. 1997. Causes and implications of species restoration in altered ecosystems. *BioScience* **47**(1):21–31.

Reiger, G. 1994. Wishful wildlife management. *Field and Stream* **99(6):12.**

Robinson, J.B. 1994. For the birds: restoring a special habitat is the only way to reverse the woodcock's steady population decline. *Field and Stream* **98**(10):24–26.

Robinson, J.B. 1995. Late season in Nova Scotia. *Field and Stream* **100**(4):22–24.

Smith, C.R., D.M. Pence, and R.J. O'Connor. 1993. Status of neotropical birds in the northeast: a preliminary assessment. pp. 172–188 *in* Finch, D.M. and P.W. Stangel (Eds.). Status and Management of Neotropical Migratory Birds, U.S. Forest Service General Technical Report RM-229. U.S. Forest Service, Washington, D.C.

Weise, T.F., W.L. Robinson, R.A. Hook, and L.D. Mech. 1975. An Experimental Translocation of the Eastern Timber Wolf, Audubon Conservation Report No. 5. Audubon Society, Washington, D.C., 28 pp.

NOVEL SOLUTIONS 20

The world's first national park was founded by the United States in 1876 when Yellowstone National Park was gazetted. Since then, many national parks in many countries have been established. The United Nations established a World Heritage Park designation, but this designation has failed to protect all such parks (see Figure 20.1). Parks so designated were meant to be the best of the best. Are national parks working today? Most people in the United States have visited a national park. Grand Canyon National Park receives 500,000 visitors a year. Great Smoky Mountains National Park receives 2.5 million people per year. Visitation is increasing. In Arches National Park in Utah, the campsites within the park are filled nightly during peak season. In Denali National Park in Alaska, visitors ride on buses because private vehicle traffic is highly regulated. The national park idea seems to work for people, so surely we can protect islands of biodiversity. The situation can be viewed in another way: If we can isolate wildlife issues in parks, maybe we can deal with them. However, for the flora and fauna of national parks, the idea that a boundary can save, preserve, and protect nature is nearly bankrupt.

Other attempts to minimize wildlife issues or prevent them altogether are more promising than national parks. The Man and the Biosphere program deals with wildlife and humans simultaneously. Humans and their activities are considered to be part of the greater landscape. Still other ways of conserving biodiversity and mitigating adverse impacts on wildlife are to ranch wildlife and to make room for wildlife by improving habitat. Private sector initiatives are also an important component of minimizing impacts on wildlife. Although each wildlife issue is ultimately unique, certain generalities can nevertheless be drawn from many examples around the world.

FIGURE 20.1   A saddlebill stork shades the water while hunting for fish in Akagera National Park, Rwanda. The park was destroyed by invading rebels in 1991.

NATIONAL PARKS

National parks and reserves will be the last refuge for many species, at least if current trends continue. Unfortunately, most parks are isolated and therefore do not permit genetic transfer between populations outside the boundaries of the park. The persistence of populations within parks and preserves is also not well understood; however, few argue that such reserves can sustain populations into the future. Newmark (1995) analyzed mammalian extinctions in several western North American national parks. Within the mammalian orders Lagomorpha (rabbits), Carnivora (carnivores), and Artiodactyla (hoofed animals), Newmark found that the number of extinctions exceeded the number of colonizations. While there were 7 colonizations of parks, there were 29 extinctions. For instance, in Bryce Canyon, Utah, white-tailed jackrabbits, red foxes, and spotted skunks have not been seen for ten years. At Crater Lake, Oregon, river otters, ermine, mink, and spotted skunks have been extirpated. The wolf has been lost from Mount Rainer, Manning Provincial Park, and Yellowstone National Park, although it has been reintroduced into Yellowstone. Pronghorn antelope and elk have disappeared from Mount Lassen in California, and the moose is gone from Manning Provincial Park in Canada.

Human activities and land use patterns near the national parks influence animal persistence patterns within the parks. For instance, if forested areas outside a heavily forested park are destroyed, the park is likely to become more like an island than a continental park. Animals within the park will most likely remain there and not venture beyond the boundaries. What Newmark's work shows is that species from rabbits and skunks to wolves and elk have difficulty persisting even in large parks and preserves. Thus, the presumption that parks and preserves will protect species is turning out to be more a dream than a reality, more a belief than a scientific fact.

Harris and Frederick (1990) noted that existing conservation strategies are inadequate to preserve the long-term biological diversity of a region. The Endangered Species Act is invoked when a species has suffered severe habitat loss and the population is reduced and declining. Harris and Frederick suggest that "old-growth forests must be conserved for many scientific, moral, and aesthetic reasons, only one of which involves an endangered species." The Endangered Species Act applies not to habitats but rather to species of plants and animals within those habitats. In an age of ecosystem management, there is no endangered habitat act.

"HOT SPOTS" OF DIVERSITY

In 1997, Princeton University and Environmental Defense Fund researchers found that collections of endangered species were clustered in several key "hot spots" across the United States. They suggested that protection efforts should focus on these areas of high biological diversity. Nationally, these areas include parts of California, Florida, Hawaii, and New Jersey. A scientist from the Environmental Defense Fund claimed that the vast majority of endangered species in the United States can be protected in the smallest possible portion of land: "If conservation efforts and funds can be expanded in these areas, we can conserve endangered species effectively and efficiently." Is this idea novel?

"Hot spot" counties included San Diego, Los Angeles, San Francisco, and Santa Cruz in California and in Hawaii the islands of Hawaii, Maui, and Kauai and Honolulu on Oahu. In Florida, both Highlands and Monroe counties ranked high. In New Jersey, Atlantic, Burlington, Cape May, and Ocean counties had high numbers of various endangered and threatened species, including nine marine animals such as the blue whale and dwarf wedge mussel.

Protecting these so-called "hot spots," however, is another form of protecting areas. National parks, game refuges, and other protected areas already exist and have been less successful than anticipated in protecting species. One reason is because animals move; most depend on movement for their existence. Blue whales do not swim in circles off the New Jersey coast, nor will they reside in protected areas, no matter how large. Some peregrine falcons summer in Alaska and winter in South America. Protecting both ends of their migration route will not protect the species because it neglects protection of their migration route and protection of their prey base, for example.

Everglades National Park was created in 1949, in part to protect the vast Everglades and its biodiversity. Even before the creation of the park, many species of wading birds were in decline. Creation of the park, contrary to expectations, did not halt the declines. This is because Everglades National Park is surrounded by agricultural areas and the vast city of Miami to the east. Thousands of miles of levees and dikes and hundreds of water control pumps were constructed to "manage the flow of water" from a central room at the South Florida Water Management District in West Palm Beach, Florida.

Creation of the park was an attempt to save the *contents* of the park while ignoring its *context*, that is, the location of the park in a human-dominated landscape. Although Central Park in New York City is the most obvious example, most parks and preserves in the United States exist in human-dominated landscapes whose impacts on the protected areas are enormous and invariably detrimental. Saving so-called "hot spots" will work about as well as saving other parks and protected areas has.

MAN AND THE BIOSPHERE

The Action Plan for Biosphere Reserves was published in 1984 by the United Nations Environment Programme (UNEP) and the United Nations Educational, Scientific and Cultural Organization (UNESCO) (Nature and Resources 1984). Governments and international organizations were invited to undertake activities to improve and expand the international biosphere reserve network, to develop basic knowledge for conserving ecosystems and biological diversity, and to make biosphere reserves more effective in linking conservation and development in fulfilling the broad objectives of the Man and the Biosphere (MAB) program. MAB was initiated in 1971 to provide the information needed to solve practical problems in resource management. The United States has 44 biosphere reserves. Everglades National Park and Grand Canyon National Park are examples.

A minimum set of activities were recommended for implementation in each biosphere reserve:

1. Collection of a baseline inventory of flora and fauna and their uses
2. Preparation of a history of research done in the reserve
3. Establishment of research facilities and research programs
4. Establishment of training and education programs

5. Preparation of a management plan that addresses biosphere reserve functions

MAB research was directed at building an understanding of the structure and function of ecosystems and the effects of different types of human activity. It was an attempt to go beyond the idea of parks and reserves existing as islands within a human-dominated landscape. Key ingredients of MAB were the involvement of local people in research projects and establishing cooperative research projects with other researchers in the social, biological, and physical sciences.

One theme of the program was the conservation of natural areas and the genetic biodiversity they contained. The concept of a biosphere reserve was conceived to be a series of protected areas, linked through a coordinated international network that would demonstrate the value of conservation and its relationship with development. The first reserves were designated in 1976. In 1984, there were 243 reserves in 65 countries. Today, there are 328 biosphere reserves in 65 countries.

The Food and Agricultural Organization (FAO) was especially interested because biosphere reserves were *in situ* reservoirs of genetic resources, especially wild crop relatives, forest species, and ancestors and close relatives of domestic stocks. The UNEP was interested in global monitoring using comparable methodologies and parameters. The International Union for Conservation of Nature and Natural Resources considered biosphere reserves to be a useful concept in regional planning because conservation was presumably linked with sustainable development.

The main characteristics of biosphere reserves were as follows:

1. They were protected areas of representative terrestrial and coastal environments that have been internationally recognized for their value in conservation and in providing the scientific knowledge, skills, and human values to support sustainable development.
2. They were united to form a worldwide network that facilitated information sharing relevant to conservation and management.
3. They were representative examples of natural or minimally disturbed landscapes within one of the world's biogeographical provinces and as many of the following areas as possible:
   a. Centers of endemism and of genetic richness or unique natural features
   b. Areas of suitable experimental manipulation to develop, assess, and demonstrate the methods of sustainable development
   c. Examples of harmonious landscapes resulting from traditional patterns of land use

     d. Examples of modified or degraded landscapes that were suitable for restoration to natural or near-natural conditions

4. Biosphere reserves were large enough to be an effective conservation unit and have value for measurements of long-term changes in the biosphere.

5. Biosphere reserves provided opportunities for ecological research, education, and training.

6. There should be buffer zones surrounding a biosphere reserve.

7. There should be effective long-term protection.

8. People should be considered part of a biosphere reserve.

9. Biosphere reserves were "harmonious marriages" of conservation and development.

10. Biosphere reserves were open systems with areas of undisturbed natural landscapes surrounded by areas of sympathetic and compatible use.

11. Because of their secure protection, large size, and the inclusion of areas free from significant human impact, biosphere reserves typically provide ideal sites for monitoring changes in the physical and biological components of the biosphere. Their protection and scientific mission make biosphere reserves particularly attractive sites for gathering scientific information. Scientists had more confidence that the integrity of their sites was respected and that collected data contributed to a growing data bank of increasing scientific significance.

12. In biosphere reserves, interdisciplinary research programs involving natural and social sciences were encouraged to develop models for sustainable conservation of a large natural region.

13. The international network provided a framework for comparative studies of similar problems in different parts of the world; for testing, standardizing, and transferring new methodologies; and for coordinating the development of information management systems.

14. Biosphere reserves served as important field centers for the education and training of scientists, resource managers, protected area administrators, visitors, and local people.

15. Biosphere reserve status provided a framework for improving cooperation at the local, regional, and international levels.

16. All biosphere reserves were part of the international network that provided a framework for communication within and among biogeographic regions. Cooperation involved sharing of technology and information and the development of coordinated monitoring and research projects, to provide better information on problems of common interest.

17. Biosphere reserves were to be used to:

a. Collect background global monitoring of biological, chemical, and physical variables

b. Carry out research in basic ecological processes that might be applied to management sciences

c. Monitor the results and effectiveness of management

d. Assemble traditional knowledge about the use of species and landscapes

e. Make knowledge gained readily available

The biosphere program is alive today. Everglades National Park in Florida has become a focal point for restoration efforts of the U.S. government. Despite the designation of Everglades National Park as a World Heritage Site, man's management of the park has nearly destroyed it. Nevertheless, the MAB program continues to be a viable alternative to unmitigated exploitation.

TIGERS

In 1989, Ranthambore National Park in northern India was home to some 30 tigers. Although the park was a lush but dry forest, large herds of gaur, chital, and sambar; groups of langur monkeys; and birds thrived. Near a lake within the park, a female tiger named Noon hunted chital and sambar. Tigers are large and brightly colored; when the tiger walks, all animals take notice. When Noon walked boldly along the lakeshore, she uttered low growls at the hoof stamping and alarm calls of the chital. She could hardly be missed. She wandered off a short distance from the lake into the high grass, where she layed down to nap. The coming and going of the chital and sambar to drink did not disturb her. She awoke around 4:00 P.M., hungry for a meal. Carefully, bringing herself up to a crouch, she raised her head slightly to view her surroundings. Several chital were drinking from the lake. As the chital browsed on the tender shore grasses, Noon's muscles began to tense and ripple.

Sensing that the chital had no way to escape, Noon launched herself from her crouch and with leaps of 15 feet at a time closed the gap between herself and her prey in a few seconds. The confused chital scattered in all directions. One unfortunate chital headed into the lake. Noon powered her way through the water with sheer brute strength and caught the chital by the back of the neck. Her enormous paws wrapped around the chital's neck and her extended claws raked across the animal's back and neck, seeking a hold. Noon's jaw locked into the chital's throat and the struggle ended quickly. Partly swimming, partly walking, she dragged her prey ashore and cleaned herself. The chital carcass lasted a day.

In 1998, seven tigers remained in Ranthambore. Surrounded by a landscape dominated by humans, with every green leaf within ten feet of the ground eaten by goats, Ranthambore was invaded by villagers. India is one of only a few places where goats feed on the trees, so desperate are they for forage. The tigers were killed one by one. First, tigers in habitats within park areas frequented less often by tourists became victims. Slowly, tour drivers noticed that fewer tigers were being seen. Village woodcutters were seen more often in the park, collecting bundles of vegetation to feed their starving goats. One by one the tigers disappeared. Noon was one of the most photographed tigers in Ranthambore. Unfortunately, her status did not save her.

Now, an estimated 5000 wild tigers (*Pantera tigris*) remain in the wild. In years past, British hunters atop elephants could kill 20 tigers in a single hunt. Those days will never return. Private breeders in Thailand want the Thai Forestry Commission to allow them to raise tigers and sell them to Chinese medicinal companies. The Chinese believe that tiger bones, penises, and other body parts can cure anything from arthritis to impotence. Of course, there is no scientific evidence that tiger penises cure impotence, but perhaps if one believes that sleeping with a tiger penis under his pillow can cure his sexual ills, it will indeed work. Currently, wild tigers are illegally taken to supply part of the demand (see Chapter 3).

There is a worldwide ban on tiger products, and tigers are an endangered species under the Convention on International Trade in Endangered Species of Wild Fauna and Flora (CITES); hence international trade is forbidden. However, Thailand has a right to sell the Asian cats and their parts domestically. The director of Siracha Farm, Mr. Somphong Temsiriphong, has invested money in the hopes that the government of Thailand will approve his scheme to raise and sell tigers. He is now raising 35 tigers in Bangkok, where a whole dead tiger can be sold illegally for $10,000. He intends to breed as many tigers as possible to meet the local demand. The potential regional demand is enormous.

Somphong has a display case with, among other items, such valuable products as dried tiger penises that can sell for $4500. Other local breeders are ready to cash in as well. Breeders' practices are less than palatable for animal rights activists. Cubs as young as six months old are separated from their mothers so that females can be made to produce more litters. Members of the Thailand Wildlife Foundation say breeders are in it for the money. "What they want is to end up selling tiger steaks and penis soup for aphrodisiac purposes, and I just don't think animals should be used like this," said one member.

A government official said that tigers could be farmed as easily as pigs and that the legal supply would reduce prices and take pressure off wild tigers. The West, the official said, is too sentimental about animals. Opponents fear that trade in wild tigers could be more easily disguised and that the last of the world's remaining wild tigers could be wiped out. Thus, the wildlife issue here seems clear. Should raising endangered species legally for fun and profit be permitted? Could some of the profits be taxed and perhaps devoted to wild tiger protection and habitat restoration?

The free market is an incredibly powerful economic force. With tigers selling for $10,000 each, there is an enormous incentive to raise them domestically. When a fishery is a stake, the tragedy-of-the-commons takes over. Everyone thinks that since the resource is going to be depleted, they should get their share before someone else does. We must remember, however, that there is a lot of middle ground between unmitigated greed and the sustainable uses of natural resources. The most common bird in the world is the chicken (*Gallus gallus*). The reason this is so is because people all over the world eat chicken eggs and meat. Even their feathers are used to fill pillows. Chickens are relatively inexpensive almost everywhere people are found. One cannot help but wonder if the California condor (*Gymnogyps californianus*) would be facing extinction if people ate the meat and eggs of this beautiful bird.

On November 17, 1994, The Associated Press reported that Asian countries at a CITES meeting agreed to work toward elimination of tiger bone and rhinoceros horn as traditional medicines. China, South Korea, and Thailand, the consuming countries, agreed to work with physicians and pharmaceutical companies to find alternative treatments.

Tiger bone was used to treat muscular problems, and rhino horn was used to reduce fever. Of course, the root problem is that 2000 years of traditional recorded medicine and belief, and not education and fact, led people to make their product decisions. These traditional cures have become increasing costly, and cost was no doubt a factor in bringing these countries to the bargaining table. Tiger bone sells for $75 to $155 a pound. Would any consumer know the difference between a vial of tiger bone and a vial of chicken bone? Rhino horn is outlandishly expensive, especially in China, where the average monthly wage is $29. Nevertheless, consumers seek minuscule amounts for treatments.

Japan was the third largest consumer of rhino horn medicines. When pharmaceutical companies discontinued use of the horn in the 1980s, consumption dropped. During that time, rhino populations dropped as well, from 70,000 in 1970 to 12,000 in 1995. Most of these are in pens or other facilities where they can be protected. Rhino horn is still used for dagger handles in Yemen.

BUTTERFLIES

Larry Orsak, director of the Christensen Research Institute in Papua New Guinea, considers selling insects such as butterflies a great way to preserve tropical rain forest. He suggests that "those who equate killing butterflies with destroying butterflies don't know much about butterflies, the tropics, or what strategies have gotten people in developing countries to save their forests. The fact is, buying tropical butterflies for your collection may be the best investment you ever made in tropical forest protection." How could this be? Perhaps insect ranching is the answer to insect conservation.

The government of Papua New Guinea set up the Insect Farming and Trading Agency (IFTA) in 1978. The agency purchases insects from local villagers and sells them to collectors around the world. Most of the insects are collected from the wild; however, the IFTA requires that common birdwing butterflies be bred by villagers. Many villagers now plant extra insect food plants and then catch and sell adult butterflies. Certainly, this is insect ranching for profit. In addition, not all butterflies are harvested, so the food plants serve to increase the available resources for all insects.

In the case of Papua New Guinea, raising and selling or collecting excess insects helps preserve wild populations. How sustainable is insect ranching? Insects generally have high reproductive rates. Obviously, they also have very high mortality rates. If more caterpillars could survive, there would be more butterflies. By collecting the caterpillars and bringing them home, villagers are increasing survival rates of the butterflies. However, caterpillars also serve as food to birds. Ultimately, by increasing the number of food plants, there will likely be an increase in the number of butterflies, and thus more can be harvested. Also, eggs can be placed on favored food plants and raised in glass containers. The emergent butterflies can be used as eggs layers or be harvested and sold.

Every legally exported Papua New Guinea lot of insects comes with an export permit. Each insect does not come with its own permit. If the lot contains birdwing butterflies (the world's largest butterflies), it must also have a CITES stamp. It is up to the buyer of the lot to then copy the permit or otherwise certify that such a permit is on file. If you are a serious collector of butterflies, your collection should certainly have a birdwing in it. Would you know the difference between a legitimate permit and a fraud? Would you be willing to risk a fine and possible jail sentence on a facsimile of a "legitimate" permit? If the "bugs" can be worked out, insect ranching might catch on.

PACIFIC SALMON

Salmon are amazing fish. After being born in the upreaches of a backwater stream, the young fry grow quickly on insect prey and then make their way to sea. Some four years later, they answer their biological clock and begin migrating back to their place of birth to spawn. The trip back is a hazardous one indeed. Presuming they have not been harvested at sea (most likely they have), the salmon head back toward the stream of their birth. Predatory seals know that the salmon are headed for freshwater streams. With a meal on the way, seals congregate in pools where rivers meet the sea. If enough salmon make the journey, the seals cannot eat all the bounty. However, when fewer salmon arrive, the seals quickly attack and eat them day after day.

A surviving salmon ceases eating when it reaches fresh water and begins to deteriorate and die. The most obvious change is its color, which turns a bright blood red. Many made-man and natural hazards lay in the salmon's path. Dams and waterfalls must be negotiated. It seems almost unbelievable that a salmon, or any fish for that matter, could possibly swim up a waterfall, but indeed salmon do. Salmon use natural pools to circle and then propel themselves upward, furiously swimming against the waterfall. Many times, they are washed back down, where they again rest and gain strength. Once again, they propel themselves upward, swimming hard against the falls. Some propel themselves not upward but sideways and land hard against rocks and boulders. Some become trapped on land and are unable to struggle free. But the drive to spawn is a powerful one, and the salmon are driven to answer this call. Many have survived a seal attack but are missing a fin or part of their tail. This severely impacts their ability to make the journey upstream, but they are compelled to try. Anyone who has stood by a waterfall and watched salmon struggle time and again to climb the falls will probably think twice before eating salmon, so incredible is their tenacity to spawn.

Fortunately, biologists know how to save the Pacific salmon populations because the biology of salmon has been studied for decades, and the causes of the decline are understood. Today, in some parts of Oregon and Washington, spawning salmon will find improved habitat in which to reproduce and die. More than any other factor, upstream habitat alteration and destruction has led to declining population numbers. Now, restoration efforts are under way to help improve these habitats. Many local and state governments are teaming with environmental groups, local volunteers, out-of-work loggers, and lumber companies to restore spawning grounds and fry habitat to more complex natural conditions the fish favor.

To reproduce, salmon require clean gravel beds, and hatchlings need a complex habitat to escape predation and feed successfully. Insects are the main prey of the young fish, and the small fish must be able to hide and seize unsuspecting insects. In a more natural setting, fallen logs, boulders, bank overhangs, and shade trees provide an ideal habitat for both insects and fish. This complex structure also creates bottom ripples, uneven current patterns, and microhabitats which increase biological complexity. In other words, increased structural complexity leads to increased biological complexity. Increased biological complexity also supports and maintains increased structural complexity. Removing any component, such as a dead tree or a boulder or any of the insects, reduces complexity and therefore is detrimental to all elements of the system.

Previously, people removed dead logs in the mistaken belief that doing so would benefit fish. Indeed, as we now appreciate, these activities had just the opposite effect. By removing a log, environmental complexity was reduced. The local complex bottom topography of the streambed was changed from alternating ripples and pools to a uniform flat bottom. Fish hiding spots were taken away. The light patterns created by the log disappeared. Insects deserted. The bank overhang that had been held in place by grass roots and a reduced flow rate created by the log collapsed into the stream. Without the cool, shady overhang, the young fish had no place to hide.

Cutting logs along the stream not only took away the patchy light patterns but also released soil into the stream, choking spawning gravel beds. Insect diversity was reduced, and more sun on the stream increased water temperatures, killing microorganisms. Long-term salmon production plummets when structural complexity decreases and requires three or four decades to recover. In an effort to jump-start the recovery process on the Upper Rogue River in Oregon, a restoration team brought in large tree trunks and anchored them in natural positions. Streambed complexity immediately began to improve. Pools and ripples developed and fish hiding places were created.

Downstream on the Rogue River, Savage Rapids Dam impedes or kills tens of thousands of migrating salmon yearly. The National Marine Fisheries Service estimates that some 45,000 more salmon and steelhead trout (20% of the population) would survive if the dam were completely removed. After a long struggle, Oregon residents decided the dam had to go. With the dam's imminent demise, restoration efforts are now under way. Certainly, the road to restoration will not be clear of obstacles. Agricultural interests fear losing hard-won irrigation rights. One question of interest that must be answered is just how far restoration efforts should proceed away from stream banks. The federal plan stipulates "a uniform zone equal to twice the height of the tallest streamside trees or 300 feet, whichever is greater"—sort of a one-size-fits-all approach to stream restoration. Nevertheless, restorers believe the cumulative impact will be great and the payoff rapid. To maintain public support and monitor restoration efforts, fish counts will be made each winter.

On March 30, 1994, the *Wall Street Journal* reported that a federal judge declared that the National Marine Fisheries Service, the U.S. Army Corps of Engineers, and other government agencies responsible for managing dams and reservoirs on the Snake and Columbia rivers had broken the law. The agencies, it seemed, had pushed through a dam-operating plan they knew would have killed millions of salmon.

PRIVATE SECTOR COOPERATION

Since habitat loss is usually the main, although not the only, problem behind a wildlife issue, acquisition of habitat is one means of protecting wildlife. However, habitat purchase is expensive and often impossible. Land use controls are also not acceptable to many landowners who feel that "big government" should not tell them what to do with their land. As a result, it is futile, almost impossible, to implement an extensive system of species-based land use controls (de Klemm 1993).

Private landowners should somehow be made more responsible for the management of their land to be consistent with species protection, while at the same time providing the necessary income a landowner expects.

Crested Caracara

Dr. Joan Morrison, of Colorado State University, is carrying out an ecological study of the life history, nesting activity, habitat associations, and demographics of the crested caracara (*Polyborus plancus*) (Figure 20.2) in cooperation with landowners in south-central Florida. The crested caracara was listed as threatened on July 6, 1987 under the Endangered Species Act (see Chapter 2).

As a cooperative effort between ranchers and scientists, the main focus of the research is to understand the biology of the crested caracara and document land management practices that are compatible with its survival. Morrison and the ranchers have forged a unique bond of trust, working together to increase awareness of the caracara among Floridians and to discuss how ranches can maintain sustainable, economically viable activities while also conserving native habitats and wildlife species.

The beginning of the year is when crested caracaras nest and raise young. In Florida, the breeding season can extend from September through June, although most of the population begins egg laying in January and February. Some young have already fledged by March. These early nesters may go on to produce another brood.

Because they nest in open habitat, caracaras are quite secretive during the nesting period. Adults do not spend much time at the nest except during incubation and while the chicks are very young. The yellow and brown pattern of

FIGURE 20.2   Crested caracara. The only population east of the Mississippi is found around Lake Okeechobee, Florida, where the species is listed as threatened.

the young provides effective camouflage, protecting them while the parents are away from the nest. Nest predators include crows, raccoons, and perhaps other raptors.

Crested caracaras reach the northern limits of their distribution in the southern United States. Populations occur in southeastern Texas, southwestern Arizona, and south-central Florida. Florida's population has been isolated since the last ice age. Crested caracaras also occur in Mexico and Cuba and in suitable habitat throughout Central and South America. They prefer open, short grassland habitats with scattered trees for nesting. In Florida, they nest primarily in palm trees.

Currently in Florida, crested caracaras are found primarily on privately owned cattle ranches. Current land management practices of the ranchers appear to keep the habitat suitable for caracaras. Nests as close together as 1 km have been found. Crested caracaras are diet generalists, feeding on a variety of vertebrate prey as well as carrion. Unfortunately, the crested caracara is threatened as its habitat falls victim to Florida's rapid urban and agricultural development.

Unless the adults have a second brood, young caracaras may remain on the territory for up to ten months, until the adults begin another breeding effort. After they leave their natal territory, juvenile caracaras are nomadic, wandering throughout the region. They often band together in groups of up to 30 individuals. If adults have a second brood, the young of the first brood are driven off after only three months. Since young birds learn to forage from their parents, perhaps there is a differential survival rate between young from the first brood that spend only a short time with their parents versus young of the second brood that spend eight months with the adults.

It is believed that caracaras, like eagles, do not breed until age three or four. Radiotelemetry is only now beginning to provide information on survival, activities, and habitat selection of young, nonbreeding caracaras. Individuals seem to be philopatric to a region of the study area not far from their natal territory, and these juveniles are most often found on cattle ranches throughout the region.

By leaving nesting trees such as palms in cattle pastures, local ranchers can help save this unique bird of prey. Habitat destruction due to conversion of cattle pastures into shopping malls, trailer parks, agriculture, and housing developments continues to threaten the crested caracara's long-term existence in Florida. With 13.7 million people in Florida and 892 people a day moving into the state, habitat destruction will likely continue. Ranchers, however, have found a way to allow the crested caracara to survive. How can ranchers be compensated for their efforts?

In 1998, one ranch decided to convert 8000 acres to sugar cane production. This particular conversion required a "take" permit from the U.S. Fish and Wildlife Service because two crested caracara nesting pairs would lose their primary habitat. Despite the Crested Caracara Habitat Conservation Plan, the U.S. Fish and Wildlife Service was about to cave in to special interests and permit critical habitat loss, the very habitat the service was required by law to protect.

In 1995, Georgia-Pacific Corporation, the largest U.S. forest products company, announced plans to protect red-cockaded woodpecker habitat on its production timberlands. Georgia-Pacific claims that some 300 to 400 red-cockaded woodpeckers live on its land. The company announced plans to stop logging on land that contains colonies of red-cockaded woodpeckers. Additionally, Georgia-Pacific will establish buffer areas around each colony, where timber will be selectively harvested. No road building will take place where colonies exist. Clearly, the private sector must steward its lands in a more ecologically responsible way. How can private companies be rewarded for protecting habitat for endangered species?

Ecotourism

Perhaps the most effective use of wildlife is ecotourism. The worldwide demand for travel to exotic places is increasing. Tourists interested in wildlife are willing to travel to remote places to get a glimpse of rare species. In Manu National Park in Peru, tourists are willing to pay more than $100 a night to see tropical birds and primates in a virgin lowland rain forest of the Peruvian Amazon. Tourist visits to these regions can be carefully controlled as well. Thus, core regions of the reserve can be carefully protected and can serve as population sources to other areas that are impacted by tourism. Areas can be placed on a rotation basis and allowed to recover.

Africa seems to be haunted by unstable governments, civil wars, droughts, extreme poverty, declining growth rates, and a rapidly increasing population. However, as the mountain gorilla project in Rwanda demonstrated, wildlife can be protected even under the most extreme conditions. The local guards in Rwanda performed admirably and went almost a year without salary. They protected the gorillas as best they could. Several of the guards lost family members to the war. Nevertheless, they protected the gorillas. There is real hope in Rwanda that the mountain gorillas can be saved and tourism will be revived. In the past, gorilla tourism was the second largest source of hard currency for Rwanda. Unfortunately, very little money went back into the local community. Perhaps this will change under the new government.

Project SAFEGUARD in Alaska

Alaska covers 656,424 square miles; protecting wildlife against poachers is therefore a formidable task. In 1984, a private, nonprofit organization formed the Alaska Fish and Wildlife SAFEGUARD program to protect wildlife. SAFEGUARD pays money to anonymous citizens who report wildlife violations that lead to arrests or citations. Several examples cited on the SAFEGUARD Web page highlight the effectiveness of the program.

On April 1, 1991, three moose were poached and wasted in Kincaid Park in Anchorage. After extensive publicity, along with advertised rewards, a SAFEGUARD call resulted in the arrest and conviction of an individual. He was sentenced to 55 days in jail, lost all hunting and fishing privileges, was ordered to perform 100 hours of community service, and was fined $7500.

In 1991, SAFEGUARD received a call about a crab fishing vessel that was stealing other vessels' crabs and crab pots in the Bering Sea. Fish and Wildlife Protection obtained and served search warrants on the skipper. As a result, the skipper was charged and prosecuted on several misdemeanor and felony counts. He was fined $40,000, forfeited all crab pots in his possession, and his fishing privileges in Alaska were revoked. In addition, his million-dollar fishing vessel was forfeited to the state of Alaska.

In 1992, a report to SAFEGUARD indicated that nine bottomfish trawlers were routinely fishing in the closed waters of Alaska near King Cove. This prompted the formation of a task force which included 40 state and federal enforcement agents. Over 1000 illegal fishing acts in Alaskan waters were documented and 26 boats were targeted for prosecution. Just one company was responsible for the illegal taking of approximately 27 million pounds of fish valued at $5.4 million. The estimated criminal and civil settlements paid to the state of Alaska resulting from this one SAFEGUARD case could exceed $15 million.

A call to SAFEGUARD resulted in the 1992 conviction of a lower Yukon fish processor for buying fish and not reporting the purchases to the Alaska Department of Fish and Game. The processor and the fishermen who sold him the illegal product were fined nearly $1.5 million, and the judge ordered 23 vessels forfeited to the state of Alaska. These included fishing boats, tenders, and several large processing barges and ships.

SAFEGUARD received a complaint in 1993 that a man from Idaho had been buying Alaska resident licenses and tags for 11 years and that he had accumulated bear, moose, walrus, and musk ox trophies. After a lengthy joint investigation by Alaska Fish and Wildlife Protection, Idaho Fish and Game, and the U.S. Fish and Wildlife Service, the suspect was convicted of multiple hunting violations as well as falsifying license documents. The judge fined him $3000, sentenced him to 30 days in jail, revoked his hunting privileges, and ordered forfeiture of all game trophies related to the case.

Interestingly, records show that approximately 75% of callers refuse to accept a reward.

SUSTAINABLE DEVELOPMENT

Most of the earth's biodiversity is in the tropics. Therefore, it is very easy for northern Europeans and North Americans to suggest that wildlife everywhere should be protected. Isn't it curious that Europeans wiped out much of their native species at least 500 years ago? Europe was once heavily forested. Now almost no forest remains. The United Kingdom had bear and moose; now there are none. It took a while longer to reach and nearly wipe out the buffalo of North America, but once the slaughter started, the end came quickly. Buffalo are only now recovering. Habitat change and loss in Europe and North America have been extensive. Still, logging companies yearn to finish off the old-growth timber of the Northwest and Alaska.

The richest nations on earth consume most of the earth's natural resources, particularly depletable resources. North America has a penchant for cheap fuel. The Japanese have an insatiable desire for rare and expensive wood products. The poorest nations generally are found in the tropics. The world's breadbaskets are in higher latitude countries. Thus, to suggest that the poorer countries on earth conserve their wildlife so that the rich nations can enjoy it is, to be generous, laughable. Why should Peruvian Indians living off the forest, as they have for thousands of years, not kill the last of this or that macaw for food? If it were not for the pet trade, there might be many more where the macaws came from. Why should an Eskimo avoid killing whales when the Japanese continue to eat whale meat?

Obviously, no nation can order another nation to conserve its wildlife or its natural resources. People everywhere want and deserve a better life for themselves and their families. Nor is it natural to assume that a nation will demand its citizens cease certain activities that have been a traditional way of life. The San peoples of the Kalahari have hunted wildlife for more than 100,000 years. And they would still be doing so today were it not for huge cattle ranches that have destroyed wildlife habitat over vast regions of southern Africa. The San are not responsible for the great loss of wildlife in southern Africa, any more than a South American forest dweller is responsible for the loss of macaws. Indeed, it is in the best interests of native peoples to ensure that no resource is consumed to extinction.

Early hunters of the post-glacial period may have hunted out the North American mammoth and the giant sloth. It is not true that native hunters have preserved every species. However, it was certainly in the best interests of native people to conserve their resources. The San people of the Kalahari Desert, for instance, eat only half of the berries on a bush and leave the remaining half for other species and for seed dispersal. Only highly efficient developed nations with modern factory ships could deplete the seas of fish, only commercial hunters could wipe out billions of passenger pigeons, and only demand for rhino horn could drive an ancient species to the brink of extinction.

Today, novel solutions are needed to address our wildlife issues. Earlier we discussed how the public of south Florida demanded an end to frog gigging. Within two days of the publication of a newspaper article educating the public about gigging in a national preserve, a public outcry ended the slaughter of frogs. What the public wants, the public gets! The novel solution to the frog gigging problem was not to call in legions of scientists but simply to educate the public to a potential problem.

What can be done about preserving biodiversity in the tropics? Can wildlife have economic as well as aesthetic value? How can aesthetic value be translated into economic value? Can humans live sustainably anywhere? Considering Easter Island, Mesopotamia, and Mesoamerica (Chapter 3), are humans even capable of living sustainably off our planet? Who must learn to live sustainably—the developed nations or the Third World nations?

The term "sustainable development" was introduced in a 1987 report entitled *Our Common Future* by the World Commission on Environment and Development of the United Nations. The commission defined sustainable development as "development that meets the needs of the present without compromising the ability of future generations to meet their own needs." Of course, meeting tomorrow's needs without meeting today's needs is irrelevant, so there is a natural priority, if only implicitly, attached to the definition. Beyond meeting basic human needs, most people agree that there is an economic meaning to the term "sustainable development."

Sustainable development was presumed to be "exploitation of resources" while ensuring that "the ability of the biosphere to absorb the effects of human activities is consistent with future as well as present needs." Certainly, "sustainable development can be pursued more easily when population size is stabilized at a level consistent with the productive capacity of the ecosystem." At the United Nations Conference on Environment and Development in Rio de Janeiro in 1992, Agenda 21, with a target date of the middle of the 21st century, was proposed to allow earthlings to adjust their needs to the finite resources of the earth.

How is sustainable development achieved? Which economies are more sustainable? Is an economy built on large capital investments and cheap energy sources paid for in paper money sustainable? Or do we have to ensure that the Native Amerindians living in the Amazon can sustain their way of life through consumptive uses of wildlife into the future? In other words, which economy is more sustainable, especially when oil prices eventually climb to new heights? The effect of a tenfold increase in fuel prices on the developed world would be catastrophic, while the Amerindian would hardly notice the effect. Indeed, the concept of sustainable development is not an Amerindian idea; it is a rich nations idea brought about by the fear that other nations want what they have— access to capital and cheap, nonrenewable resources. This concern has been brought about by increasing population densities, species extinctions, and loss of environmental quality.

In Robinson and Redford (1991), the question is asked, "Do we, as human beings, have the right to use wildlife for our own purposes and benefits, or do wild species have inalienable rights of their own?" If wildlife has no value, utilitarian or otherwise, then why should it be worth saving? If all people agreed that wildlife has no value, there would never be another wildlife issue.

Apparently, judging by the number of wildlife issues, some fraction of people think wildlife has value. This value might be aesthetic, scientific, spiritual, consumptive, or for some other reason. For instance, you will probably never see a wild snow leopard and maybe you will never see one in a zoo, but you might be glad that such an animal as ghostlike as the wind exists in the high Himalayas. You may not know why hawk-moths exist, but you might understand the fact that because a hawk-moth's tongue is as long as its body, there is something special going on here. In fact, the hawk-moth pollinates certain plants. Without the hawk-moth, these plants could not reproduce.

Perhaps you are the kind of person who does not need to have an immediate reason to believe that everything is connected to everything else and that a creature's existence did not come about by chance alone. For instance, Emmons (1987) claims that jaguars, panthers, and ocelots eat prey in relation to its abundance in the environment. If one prey becomes more common, it is more likely to be eaten. If something is rare, it is less likely to end up as a meal. Thus, biodiversity is higher where these top carnivores knock back any species that becomes too common. The same might be true for birds of prey, seals in the sea, and sharks as well. It just might be that all top predators help maintain a healthy environment and increase diversity by their food habits. But if you did not understand this and you hunted jaguars and panthers to extinction because they are big cats and you think it is fun to shoot big cats, you might destroy biodiversity in the process of having your fun.

Wildlife has many uses to humankind, one of which goes well beyond the obvious uses suggested in Robinson and Redford (1991). Wildlife is important in maintaining a healthy environment because of the so-called ecosystem services, the work that it does for humans every day free of charge.

Think of insectivorous birds gleaning insects from trees. One famous ecological experiment compared the success of trees with and without services provided by insectivorous birds. In side-by-side experiments, some trees were covered with a mesh that let bugs in and kept birds out. Other trees were left alone. Net-covered trees failed to produce healthy leaves. Those leaves that opened were attacked by hordes of insect pests. Leaves of trees not covered by nets did fine, largely because birds consumed the leaf-eating insects. The birds performed a service—they ate insects. And they did it without sending us a bill.

If the birds were gone, our use of chemicals and pesticides would have to increase if we wanted to have trees.

There is no doubt that wildlife can be of economic value and can be sustainably harvested commercially. Alligator farms in Florida are doing a booming business in meat and hides. Wild alligator populations have responded to protection and there is now a limited hunting season on alligators. Where people depend on wildlife for food, the question is one of sustainably harvesting wildlife. Can wildlife be harvested sustainably to feed a family? How about two families? Can wild animals be harvested commercially?

Notice that we are not asking whether or not wildlife should be harvested. The fact of the matter is that people across most of the world depend on bush meat to live. Perhaps people have replaced jaguars, panthers, lions, tigers, and leopards as the world's top predators.

South American Vicunas and Guanacos

Overhunting and habitat destruction continue to be the most severe and widespread human activities that threaten wildlife. In marginal lands, such as semi-arid and arid lands, human activities are especially detrimental. The high Andes of South America is important habitat for the South American vicuna (*Vicugna vicugna*) and the guanaco (*Lama guanicoe*). Both these animals have high economic potential for their wool; however, both have experienced dramatic population declines over the last century (Franklin and Fritz 1991).

Sustainable yield of wildlife products, especially commercially sustainable yields as compared to subsistence yields, is a difficult problem, especially in a semi-arid region where wildlife populations are scattered across a vast tractless region. Vicunas and guanacos might have a large economic impact if they could, in some way, be managed. The same is true for the other two members of the South American camelid family, the llama (*Lama glama*) and the alpaca (*L. pacos*). Vicunas produce "silky fine wool," followed by guanacos in terms of quality of wool. The llama is a beast of burden in South America, and the alpaca is a domestic wool producer. There is also a market for the meat and hides of these animals. Certainly, harvesting wild populations of these animals might be possible if their basic biology were understood. Can wild populations of these animals be harvested for their meat and hides and nondestructively for their wool—on a sustainable basis? Is it necessary to allow free-ranging populations to exist and harvest them carefully, or must a ranch be established for a sustained yield to be economically viable? The answers depend critically on knowing the animal's basic biology and ecology.

Green Iguanas

In many countries, slash-and-burn agricultural practices are the means by which land is converted from forest to pasture. Roughly 2 to 3 ha of forest is burned and cultivated by local people and used for three years, after which time the land is sold to a cattle rancher. The people move on to another patch of forest. Typically, beef cattle are raised for the export market and, because developed nations are willing to pay a higher price, local meat shortages develop. In some regions, armadillos, agouti, paca, iguanas, deer, and other small animals make up the meat people consume. Werner (1991) undertook research on green iguanas (*Iguana iguana*) as part of the Iguana Management Project sponsored by the Smithsonian Tropical Research Institute and the Pro Iguana Verde Foundation.

Iguana meat is a valuable protein source for rural people, and the eggs are considered an aphrodisiac and a delicacy. In Panama, some 70% of the people would eat iguana if it was readily available. However, the iguana is considered endangered in many countries where it was once abundant. Once again, the wildlife issue is clear. Humans have overexploited the iguana so that now it is endangered. The solution is to sustainably manage iguana populations to increase their numbers and provide an inexpensive protein source for humans.

Iguanas are distributed from southern Mexico to Paraguay in South America. They are found on some islands as well. Iguanas are prolific vegetarians and live in a diversity of habitats from tropical to subtropical forests. Much is known about their breeding biology and ecology, and managing wild populations of iguanas is considered difficult. Ranching iguanas, however, might be a viable alternative, but experience shows that captive management of reptiles is "not easy." Werner (1991) claims that populations of iguanas can be made artificially high if their food supply can be maintained at a constant level. Artificial nests can be built, and trees favored by iguanas can be planted. All these management practices are predicated on creating and maintaining artificially high populations so that harvests can be sustained at particular levels.

Howler Monkey Sanctuary

The black howler monkey (*Alouatta pigra*) is known in north-central Belize as the baboon. The baboon is locally abundant and is not hunted by the local people. The Community Baboon Sanctuary in Belize was established on private lands where small-scale cattle ranching and subsistence farming were taking place. Horwich and Lyon (1995) described a "community-based plan" that paralleled the Man and the Biosphere program. Land management plans were jointly developed with local stakeholders because these people were the most knowledgeable about both local conditions and the baboon's requirements.

A tourism program was set up for baboon monkey viewing, and a community education program was established in the form of a museum exhibiting important ecological, conservation, and cultural themes. There is also an ecotourist education program to describe the entire project to tourists interested in visiting the sanctuary. So successful was the program that two other sanctuaries have been started—a sea turtle program to protect beaches where the turtles nest and a manatee community reserve that offers tourism and multiple use of nearby forests. These projects are community based and locally fine-tuned. They offer a substantial improvement over the declaration of national parks as the last refuges of our wildlife.

New Initiatives

Common species of wildlife throughout the United States are declining. Even species we take for granted are diminishing in number because of loss of habitat. More than 89% of all Americans participate in outdoor activities. Americans, it seems, have a love affair with the outdoors and with wildlife. More than 31 million people photographed wildlife in 1991. Between 1980 and 1990, the number of Americans who took wildlife-watching trips increased 63%. More than 76.5 million Americans watch wildlife, 35.6 million fish, and 14.1 million hunt. A proposal that was to be introduced before the U.S. Congress in 1996 called for a small tax to be levied on recreational products to support state-based wildlife conservation, education, and recreation. The Wildlife Diversity Funding Initiative started by the International Association of Fish and Wildlife Agencies might provide as much as $350 million annually to support conservation. More than 500 well-known companies and 700 conservation, outdoor education, and recreation groups supported the proposal.

Previous support programs have been successful. Populations of many species of duck, turkey, pronghorn antelope, and striped bass among others have increased through such programs.

Animal Rights Activists and Endangered Species

The piping plover (*Charadrius melodus*) is endangered. Only 14 nesting pairs of piping plovers were found at the Monomoy National Wildlife Refuge near Chatham, Massachusetts, in 1996. According to the U.S. Fish and Wildlife Service, over 900 nested there in 1966. In 1996, however, about 5700 great black-backed gulls (*Larus marinus*) and herring gulls (*L. argentatus*) were breeding in the refuge, whereas no gulls could be found in 1961. Gulls are predatory and will readily eat the young of other birds. On May 19, 1996, The Associated Press reported that the U.S. Fish and Wildlife Service planned to poison the gulls to protect the piping plover and other birds. The gulls now number "in the tens of thousands."

Poison-laced margarine on white bread was placed in the nests of 5700 breeding gulls. Gulls seem to do well wherever humans do well. Large numbers of gulls are seen at garbage dumps and refuse stations. There is little doubt the gulls have increased in number as a result of their association with humans. Animal rights activists, having lost a court appeal to halt the poisonings, placed tuna and charcoal filter sandwiches in 2400 gull nests in an attempt to save them from being poisoned.

The Endangered Species Act obligates the federal government to act within its power to protected listed species. Greater black-backed and herring gulls need no form of similar protection. There is little doubt that the larger and more aggressive gulls would eat the young piping plovers and drive away the adults. Indeed, the result is already clear: without constant vigilance, there will be no piping plovers.

On May 20, 1996, 400 dead gulls were found. The poison was only partially successful. Obviously, this action must be continued. The issue here is quite simple and was decided by the courts. In a democracy, a solution rarely pleases both sides.

TEAMING WITH WILDLIFE

As we saw earlier, the Pittman–Robertson Act of 1937 and Dingel–Johnson/ Wallop–Breaux Acts of 1950 played a significant role in providing sustained funding to support state efforts to restore games species, conduct hunter-education programs, and make critical land acquisitions. These acts were supported by federal excise taxes on hunting supplies. Teaming with Wildlife, also called the Fish and Wildlife Diversity Funding Initiative, is a national effort to provide sustained, long-term funding for nongame fish and wildlife conservation, recreation, and education programs, which have historically operated with little or no funding. The conservation of most species of fish and wildlife and countless invertebrate species typically received less than 5% of all funding for wildlife. Because most species are not listed as endangered or threatened and because they are not hunted or fished, many remain unstudied, unmonitored, and unprotected. State efforts to raise money to support programs for these species include tax form checkoffs, specialty automobile license plates, private donations, and the use of lottery proceeds (as in Arizona); however, most efforts have fallen far short of what is needed (as in New Mexico).

Funding to support the Teaming with Wildlife program will come from a small excise tax on outdoor recreation equipment, including camping equipment such as backpacks; outdoor recreation equipment such as canoes; optical equipment including binoculars and spotting scopes; photographic equipment including cameras, film, and carrying bags; and even backyard wildlife supplies such as birdseed, feeders, houses, and birdbaths. Outdoor and recreational books include field identification guides and where-to-go and what-to-do books. Recreational vehicles, including large, so-called RVs and sport utility vehicles, would also be taxed. Because the tax is broad based, the amount is comparatively small. The user fee depends on the manufacturer's price (not the retail price) and is envisioned to be between 0.25% and 5%. Naturally, the consumer ultimately bears the increased cost. For example, a consumer would pay a 5% tax or an additional ten cents for a guidebook that retails for ten dollars and costs the publisher two dollars to produce. For an RV that cost $10,000 to manufacture and sold for $25,000, the consumer would pay a tax of $25 to $500. The fund's green logo will be displayed on each product, along with an explanation of how the funds are dedicated for wildlife conservation, recreation, and education.

Similar to hunting and angling user fees, the fees will be collected by the U.S. Treasury from manufacturers or from import duties and given to the U.S. Fish and Wildlife Service for distribution. An administrative cap of 4% will ensure against bureaucratic excesses. States will be expected to provide 25% matching funding. That is, if a project is expected to cost $100, the state must provide $25. A formula based on population (two-thirds) and land area (one-third) will be used to calculate each state's share. No state or territory will receive less than 0.5% or more than 5% of the funds. No diversion of funds for purposes other than wildlife projects focused on conservation, recreation, or education will be permitted. Based on the matching funding rule, those states that have viable nongame wildlife programs will likely receive much more money (as will Missouri), while those states that have nonviable nongame programs (such as New Mexico) will have a hard time providing the 25% required matching funding. What kind of money is at stake?

The Pittman–Robertson Act alone has raised more than $2.7 billion for wildlife research, habitat acquisition, and other wildlife programs as of 1996. Estimates are that Teaming with Wildlife fees will raise more than $350 million annually. Split evenly, this would mean $7 million annually to each state. Under the formula for funding, large, highly populated states (like California and Florida) would receive more money. Pennsylvania's share of Teaming with

Wildlife funding could be up to $13.4 million, if the state is willing to dedicate at least $4.4 million in matching funds for nongame fish and wildlife programs. How likely is it that Teaming with Wildlife will pass both houses of Congress and be signed into law by the president?

A recent study concluded that 40% of all Americans 16 years of age or older participate in some form of wildlife-associated recreation, such as fishing or hunting, and other nonconsumptive outdoor activities, such as hiking, wildlife viewing, and photography (see the section on Americans and Wildlife later in this chapter). The American Ornithologists Union, American Fisheries Societies, Ruffed Grouse Society, Archery Manufacturers and Merchants Organization, Bat Conservation International, Colorado Bow Hunters Association, Conservation Federation of Missouri, Florida Wildlife Federation, J.N. "Ding" Darling Foundation, Partners in Flight, Quail Unlimited, Society of American Foresters, Society of Conservation Biology, Sportsmen Conservationists of Texas, Wildfowl Trust of North America, Rivers Council of Washington, The Wildlife Society, National Audubon Society, Defenders of Wildlife, National Wildlife Federation, Wildlife Management Institute, Boone and Crockett Club, Sierra Club, Trout Unlimited, American Birding Association, and World Wildlife Fund are only a few of the more than 100 organizations that have urged the U.S. Congress to pass the act. Indeed, as of February 13, 1997, 1609 groups and businesses offered to support the passage of Teaming with Wildlife. Although passage of a Teaming with Wildlife act is by no means a forgone conclusion, the initiative does have broad-based industry support.

With the influence of government declining, the "pay-to-play" fee is politically acceptable. The number of both consumptive (hunters and anglers) and nonconsumptive users of wildlife has grown significantly. A Forest Service Recreation Executive Report released in May 1994 named wildlife viewing as the number one outdoor recreational activity in the United States, with 76.5 million participants, followed by fitness walking and camping. The same report predicted a 16% increase in wildlife watching and a 23% increase in outdoor photography through the year 2000. The importance of seeing wildlife as part of an outdoor experience—from backyard bird observation to hiking and canoeing—also translates into dollars for the outdoor recreation industry. More than $18 billion is spent annually on recreation related to wildlife viewing. For example, a survey in Yellowstone National Park revealed that 95% of the visitors consider seeing wildlife of highest importance.

Naturally, not all organizations support the Teaming with Wildlife initiative. As of March 1997, such notable exceptions were the Wild Bird Feeding Institute, National Bird Feeding Society, American Whitewater Association, and American Canoe Association, along with over 40 river- and wilderness-oriented environmental groups.

The American Canoe Association questioned the spending priorities and the agencies that were to receive the funding set forth by the initiative. They argued that the broad spectrum of possible projects that could be funded under the initiative created a risk that critical needs such as land acquisition would be usurped to fund building and maintenance of wildlife-viewing areas; observation towers, blinds, shelters, parking lots, road signage, and other facilities for fish and wildlife interpretation; production and distribution of fish and wildlife educational materials for schools, community groups, and other interested parties; and research. The American Canoe Association feared that the broad array of projects permissible under the proposal would compete with more pressing resource needs such as land acquisition and the operation of existing parks and forests (which is already covered under separate funding from the Department of the Interior). While the American Canoe Association noted that the projects were worthwhile, they had high administrative costs, offered more opportunity for waste, and results would be difficult to track. The association stated that the cost of education consistently increased faster than inflation by a significant margin, building projects routinely exceeded their cost estimates, and the cost of a single research project could easily run millions of dollars, thus diluting the intent of the proposal. Furthermore, most of the projects proposed for support under the Teaming with Wildlife program targeted primarily the general public and not the outdoor enthusiasts, who would pay the bill.

U.S. FISH AND WILDLIFE SERVICE
CORPORATE WILDLIFE STEWARDSHIP AWARD

On December 5, 1996, Secretary of the Interior Bruce Babbitt presented natural resources company Champion International Corporation of Stamford, Connecticut, with the U.S. Fish and Wildlife Service Corporate Wildlife Stewardship Award in recognition of the company's outstanding contributions to fish and wildlife conservation.

In presenting the award, Mr. Babbitt stated that "Champion International Corporation was among the first to provide practical solutions that allow us to enjoy a healthy environment while promoting economic growth. Wherever Champion has a presence, it has worked with the U.S. Fish and Wildlife Service and its counterpart state agencies to develop solutions that protect wildlife and allow land use. Champion has shown that we can use our lands while protecting our natural heritage."

Champion developed a comprehensive approach to protecting forest resources as part of the company's ongoing sustainability and stewardship effort and its commitment to the American Forest and Paper Association's Sustainable Forestry Initiative. Champion has supported a range of activities designed to protect endangered and threatened species and their habitats over the past several years. The corporation's efforts include adopting specialized land management techniques that benefit birds and fish and educational programs for its employees and contractors.

In 1994, Champion signed a cooperative agreement with the U.S. Fish and Wildlife Service, the U.S. Department of Agriculture Forest Service, and the state of Texas to advance restoration goals for the endangered red-cockaded woodpecker. Under the agreement, Champion manages 2000 acres of its Brushy Creek Wildlife Management Area in east Texas to protect existing woodpecker colonies and to provide additional nesting habitat for transplanted woodpeckers. The corporation has also built and installed artificial nesting cavities, conducted prescribed burns, and established open stands of longleaf pine to benefit the woodpecker.

In its role as one of Maine's major forest landowners, Champion helped establish the Salmon Habitat and River Enhancement Project, or SHARE. SHARE has become a focal point in developing cooperative solutions to conserving Atlantic salmon. As part of this effort, Champion contributed funds, personnel, and equipment to map salmon habitat, clear obstacles to spawning, repair water control structures, and build and tend weirs to track returning fish.

Champion has also worked with the U.S. Fish and Wildlife Service to arrange or fund endangered species training for its contract loggers. It produced an illustrated guidebook to endangered species in Alabama. The corporation also developed a series of educational videotapes about endangered species in the South. Because of the success of these ongoing efforts, Champion is developing endangered species guidebooks for each of the 17 states in which it operates.

Champion is the eighth company to receive the Corporate Wildlife Stewardship Award since it was created in 1990.

AMERICANS AND WILDLIFE

On February 20, 1997, the U.S. Fish and Wildlife Service issued a press release summarizing the economic impact of Americans and their passion for outdoor activities. Although the figures are for 1991, they are a valuable indication that outdoor activities are of growing importance to people. The report claimed that more than 76 million Americans spent $18.1 billion watching, photographing, and feeding birds and other wildlife in 1991. The spending generated nearly $40 billion in total economic activity across the country, supported 766,000 jobs, and resulted in $3 billion in state and federal tax revenues.

Significantly, equipment and other expenditures accounted for $10.6 billion of the $18.1 billion in direct expenditures. Of this amount, nearly a third was for off-road vehicles, tent trailers, motor homes, and pick-up trucks. The Teaming with Wildlife initiative, whose excise fee would be between 0.25 and 5% and might average 3%, could thus be expected to raise $300 million in 1991 dollars.

Wildlife watchers spent $7.5 billion on travel-related goods and services. Of this amount, 40% was for food and drink, 35% was for transportation, and 19% was for lodging. They also spent $2.2 billion on cameras, film, and developing pictures and $1.5 billion on wild bird food.

The report noted that 109 million Americans participated in wildlife-related recreation, including hunting and fishing. By comparison, 105 million attended major league football, basketball, hockey, and baseball games in 1991.

THE NEXT 50 YEARS

In the end, humans will have to decide whether they want to live with wildlife or without it. While so much has been lost, ecotourism to exotic places is the fastest growing segment of the tourism industry. People need wild places and wildlife, and both are becoming increasingly rare and expensive.

As Mathews (1994) pointed out, the next 50 years of human existence will probably be the most important in human history. For the first time, humans have the power and technology to alter vast areas of the planet. The population will swell to ten billion people by 2040, and economic output will grow about ninefold. Energy consumption and waste production will continue to grow. Nearly three-fourths of the earth's present population uses as much energy as the other quarter. These fractions will grow closer as other nations develop. Humans will require even more of the earth's primary production to exist. Yet, human understanding and appreciation of ecology are lagging behind our technological prowess. Humans wield powerful machinery and capital, but we lack a fundamental understanding of the consequences of our actions. In many cases, Humpty Dumpty cannot be put back together again.

Wildlife becomes an issue when humans try to live outside nature. Human agricultural schemes attempt to defy nature; dams are built to tame rivers, swamps are drained to bring more land into production, parks are created to "preserve nature," trees are harvested in the name of progress, and fossil fuels are used with abandon to power the "modern" society. Species introductions are altering ecological relationships that evolved through millions of years. Increasingly often, we read, see, or hear about natural disasters. Indeed, the very process that created and maintained our beautiful forests is seen by humans as a destroyer of forests. But humans have prevented fires for so long that the accumulation of debris on the forest floor is indeed a disaster waiting to happen. In our infinite wisdom to protect our forests, we have condemned them to destruction.

The reason the fire in Yellowstone National Park during the summer of 1988 was such a "disaster" is not because the fire was too big but because the park is too small and humans have suppressed fires for too long. We rarely hear that fire in Yosemite National Park acts to increase diversity or a fire in the Great Smokies acted to increase the health of the forest, or a river flooded and fertile silt was spread over 100,000 ha, resulting in one million new cottonwood seedlings springing forth from the floodplain. Instead, it's always something like "a flood destroyed a mobile home park." Human free-water schemes are almost always disasters. Dams are ultimately failures because of their long-term, and unforeseen, deleterious effects.

Many problems arise because of government subsidies. Schemes that are not possible for whatever reason become possible because of a subsidy. Thus, there should be no more government subsidies. There should be no more grazing on public lands, no more harvesting of timber on public lands, no more sugar subsidies, no more dams, no more water projects, no more animal damage control. Habitat destruction must be stopped now, today. Eradication of nonnative species, including those introduced for sport hunting, should begin at once. Without government subsidies, private initiatives that preserve habitat and permit living in harmony with nature will become profitable ventures. Humans must live with nature, as part of nature and not apart from it. Most recreationists would gladly pay modest access fees for use of public forests, especially if they did not see clear-cuts or cattle. Increased fees for the national parks would allow land connecting the parks to be purchased. More land, not less, should be put into the public sector. Massive habitat restoration projects, supported by businesses, should begin at once. Many private landowners have already decided to preserve natural habitat remnants. These efforts must be encouraged through tax reductions and other incentives. Grants for public education programs must be made available. Community assistance grants to support wildlife programs in grammar schools across the country would act to increase awareness of wildlife and habitat issues. There must be a new ethic of responsibility toward wildlife and habitat.

Living with nature and realizing that nature most often acts to benefit species increases human quality of life. Quality of life means living closer to nature, not farther from it. People everywhere have already lost enough. Most do not accept that enough is enough. Many think enough has been too much. Now they want it back.

EXERCISES

20.1 Visit the Crested Caracara Conservation Project Web page. What incentives do private property landowners have to protect threatened or endangered species?

20.2 The red-cockaded woodpecker (*Picoides borealis*) is endangered. Recently, several forest product companies have established landmark conservation partnerships to help protect the bird's habitat. Do a Web search to learn more about the red-cockaded woodpecker. What makes this bird unique, and why is it important that it be protected?

20.3 Private property owners object to "big government" telling them how to manage their property. Wildlife does not belong to the property owner, however. Since protecting habitat is important, how can government help small private landowners to do so?

20.4 Steelhead trout are amazing fish. Unlike salmon, they return to the sea after spawning. Find out more about steelhead trout in the library or on the Web.

20.5 Use a World Wide Web search engine to locate any interesting approaches to conservation.

20.6 Make a list of the national parks, their areas, and the number of species in each national park.

20.7 Make a list of the World Heritage Parks, their areas, and locations. How many of them are in the tropics? Make use of the library and the World Wide Web.

20.8 Ranchers in Florida are providing excellent habitat for crested caracaras, as well as Florida panthers and black bears. In the Rocky Mountains, ranchers provide habitat for deer and elk. Perhaps tax incentives would encourage more landowners to preserve habitat for more species. What other incentives might ranchers be offered? Present such a proposal to your state representatives. What the public wants, the public gets.

20.9 Do a search on "sustainable development" and decide what "sustainable" and "development" mean. Is this term an oxymoron?

20.10 Do a search on "fragmented forest" and summarize what happens to biodiversity when forests are continually fragmented. Isn't it time we started reconnecting forests?

20.11 What species are currently harvested commercially? Can wild animals be sustainably harvested? Read Part 4, Wildlife Farming and Ranching, in Robinson and Redford (1991) and update their examples using the World Wide Web.

20.12 In 1991, Werner's research was "in progress." What has happened to iguana ranching since then?

20.13 Create a SAFEGUARD program for your state. What are the essential ingredients of such a program?

20.14 Organize a campaign to restore a critical wildlife habitat in your state. How can you get people interested?

20.15 Name the other seven companies that have received the U.S. Fish and Wildlife Service's Corporate Wildlife Stewardship Award. How did each company earn its award?

20.16 Compare the objections of the American Canoe Association with the benefit of hindsight gained from the Pittman–Robertson Act. Who benefited from the excise tax paid mostly by hunters and anglers? Answer the remaining charges made by the organization.

20.17 How might the Teaming with Wildlife initiative be broadened to include groups, such as river enthusiasts, which feel they have been left out of the initiative?

LITERATURE CITED

de Klemm, C. 1993. *Biological Diversity Conservation and the Law.* IUCN, Gland, Switzerland.

Emmons, L.H. 1987. Comparative feeding ecology of felids in a neotropical rainforest. *Behavioral Ecology and Sociobiology* **20**:271–283.

Franklin, W.L. and M.A. Fritz. 1991. Sustained harvesting of Patagonia guanaco: is it possible or too late? pp. 317–336 *in* Robinson, J.G. and K.H. Redford (Eds.). *Neotropical Wildlife Use and Conservation.* University of Chicago Press, Chicago.

Harris, L.D. and P.C. Frederick. 1990. The role of the Endangered Species Act in the conservation of biological diversity. pp. 99–117 *in* Cairns, J. and T. Crawford (Eds.). *An Assessment: Integrated Environmental Management.* Lewis Publishers, Boca Raton, Florida.

Horwich, R.H. and J. Lyon. 1995. Multilevel conservation and education at the community baboon sanctuary, Belize. pp. 235–253 *in* Jacobson, S.K. (Ed.). *Conserving Wildlife.* Columbia University Press, New York.

Mathews, J. 1994. Hard lessons from the death of the Aral Sea. *St. Petersburg Times* October 19.

Nature and Resources. 1984. *UNESCO* **20**(4):1–12.

Newmark, W.D. 1995. Extinction of mammal populations in western North America national parks. *Conservation Biology* **9**(3):512–526.

Robinson, J.G. and K.H. Redford (Eds.). 1991. *Neotropical Wildlife Use and Conservation.* University of Chicago Press, Chicago.

U.S. Fish and Wildlife Service. 1997. 1991 Economic Impacts of Nonconsumptive Wildlife-Related Recreation. U.S. Fish and Wildlife Service, Washington, D.C.

Werner, D.I. 1991. The rational use of green iguanas. pp. 181–201 *in* Robinson, J.G. and K.H. Redford (Eds.). *Neotropical Wildlife Use and Conservation.* University of Chicago Press, Chicago.

EPILOGUE

e cannot help but feel somewhat like anxious parents preparing to send a child on a long journey. In this spirit, we offer some last-minute thoughts about wildlife issues to carry with you along the way. Several important points regarding wildlife issues merit further consideration. Throughout the book we have noted repeatedly the importance of citizen participation in resolving wildlife issues. We sincerely hope that you have been empowered by the numerous stories of citizens enacting substantial change in the way we deal with wildlife, whether rare and facing extinction or abundant and pestiferous. Some of the main themes in the book are summarized here.

WILDLIFE RESTORATION WORKS

Each day, more and more citizens become interested in wildlife restoration. We believe that many would like to see walrus on the rocky coasts of Maine, just as we would. Scientists have the knowledge and expertise to restore such populations. What they need is the political will to make it happen. Who can give them this will? You can. We are optimistic that you, as a voter, will demand such restoration efforts in Maine and elsewhere.

The evidence to support our positivism is overwhelming. In the early 1900s, deer were consumed for food in the United States, and their numbers dwindled severely. Conservation efforts were financed chiefly by the Federal Aid in Wildlife Restoration Act of 1937 (Pittman–Robertson Act) and by hunters' licenses and fees. These funds played an important role in the restoration of the wild turkey (*Meleagris gallopavo*). Kennamer et al. (1992) detailed the history of the wild turkey in the United States. Wild turkey populations reached their lowest levels just before the end of the 19th century. They persisted only in the most inaccessible areas. Since then, through wildlife management efforts, wild turkey now appear in 49 states, absent only from Alaska. Unfortunately, wild turkeys were successfully introduced in Hawaii.

As pointed out by Kennamer et al. (1992), Pittman–Robertson funds increased substantially after large numbers of soldiers returned from World War II. These funds were then available to finance trapping and release programs, as well as studies of turkey dietary and habitat needs.

A similar story holds for white-tailed deer. McCabe and McCabe (1984) argued that the white-tailed deer population went through three stages historically between the years 1500 and 1900. The first stage extended from 1500 to 1800 and mostly was a period of intensive slaughter by Native Americans, who may have reduced the total number of white-tailed deer by as much as 50% of its pre-1500 size. During the second stage (1800–1865), the population increased somewhat as the impacts by Native Americans were terminated by colonizing European Americans. The third phase (1850–1900), saw the most intensive slaughter ever by market and subsistence hunters. Again, through financing from the Pittman–Robertson Act, the white-tailed deer made a comeback throughout most of its former range.

There is no doubt that wildlife biologists and managers know, in most cases, exactly what is needed to restore populations headed for extinction. The main drawbacks to any restoration are economic or political. Wildlife biologists and conservation biologists are more knowledgeable than politicians regarding the conservation of species. We live in a world where scientists can clone genes from a bacterium and insert them into corn seed, so that the corn plant can unleash the defenses of the bacterium onto its insect pests. In the field of conservation biology and wildlife ecology, technological tools such as in vitro fertilization are commonplace. Studies of small mammal populations once involved the amputation of various combinations of an animal's toes to facilitate future identification. These amputations were often detrimental to the animal. Today, tiny transponders can be implanted just below the skin of these animals so they can be identified when recaptured. Each transponder has a unique bar code that can be read with a wand in a fashion very similar to the way a cashier scans grocery items in a supermarket.

We now have tiny radio transmitters that can be placed on birds as small as the Cape Sable seaside sparrow so that individuals can be tracked and monitored. We have the so-called Geographic Information System (GIS) software so that habitats can be mapped more accurately than ever before. In short, the tools available to wildlife ecologists today are staggering in their capabilities. But the list of tools is not limited to the laboratory or the field. Consider also the tools available for acquiring information. The Internet and World Wide Web put information at our fingertips in an instant. Searches for information that took weeks or months a mere ten years ago now take seconds.

Given the state of the science and technology in use today, we hope that in the future you will be astonished if a politician, corporate agriculturist, or representative of a timber company ever says that "more data are needed" before acting to resolve any particular wildlife issue. But what study can be done, for instance, to prove that the northern spotted owl will vanish if the chain saws of the timber industry succeed in destroying all the remaining old-growth forests? The answer is none, because proof is not part of the scientific method. The same can be said for the cleanup of the Everglades, the effects of acid rain on eastern deciduous forests, and the impacts of grazing on public lands. Thus, the barriers to solutions to many wildlife issues are mostly political. Conservation scientists generally know exactly what needs to be done to resolve the majority of (if not all) wildlife issues if the only concerns are biological.

NOT ALL WILDLIFE ISSUES INVOLVE POPULATION DECLINES—SOME INVOLVE POPULATION EXPLOSIONS

As seen in Chapter 19 and the successes in wildlife restoration noted above, Americans are now faced with several species that have become very abundant. Some of these species were introduced from foreign environments and some were native. Several of these species were serious pests either to agriculture or in a more general way.

In 1997, The Wildlife Society (the national organization for professional wildlife biologists) devoted an entire issue of the *Wildlife Society Bulletin* (one of its publications) to the problem of deer overabundance. Deer and numerous other species threaten motorists at night as the animals are caught in the headlights of cars.

The exploding snow goose population now seriously threatens fragile arctic vegetation on its nesting grounds. Surely, one might think, a population explosion in the arctic, far from any human activities, must be a purely natural phenomenon. But many wildlife scientists argue that this is an example of how human activities in one place (the Great Plains) can impact wildlife populations hundreds of miles away. To see this, we must consider that in pre-European times, snow geese nested in the arctic and wintered along the coast of the Gulf of Mexico. Nesting success depended on nutrient stores (fat) the birds accumulated during the winter in nutrient-rich coastal marshes (Baldassarre and Bolen 1994). Of course, the journey from the coast to the arctic required the consumption of some of the stored energy. Birds with inadequate energy stores burned too much of their energy on the return trip and suffered lower nesting success. Thus, energy "deposits" were made during winter feeding, and "withdrawals" were made during the trip to the nesting grounds. Not surprisingly, a balance between energy deposits and withdrawals evolved. Just enough energy was left over after the migration to produce a few young.

Then wide-scale agriculture across the Great Plains entered the picture. Migrating snow geese no longer fly over vast stretches of prairie. Instead, they fly over fields of corn and grain, a veritable cornucopia. This availability of food meant that it no longer was necessary for the geese to travel all the way to the Gulf coast. Individuals thus were able to add to their winter deposits and reduce withdrawals by shortening the trip. This has drastically increased nesting success, and now there is a population explosion caused by human activities hundreds of miles from the nesting grounds.

Introduced nutria are devastating wetland vegetation in Louisiana. Nutria were introduced in the United States to develop a fur industry. When fur was popular in the fashion trade, people in the southeastern United States harvested many nutria for their skins. Then attitudes changed. People began to protest the killing of animals for fur. The market for nutria pelts declined to the point that it was no longer profitable for people to harvest them. The high reproductive capacity of nutria (they replace individuals that are harvested in a short period of time) was one of the attractive features of the species for a commercial harvesting enterprise.

Unfortunately, with reduced harvests came the inevitable population explosion. And what do the nutria eat? They eat wetland vegetation. They also tunnel through levees that hold back water and in general have become a major nuisance in Louisiana. Nutria are such a problem that state officials offer free samples of cooked nutria meat from stands set up in shopping malls. The idea is to convince people that nutria are good to eat and people should go out and harvest them for meat instead of skins.

There are numerous other examples of populations that have increased and become serious pests. What possible solutions are there to these increases?

Without question, managed hunting offers one potential solution to some of these problems. Unfortunately, many citizens are vehemently opposed to hunting of any type. Keep in mind that citizens are also registered voters. In some states, various types of hunting have already been banned by voters.

Few issues have simple solutions, however. We simply want to point out that human activities can set off a chain of events that impact populations of wildlife in highly indirect and often unpredictable ways.

OVERKILL, AGRICULTURE, AND WILDLIFE ISSUES

In the early history of the United States and Canada, commercial hunting and the resultant overkill were the most negative human activities as far as wildlife were concerned. We know all too well what happened to the great bison herds of the Great Plains. The story of the bison is similar to the story told by Farley Mowatt in his book *Sea of Slaughter* (Mowatt 1996). Mowatt describes the extermination of species such as the great auk (the symbol of the American Ornithologists Union). The great auk was a seabird. Like all seabirds, auks spent most of their lives at sea, coming to shore only to breed. Early seafarers from Europe harvested the auk for meat and later for oil and feathers. On shore, the auks were easy prey for hunters who could kill the hapless birds with paddles. But these European hunters did not limit the slaughter to adult birds. They also took the eggs.

Auk eggs, and those of other seabirds for that matter, were laid on bare rock on islets in the North Atlantic. Early European colonists soon learned that the eggs, if fresh, were a delicacy. But only the freshest eggs were desirable; a more mature egg might contain a partially developed auk chick, which apparently detracted from its palatability. But how does a hunter know if an egg is fresh and desirable or mature and undesirable? To guarantee that the eggs were fresh, hunters would find an auk nesting colony and smash all the eggs they could find. Since the eggs were laid on bare rock, we can be reasonably certain that few, if any, eggs on a given islet escaped the hunters' boots. The hunters would then return after a few days and collect all the replacement "fresh" eggs. The population of the great auk vanished by the mid-1800s under the boots of greedy commercial egg harvesters. We regret the loss of the auk and feel cheated that we will never see this great bird.

Extinction also befell several other species that were commercially harvested. With no controls and no management, greed inevitably led to extinction. Several wildlife populations today also face extinction, but now from more indirect causes such as habitat loss and fragmentation and also from introduced species.

Habitat loss due to agricultural practices has now replaced overkill as the most serious negative impact on wildlife. In fact, Aldo Leopold (one of the great early wildlife biologists) once remarked that what remains of natural biodiversity (animals and plants) only remains because "...agriculture has not got around to destroying it." There are two main types of agriculture: farming and ranching. Both can have negative and positive impacts on wildlife.

Farming, by definition, requires the replacement of native plant communities with monocultures of what are often introduced species of crops. This leads to the loss of native habitat, fragmentation of habitat, and production of an artificial food supply for many undesirable and often introduced species. The loss of native species of plants due to farming can lead to chains of extinctions. And, as the great British ecologist Charles Elton once observed, introduced species might have higher success rates in agricultural as opposed to natural habitats. Thus farming can be linked to all four members of Diamond's Evil Quartet.

Many types of farming also require the use of large quantities of water, as well as pesticides, herbicides, fungicides, and fertilizer. As an example of the effects of overuse of water in farming and its environmental effects, recall the impacts on the Aral Sea. Pesticides are needed because native predators of pests either have been eliminated (overkill) or, in the case of introduced species, were never present. Fertilizers are needed because local biogeochemical cycles have been disrupted and the balance in the cycles is gone. Instead, there is a net loss of some nutrients and a net gain of others. Fertilizers are designed to make up for these losses due to runoff and leaching.

By definition, chemicals that leave farms as runoff enter the local watershed (they "run off" the fields into ditches that feed into creeks that in turn feed rivers and so on). These pollutants can wreak havoc in aquatic systems, and the effects can be felt in estuaries and coastal environments that are hundreds of miles from the farms where they originated. Some chemicals also have serious impacts on the atmosphere. For example, consider methyl bromide, which is used as an insecticide. Each molecule of methyl bromide is said to have 50 times the negative impact of a chlorofluorocarbon molecule on the ozone layer.

As negative as these impacts are, there are more. Croplands produce huge amounts of food for several species, among which are such species as crows and brown-headed cowbirds, which harm other species. In the eastern United States, forests were cleared for farming. This created more edge habitat (the edge is the border between field and woodlot). The crops also produced food in the form of waste grain and weeds that colonized the borders of the fields. These landscape changes enabled brown-headed cowbirds to spread into the eastern United States (Robinson et al. 1995).

The brown-headed cowbird is a brood parasite. Rather than building a nest of its own, the cowbird lays its eggs in the nest of another species. The cowbird egg is often much larger than that of the host. But the host is conditioned, through evolution, to feed the nestling in its nest. It does not consider that the nestling is not its own. Indeed, the victim incubates the cowbird egg right along with its own eggs. The cowbird nestling pushes any eggs and nestlings from the nest. Even if the cowbird hatches somewhat after the host nestlings, it might still outcompete them for food from the parents, who typically feed the largest mouth in the brood. The victimized parents feed the cowbird until they are exhausted and so depleted physically that their future reproductive efforts become jeopardized. In the end, the unsuspecting parents may raise a cowbird that, ironically, will parasitize their nest again the next year.

Not all species are equally vulnerable to cowbird parasitism. Some species are considered to be "acceptors" whereas others are clearly "rejecters" of cowbird eggs. Each female cowbird has the potential to destroy 40 nests per year (one per egg [Ehrlich et al. 1988]). And each cowbird may live for more than ten years.

The brown-headed cowbird has been in North America since long before there was any sort of farming, yet farming has tipped the balance in its favor. Because cowbirds have co-existed with many species of birds on an evolutionary time scale, we might assume that not all species are equally vulnerable to cowbird nest parasitism, regardless of the changing landscape. Indeed, that is the case. Brown-headed cowbirds are known to have successfully parasitized the nests of 144 species and have been unsuccessful in the nests of perhaps an additional 100 species (Ehrlich et al. 1988). Some species, such as the yellow warbler (*Dendroica petechia*), may respond to cowbird parasitism by abandoning the nest or by burying the cowbird egg (Robinson et al. 1995). Other species, such as the blue jay and brown thrasher, eject the cowbird egg (Ehrlich et al. 1988). But other species simply feed the growing cowbird nestling even after it has ejected their own eggs or nestlings from the nest. As eastern forests become increasingly fragmented, remaining woodlots become smaller and smaller. Scientists have found that some hosts have higher rates of cowbird parasitism in smaller fragments of forest (Robinson et al. 1995).

But habitat alteration at the landscape level is just one of the numerous negative effects of farming on wildlife. In an article in the September–October 1996 issue of *Audubon* magazine, Les Line discussed the effects of American manufacturers exporting to other nations chemicals that have been outlawed in the United States. A case in point is Argentina, where farmers sprayed the chemical monocrotophos, banned in the United States, to kill grasshoppers that were ravaging their crops. Grasshoppers were a staple food of wintering Swainson's hawks, which nest in North America. A massive kill of as many as 20,000 Swainson's hawks in Argentina one winter was attributed to the use of this pesticide. The farmers were also victims in this case, as they had not been educated as to the effects of monocrotophos. Several farmers in Argentina have worked to solve the problem by reporting hawk kills and by searching for alternative pesticides, according to an article by Les Line in the *New York Times* (October 15, 1996).

As pointed out by O'Connor and Shrubb (1986), who studied farming and birds in Great Britain, pesticides can have three negative effects on populations of wild birds. First, they kill both plants and animals and thus have an impact on the local ecological relationships in an area. Second, they can have the indirect effect of altering the local food supply for birds by eliminating some weeds and thus their seeds. Third, chemicals allow farmers greater freedom in how they farm and what they farm.

Ranching also has potentially severe effects on wildlife. In the western United States, millions of acres of public land are subjected to grazing by private livestock. Jacobs (1991) reported that of the 750 million acres of land in the 11 contiguous western states (Montana, Wyoming, Colorado, New Mexico, Arizona, Utah, Idaho, Nevada, Washington, Oregon, and California), about 525 million acres are subject to grazing. However, 260 million of these acres are federal lands managed by the Bureau of Land Management and the Forest Service. These 260 million acres represent 85% of the public lands in the 11 western states (Jacobs 1991). Unfortunately, according to Jacobs (1991), this grazing produces very little livestock; only about 2% of the livestock produced in the United States comes from grazing on public lands in the 11 western states. Ranchers who use these public lands have for years benefited from low-cost grazing, and environmentalists have recently made a concerted effort to raise grazing fees on public lands. Incredibly, politicians from western states, such as Senator Peter Domenici from New Mexico, have argued that public lands should simply be turned over to grazers! In 1996, Senator Domenici sponsored a bill (SB 1459) called the Public Rangelands Management Act. An editorial in the *Tampa Tribune* (May 16, 1996) claimed that this bill would give ranchers priority use of national forests, national grasslands, and Bureau of Land Management lands. As pointed out by the columnist, this is an outrage, as the lands Domenici was apparently willing to turn over to his special interest group belong to the American public, not Congress or western ranchers. Supporters of this bill argued that 22,000 ranching families would be hurt if public lands were not open for grazing (104th Congress, 2nd Session, March 21, 1996, 5:51 P.M., page s-2622 Temp. Record, Vote No. 50). Without a doubt, this is the image that supporters of grazing on public lands hope the public will swallow: small family businesses struggling to eke out a living in the vast and otherwise unused West. But this argument is more style than substance. According to an editorial in the *St. Petersburg Times* (December 23, 1994), the owners of the 500 largest grazing leases are all "giant corporations."

The negative effects of heavy grazing are legion and include loss of topsoil, siltation of riparian systems, and loss of native vegetation. One species that suffered severely from the effect of grazing is the masked bobwhite of the American southwest. In Arizona, one species of gamebird that has had a bizarre history is the masked bobwhite. Early naturalists in Arizona observed bobwhites as early 1864. Specimens were obtained in 1884, and Herbert Brown published a note on these quail which was republished in the magazine *Forest and Stream*. Ornithologist R.R. Ridgway of the U.S. National Museum in

Washington, D.C. disputed the claim that bobwhites occurred in Arizona. There was little that Herbert Brown could do to counter this rejection, as the original specimens were spoiled. However, fortune smiled on Brown when he later obtained additional specimens and sent them to the American Museum of Natural History in New York, where the species was described as *Oreotyx ridgwayi*—in honor of the very person who had apparently denied its existence in Arizona (Brown 1989). Thus the masked bobwhite was discovered fairly late in the history of American ornithology, and by the time it was recognized it was already disappearing, along with its habitat, under the hooves of grazing livestock. Restoration efforts have focused on reintroductions of masked bobwhites from Sonora, Mexico, into southeastern Arizona and southern New Mexico (Brown 1989, Campbell 1976). Releases into New Mexico apparently have all failed (Campbell 1976).

It must be noted that although many (but not all) agricultural practices have negative effects on wildlife, as discussed above, agriculture is a critical component in our life support system. Most of us would no doubt soon starve without the efforts of agriculture. However, let's not simply accept the notion that the end (groceries) justifies the means (e.g., unrestricted use of pollutants) in every case. As badly as people need food, we consider unacceptable the complete extermination of wildlife to produce food for humans.

THERE ARE AT LEAST TWO SIDES TO EVERY ISSUE AND THERE MUST BE A BALANCE IN THE RESOLUTION

We might be able to survive without agriculture, but who would want to? Many people derive tremendous pleasure from tending gardens and producing vegetables for their own consumption. Fewer people raise livestock for meat. And still fewer of us put meat on the table by hunting.

It is easy for those interested in wildlife conservation to simply condemn commercial fishers, farmers and ranchers, and land developers. We all rely, to some extent, on some, or all, of the activities of these people. Here, we only want to point out that such reliance does not justify the carte blanche destruction of the environment to produce various products. We hope that you, as a citizen, will consider future issues by looking at both perspectives. For instance, consider the tomato farmers on the outskirts of Everglades National Park. To bring in their crops, they spray tons of methyl bromide into the atmosphere. This kills insect pests but also has a highly negative impact on the ozone layer. What should be done? The Environmental Protection Agency has banned production of methyl bromide by the year 2000. Some legislative representatives are fighting the ban. Is it worth damaging the ozone layer to have juicy red tomatoes? In the end, you, as an educated voter, will decide this issue. We hope you will draw upon your intellect to make your decision and not be blinded by the rhetoric and hype that so often shroud wildlife issues in a fog that only creates distrust and widens the gulf between sides. In the end, your vote counts!

LITERATURE CITED

Baldassarre, G.A. and E.G. Bolen. 1994. *Waterfowl Ecology and Management.* John Wiley & Sons, New York.

Brown, D.E. 1989. *Arizona Game Birds.* University of Arizona Press and Arizona Game and Fish Department, Phoenix.

Campbell, H. 1976. Foreign Game Birds in New Mexico, Bulletin 15. New Mexico Department of Game and Fish.

Ehrlich, P.R. D.S. Dobkin, and D. Wheye. 1988. *The Birder's Handbook.* Simon and Schuster, New York.

Jacobs, L. 1991. *Waste of the West: Public Lands Ranching.* Lyn Jacobs, Tucson, Arizona.

Kennamer, J.E., M. Kennamer, and R. Brenneman. 1992. History. pp. 6–17 *in* Dickson, J.G. (Ed.). *The Wild Turkey: Biology & Management.* Stackpole Books, Harrisburg, Pennsylvania.

McCabe, R.E. and T.R. McCabe. 1984. Of slings and arrows: an historical retrospection. pp. 19–72 *in* Halls, L.K. (Ed.). *The White-Tailed Deer: Ecology and Management.* Stackpole Books, Harrisburg, Pennsylvania.

Mowatt, F. 1996. *Sea of Slaughter.* Chapters Publishing, Shelburne, Vermont.

O'Connor, R.J. and M. Shrubb. 1996. *Farming and Birds.* Cambridge University Press, Cambridge, U.K.

Robinson, S.K., S.I. Rothstein, M.C. Brittingham, L.J. Petit, and J.A. Gryzbowski. 1995. Ecology and behavior of cowbirds and their impact on host populations. pp. 428–460 *in* Martin, T.E. and D.M. Finch (Eds.). *Ecology and Management of Neotropical Migratory Birds.* Oxford University Press, New York.

INDEX